Vertebrate blood cells

Vertebrate blood cells

EDITED BY

A.F. Rowley & N.A. Ratcliffe
Biomedical and Physiological Research Group,
School of Biological Sciences, University College of Swansea

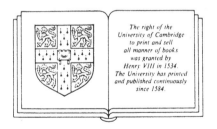

The right of the
University of Cambridge
to print and sell
all manner of books
was granted by
Henry VIII in 1534.
The University has printed
and published continuously
since 1584.

CAMBRIDGE UNIVERSITY PRESS

Cambridge

New York New Rochelle

Melbourne Sydney

CAMBRIDGE UNIVERSITY PRESS
Cambridge, New York, Melbourne, Madrid, Cape Town, Singapore, São Paulo, Delhi

Cambridge University Press
The Edinburgh Building, Cambridge CB2 8RU, UK

Published in the United States of America by Cambridge University Press, New York

www.cambridge.org
Information on this title: www.cambridge.org/9780521103664

First published 1988
This digitally printed version 2009

A catalogue record for this publication is available from the British Library

ISBN 978-0-521-26032-9 hardback
ISBN 978-0-521-10366-4 paperback

Contents

Contributors

BORYSENKO, M. PhD, Associate Professor, Department of Anatomy and Cellular Biology, Tufts University School of Medicine, Boston, MA 02111, USA

DIETERLEN-LIEVRE, F. PhD, Directeur de Recherche au CNRS, Institut d'Embryologie, 49 av de la Belle Gabrielle, 94130 Nogent-Sur-Marne, France

HUNT, T.C. PhD, Biomedical and Physiological Research Group, School of Biological Sciences, University College of Swansea, Singleton Park, Swansea SA2 8PP, UK

MAINWARING, G. PhD, Biomedical and Physiological Research Group, School of Biological Sciences, University College of Swansea, Singleton Park, Swansea SA2 8PP, UK

MILLAR, D.A. M.Sc., Senior Research Assistant, Biomedical and Physiological Research Group, School of Biological Sciences, University College of Swansea, Singleton Park, Swansea SA2 8PP, UK

PAGE, M. PhD, Postdoctoral Research Fellow, Department of Anatomy and Embryology, University College, Gower Street, London, UK

PARMLEY, R.T. MD, Associate Professor/Division Chief, Department of Pediatrics, The University of Texas, Health Center at San Antonio, 7703 Floyd Curl Drive, San Antonio, Texas 78284, USA

RATCLIFFE, N.A. PhD, DSc, Professor and Leader of the Biomedical & Physiology Research Group, School of Biological Sciences, University College of Swansea, Singleton Park, Swansea SA2 8PP, UK

ROWLEY, A.F. PhD, Lecturer in Immunology, Biomedical and Physiological Research Group, School of Biological Sciences, University College of Swansea, Singleton Park, Swansea SA2 8PP, UK

SYPEK, J.P. PhD, Assistant Professor, Division of Geographic Medicine & Infectious Diseases, Tufts University – New England Medical Center Hospitals, Boston, MA 02111, USA

TURNER, R.J. PhD, Lecturer in Immunobiology, Department of Zoology, The University College of Wales, Penglais, Aberystwyth, SY23 3DA, Wales, UK

Preface

Over two decades have elapsed since the publication of *Comparative Haematology* by Warren Andrew (Grune & Stratton, 1965) in which the haematology of all major animal groups was elegantly reviewed. Since then, there have been several volumes dealing with invertebrate haematology such as *Invertebrate Blood Cells* Volumes I & II (N.A. Ratcliffe & A.F. Rowley, 1981), *The Reticuloendothelial System: A Comprehensive Treatise* Volume III, *Phylogeny & Ontogeny* (N. Cohen & M.M. Sigel, Plenum Press, 1982) and *Blood Cells of Marine Invertebrates* (W.D. Cohen, Alan R. Liss, 1985) but no collected review of vertebrate haematology has appeared. Tremendous advances, however, have been made in vertebrate haematology and immunology since *Comparative Haematology* was published in 1965 and the present volume attempts to bring much of the more recent literature together in a comprehensive form.

The book is divided into six chapters. The first, reviews the possible stages in the evolution of the blood cells typical of all vertebrate classes and compares and contrasts some aspects of invertebrate and vertebrate haematology. Subsequent chapters cover the different vertebrate classes in depth. Each chapter, although written by different authors, has a uniform approach covering fundamental topics such as the structure of the circulatory and lymphatic systems, and blood cell ontogeny, structure and functions. The chapters are copiously illustrated with light and electron micrographs of different blood cell types.

The book should find a wide scientific audience including immunologists, clinical and veterinary haematologists, and evolutionary and developmental biologists. The main aim of *Vertebrate Blood Cells* is to stimulate research into comparative haematology in which so much still remains to be discovered.

AFR
NAR

1 Comparative aspects and possible phylogenetic affinities of vertebrate and invertebrate blood cells

N.A. Ratcliffe & D.A. Millar
Biomedical and Physiological Research Group, School of Biological
Sciences, University College of Swansea, Singleton Park, Swansea
SA2 8PP, UK.

Introduction

All vertebrates possess blood cells responsible for
respiratory, immune and haemostatic processes but, of these functions,
only cellular and humoral immunity are mediated by complex interactions
between subpopulations of sessile and freely circulating cells.
Erythrocytes transport oxygen, thrombocytes or platelets initiate
haemostasis while a heterogeneous array of T- and B-lymphocytes,
macrophages, monocytes, basophils, mast cells, eosinophils and
neutrophils are components of the vertebrate immune system. This
system is highly evolved and extremely sophisticated so that most
vertebrates respond to antigenic stimulation utilizing finely
discriminative humoral and cellular processes involving the production
of specific antibodies, lymphokines and memory cells. Even the
agnathans, at the base of the vertebrate phylogenetic tree, synthesize
antibody, exhibit allogeneic recognition, possess lymphocyte
heterogeneity and plasma cells. The question thus arises as to the
origin of vertebrate blood cells responsible for these immune and other
reactions.

In this brief consideration of vertebrate blood cell origins, the
problems of such a study are indicated, then an outline of the possible
steps or factors involved in the phylogeny of the immune system is
given, and, finally, comparisons are made between vertebrate and
invertebrate blood cell types. Most of which is presented owes much to
original concepts published by Burnet (1968, 1971), Marchalonis & Cone
(1973), Manning (1975, 1979, 1980), Cooper (1976a,b, 1977, 1982),
Marchalonis (1977), Hildemann et al. (1981), and by others, detailed
below, to whom the reader is referred.

PROBLEMS IN TRACING THE ORIGINS AND EVOLUTION OF VERTEBRATE BLOOD CELLS

Unfortunately, since the ancestors of vertebrates are now
extinct, there has been considerable controversy concerning their
phylogenetic origins. It is often stated that echinoderm- or tunicate-
like animals gave rise to the vertebrates, (Fig. 1) by way of a filter
feeding larval form (Garstang 1928; Berrill 1955; Romer 1967), although
the adult stage of these animals bears little resemblance to
vertebrates. The idea of a prototunicate larva as the vertebrate
ancestor has, however, more recently been challenged by Jollie (1973)
since it seems likely to him that a filter feeder would lead to a

sessile or inactive life rather than an active protovertebrate with
highly developed sensory, neural and locomotor systems. Thus, attempts
to trace the origin of vertebrate blood cells within the invertebrates
will at best be speculative and based on the assumption that
comparisons with living species are valid since the enigmatic ancestral
forms are no longer available.

Another problem emanating from the uncertainty of vertebrate ancestry
is associated with the identification of truly homologous cells and
structures (Cooper 1976a, 1977, 1982; Warr & Marchalonis 1978). By
definition, homologous blood cells would have the same phylogenetic
origins whereas analogous cells may resemble each other structurally or
functionally but would not share a common ancestor. As we shall detail
below, the coelomocytes of some annelids may resemble vertebrate
lymphocytes in their response to certain mitogens and functions during
graft rejection (see refs in Cooper 1981, 1982) but the cells involved
are probably analogous due to the relatively distant phylogenetic
positions of the annelids and the vertebrates. The annelids, together
with the flatworms, molluscs and arthropods, are protostomes (Fig. 1)
and characterised during development by determinate cleavage and a
blastopore which divides to form both the mouth and anus. In contrast,
vertebrates, together with echinoderms, protochordates and
hemichordates, are deuterostomes (Fig. 1), undergoing indeterminate
cleavage with the mouth forming some distance from the blastopore which
gives rise to the anus (Barnes 1980). The similarities noted in the
blood cells of these two groups may represent an example of convergent
evolution in which cells from different evolutionary stocks evolve to
resemble each other in response to similar evolutionary pressures.

Another example, which clearly illustrates the problem in identifying
homologous structures, concerns the lymphocyte- and erythrocyte-like
cells of echinoderms. Lymphocyte-like cells of echinoderms not only
closely resemble vertebrate lymphocytes ultrastructurally, but they
also respond to mitogens, produce antibody-like molecules, interact
during immunity and may be produced in lymphoid-like organs (see
summary in Ratcliffe et al. 1985 and original refs cited below).
Certain blood cells of echinoderms also closely resemble the nucleated
erythrocytes of lower vertebrates and the early erythroblast stages of
mammals (Fontaine & Lambert 1973). In this case, since both the
echinoderms and vertebrates are deuterostomes, it is much more
difficult to decide whether echinoderm blood cells and associated
lymphoid-like organs (Smith 1978, 1980) are truly homologous or
analogous with those of lower vertebrates. Not surprisingly, there is
disagreement on this point in the literature. Some authors advocate
homology (Smith 1978, 1980), others supporting analogy and convergent
evolution (Fontaine and Lambert 1973), while others are more cautious
and just state that the echinoderm immune system has characteristics
resembling those of vertebrate immunity (Leclerc et al. 1986). An
identical situation is found with the tunicate blood cells, presumably
also as a result of their phylogenetic position, with the
lymphocyte-like cells regarded as homologous by some (e.g. Wright 1976)

while others believe that such a conclusion is highly questionable (e.g. Warr & Marchalonis 1978). No doubt additional biochemical and molecular analyses will help to clarify matters.

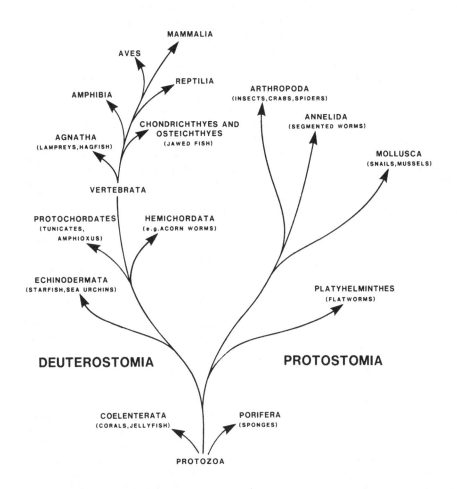

Fig. 1   Phylogenetic scheme for the main groups in the animal kingdom.

A final problem which, as pointed out by Cooper (1976a), continues to frustrate attempts to trace the origins of blood cells concerns the confusing terminology still adopted by many invertebrate haematologists. There is the continuing insistence of using vertebrate terms such as lymphocyte, granulocyte and macrophage to describe invertebrate leucocytes when in actual fact there is little evidence for true homology. Therefore, any vertebrate immunologist hoping to work on the origin and evolution of blood cells will undoubtedly be totally confused by the use of familiar names, such as lymphocyte, to describe cells that are little more than progenitor or stem cells and which occur throughout many of the invertebrate phyla. The situation is further aggravated by the sheer diversity of the invertebrates, together comprising over 95% of all known species in the animal kingdom, and which has resulted in a multitude of names for their blood cells. For example, the invertebrate phagocyte has been variously called an amoebocyte, granulocyte, macrophage, monocyte, plasmatocyte and granular cell (Ratcliffe & Rowley 1979). More recent publications have hopefully clarified invertebrate haematology (e.g. Ratcliffe & Rowley 1981) while, until such time as true homology can be proven, a functional rather than a morphological classification scheme might prove to be most useful (Ratcliffe et al. 1985).

POSSIBLE STAGES AND FACTORS INVOLVED IN THE EVOLUTION OF BLOOD CELLS AND THE IMMUNE SYSTEM

Since the majority of blood cell types are involved in the immune system then consideration of the main steps in the evolution of this system may provide clues as to the origin of the various cell types (Table I).

The first blood cells

All invertebrates are capable of recognising foreignness and effecting phagocytosis. Even the protozoans possess cell surface determinants, exhibit transplantation incompatibilities (so that they can distinguish self from non-self)(Hildemann et al. 1981), and contain lysosomes within their cytoplasm. They thus appear to be independent, wandering, phagocytes, well-suited as potential immunocytes. With the evolution of the Metazoa, possibly from some form of colonial protozoan (Haeckel 1874), organised cell layers first appeared incorporating wandering phagocytic cells, termed archeocytes in sponges. Most, if not all, subsequent animal phyla contain wandering phagocytes which may have arisen from those of the lowly invertebrates by adaptive radiation (Cooper 1976a). Once the third body layer or mesoderm evolved, a cavity (the coelom) developed between the body wall and the gut so that a circulatory system was required to transport trophic and waste substances around the body. At this point in evolution, the blood cells probably migrated from the surrounding connective tissues into the haemolymph so that phagocytes were freed from their food-scavenging role and evolved into an array of leucocyte types with more complex immune reactivity (Cooper 1976b). This stage in evolution can be seen in certain Platyhelminthes where, in some advanced forms, the phagocytes move out of the parenchymatous connective tissue into

TABLE 1   Evolutionary Steps of Possible Significance in the
           Phylogeny of Blood Cells and the Immune System [a]

EVOLUTIONARY STEP/PRESSURE	IMMUNOLOGICAL IMPLICATIONS
1. Protozoa	Recognition and Discrimination
2. Metazoa Including Colonial Forms	Histocompatibility System, Allogeneic Recognition and Short Term Memory
3. Mesoderm and Circulatory System. Nutrition and Defence as Separate Functions	Freely Circulating and More Diverse Blood Cell Types. Cellular Immunity and Erythrocytes
4. Cancer Associated with Increasing Complexity and Longevity	Immunosurveillance
5. Ancestral Protovertebrate	Increased Recognition and Discriminatory Powers?
6. Lower Vertebrates with Increased Body Size, Longer Life Span and Reduced Reproductive Potential	True Lymphocytes, Lymphoid Tissue and Antibody Production (IgM), Longer Term Memory
7. Emergence on to Land, Exposure to Irradiation and Development of High Pressure Blood Vascular Systems	Bone Marrow, Additional Antibody Classes, T- and B- Lymphocytes, Lymphoid Organs with Increased Complexity, GALT
8. Amniotes (reptiles, birds, mammals) with Loss of Free-living Larval Form	Advanced Differentiation of Immuno-competent Cells Allowing Increased Diversity and Efficiency of Immune System
9. Homoiothermy with More Favorable Environment for Pathogens	Increased Efficiency of Immune System, Integrated Cellular and Humoral Responses, Germinal Centres in Secondary Lymphoid Organs, Lymph Nodes
10. Viviparity with Maternal-Foetal Interactions	Additional Fine Tuning of Immune System

[a] Information from Burnet (1968), Marchalonis and Cone (1973), Marchalonis (1977)
Manning (1975, 1979, 1980), Manning and Turner (1976), Cooper (1976a, 1982)
Cooper et al. (1980), Hildemann et al. (1981), Warr (1981).

the haemolymph (Stang-Voss 1974).

The MHC system
Another significant development in the evolution of the
immune system also appears in primitive and other invertebrates. Thus
transplantation experiments with sponges, coelenterates, annelids,
echinoderms and tunicates have shown the presence of allogeneic
recognition with a short-term memory component in many of these animals
(reviewed in Ratcliffe et al. 1985). It is surprising that even the
lowly sponges exhibit alloimmune rejection with specific short-term
memory (Hildemann et al. 1979) and that the actual cytotoxic reactivity
during allograft rejection involves cytotoxic or "killer" cells
crossing over via tissue bridges to destroy the allogeneic cells
(Bigger et al. 1981) These results have been interpreted to indicate
the presence, in these invertebrates, of an ancestral MHC system with
considerable polymorphism at the locus or loci controlling graft
acceptance and rejection. Hildemann (1977) therefore proposed that
the histocompatibility (H) system underlying cell-mediated immunity is
ancestral to and also separate from the immunoglobulin (Ig) system
found in vertebrates. At the level of the vertebrates, the Ig system
was possibly added to provide a more finely tuned recognition ability
in the physiologically complex vertebrates. Such additional regulation
would be furnished by the Ig system of circulating antibodies and cell
surface receptors (Hildemann et al. 1981). The acquisition of both Ig
and H systems has important implications as far as the evolution of
blood cells is concerned. In the 'higher' vertebrates, as these two
systems become more closely integrated then immune reactivity was more
exquisitely controlled and the interactive T- and B-lymphocytes
evolved. Indeed, the evolution of the MHC in the vertebrates was an
essential pre-requisite to the development of the various T-cell
subpopulations with surface receptors for class I (cytotoxic T-cells)
or class II (helper T-cells) MHC-encoded proteins.

Further evidence for the emergence of lymphocytes with T-cell
properties before those with B-cell properties has been provided by
studies on naturally-occurring, dinitrophenyl hapten-binding proteins
in the tunicate, Pyura stolonifera. Marchalonis & Warr (1978) showed
that these molecules consist of a single subunit with a molecular
weight of 65,000- 70,000 daltons and upon electrophoresis resemble the
heavy (μ) chains of IgM which is the most primitive immunoglobulin.
Further analysis showed that these subunits are of similar mass to the
T-cell receptor so that specific cellular immunity may well have
evolved before the humoral system.

Various reasons have been postulated for the presence of alloreactivity
at such an early stage during evolution in animals as primitive as
sponges and corals. Since allogeneic recognition is apparently absent
in both molluscs and insects, alloreactivity seems to have been
selected for in colonial invertebrates in which transplant rejections
may be essential for survival (Lackie 1980; Scofield et al, 1982). The
lack of allogeneic and xenogeneic incompatibility in such colonial

forms would soon result in colony-colony overgrowth and fusion with
resultant loss of individual integrity. In keeping with this
hypothesis, Buss and Green (1985) have suggested that any fusion of
allogeneic colonies would cause loss of fitness of one of the partners
in the chimaera unless each partner has precisely the same commitment
in gamete production. In other words, one partner would be
effectively parasitized by the other component of the chimaera.

Cancer
Another force favouring development of the cellular elements
of immunity may have been somatic mutation. There is some dispute as
to whether neoplasms identical to those of mammals occur in
invertebrates although uncontrolled cancer-like growths have been
reported in insects and molluscs (Farley 1969; Gateff 1978). It may be
significant that of all the invertebrates, these two groups apparently
lack allogeneic recognition and therefore the ancestral H system may be
poorly developed. With the increase in complexity and longevity of
animals during evolution, somatic mutations would constantly occur and,
unless a competent immunosurveillance system is present, cancer and
premature death will result. As pointed out by Burnet (1968), this
would be disastrous if it occurred in the prereproductive period which
would be possible if cancer was contagious and could be passed on from
old to young individuals. There must, therefore, have been great
pressure to develop a surveillance system based upon the recognition of
foreign histocompatibility antigens on the surface of a mutated cell by
circulating effector cells. The implications of this are obvious as
the recognition system involved would have to be extremely finely
tuned, to distinguish abnormal from normal body cells, and mediated by
specific cell surface receptors. There would thus be considerable
evolutionary forces favouring retention and development of the MHC
system and of the cellular rather than the humoral components of
immunity. This again would favour the diversification of blood cell
types, for example, the cytotoxic T-cells involved in
immunosurveillance.

Evolution of Vertebrates
Two facts are immediately evident from a comparison of blood
cells and immune systems from different vertebrate groups. First,
since cyclostomes (hagfish and lampreys) exhibit lymphocyte
heterogeneity, possess immunoglobulin-bearing lymphocytes and plasma
cells, and produce specific antibodies (Hildemann & Thoenes 1969;
Kilarski & Plytycz 1981; Raison & Hildemann 1984; Zapata et al. 1984),
then the blood cells and immune system of the most primitive
vertebrates are considerably more advanced than those of the most
highly evolved invertebrates (see below, however, 'lymphocyte-like
cells', for the more recent work on echinoderms). Second, comparison
between the cyclostomes and mammals shows just how much the blood cells
and associated immune processes have evolved within the vertebrates.
Thus, mammals exhibit accelerated allograft rejection and specific
memory, interactive T- and B-lymphocytes including T-helpers,
T-suppressors and cytotoxic T-cells, as well as killer and natural

killer cells, diversity of antibody classes and well defined primary
and secondary lymphoid organs.

The main events and forces probably responsible for evolutionary
changes in the blood cells and immune potential of vertebrates have
been discussed previously (e.g. Manning & Turner 1976; Cooper 1976a;
Cohen 1977; Marchalonis 1977). The spleen and thymus first emerge as
discrete organs during evolution from cyclostomes to primitive fish,
while accelerated allograft rejection first appears in the teleosts,
indicating the presence of a modern-type MHC system. Despite these
advances, there is still doubt about the presence in agnathous fish of
typical mammalian type B- and T-cells although lymphocyte heterogeneity
seems to occur in gnathostomate fish (e.g. Miller et al. 1985; see also
Chapter 2).

Evolution of the lymphocyte into a number of well-defined cell types
seemed to await the emergence of vertebrates from water to land. This
single evolutionary step heralded major changes in the blood cells and
immune systems of vertebrates. Among the anurans (frogs and toads) we
find, mainly for the first time, bone marrow, lymph nodes,
gut-associated lymphoid tissues, antibody diversity with both IgM and
IgG, and, although debatable, true B- and T-cell dichotomy (eg
Borysenko 1976; Du Pasquier 1976; Cooper et al. 1980; Manning & Horton
1982). The appearance of bone marrow is presumably related to the
hollow bone structure for locomotion on land and is an ideal site for
blood cells which would be protected from harmful ionizing radiation by
the hydroxyapatite of bone (Cooper et al. 1980). The reason for the
diversity in antibody classes at the amphibian level may be related to
the progressive evolution of high-pressure blood vascular systems in
terrestrial animals. Changes in the rate of circulation and
hydrostatic and osmotic pressures of the blood may make it difficult
for the movement of high molecular weight antibody (IgM) into
extravascular compartments to reach infectious agents, so that a lower
molecular weight antibody (IgG) evolved (Manning & Turner 1976). In
'higher' vertebrates, such as mammals, although as in most vertebrates,
IgM acts as the antigen receptor on B-lymphocytes, it is IgG which is
capable of binding to monocytes and neutrophils via specific Fc
receptors to enhance both the specificity and activities of these cells
during phagocytosis and cytotoxicity (Atassi et al. 1984). Thus, many
of the activities of the blood cells and of the immune systems of
higher vertebrates are controlled at least partially by the
phylogenetically recently acquired IgG and IgE antibody classes. The
acquisition of a variety of new antibody classes may well have
encouraged the diversification of the lymphoid cells of higher
vertebrates.

Other important evolutionary steps influencing the increase in
efficiency and sophistication of vertebrate immunity and blood cells,
include the loss of the free-living larva with protection of the young,
and development of homiothermy and viviparity. In amniotes, with the
loss of free-living larvae, there is no longer a pressing need to

provide an effective immune system during early stages of development and this may allow a build-up of immunocompetent cell populations before functional commitments are necessary (Manning 1975). The development of homiothermy in birds may also have provided an evolutionary incentive for promptness and diversification in the immune system in order to protect the young from invading microorganisms which would multiply rapidly in the warm and nutritionally rich environment (Manning & Turner 1976). Finally, the development of viviparity in most mammals, during which maternal/foetal interactions occur over long periods, would also have necessitated refinements in the control and feedback mechanisms of the immune system in order to avoid potential rejection processes. The protected environment would also have allowed long periods for the production of immunocyte populations and may have aided in the acquisition of functional heterogeneity of the cells (Manning & Turner 1976).

A COMPARISON OF INVERTEBRATE WITH VERTEBRATE BLOOD CELLS

In this final section, a basic outline of invertebrate blood cell types is given and a brief comparison with vertebrate blood cells is made. The latter are described in detail in the remaining chapters of this book.

Macrophage- and granulocyte-like cells

Since all invertebrates have macrophage- and granulocyte-like cells then phagocytic cells have been conserved throughout phylogeny (Ratcliffe & Rowley 1981; Cooper 1982). In invertebrates, these cells are responsible for phagocytosis and entrap invading metazoan parasites within multicellular capsules too (Ratcliffe et al. 1985). Descriptions of encapsulation by granulocytes of metazoan parasites injected into the peritoneal cavity of mammalian species are almost identical to those reported for invertebrate species. For example, when newly encysted juvenile Fasciola hepatica are injected intraperitoneally into sensitized rats they rapidly become encapsulated by interactive populations of mast cells, eosinophils and neutrophils (Davies & Goose 1981). This process appears, super- ficially, almost identical to the granular cell/plasmatocyte co-operation occurring during the encapsulation process in insects (Schmit & Ratcliffe 1977).

Anderson (1981) undertook an interesting comparison of invertebrate and vertebrate leucocytes in which he pointed out that invertebrate leucocytes contain many different enzymes, including acid and alkaline phosphatase, β-glucosaminidase, indoxyl esterase and non-specific esterase. Some of these occur in discrete granules which must therefore be lysosomal and capable of killing and degrading ingested materials such as bacteria. In addition, enzymes are released from these leucocytes and must serve as mediators of protective processes. He concluded, however, that based on lysosomal hydrolases it would be difficult to decide whether invertebrate phagocytes were more like mammalian granulocytes or macrophages although the absence of myeloperoxidase in the invertebrate cells did suggest a closer affinity to the macrophage cell types. Subsequent work by Sminia et al. (1982), Nakamura et al (1985) and others does indicate, however, that some invertebrate phagocytes do contain the myeloperoxidase system. Anderson (1981) has shown too that, in contrast to vertebrate phagocytes, no Fc or C3 receptors can be detected on invertebrate blood cells, and the metabolic events accompanying phagocytosis by invertebrate leucocytes are fundamentally different from those characteristic of granulocytes and macrophages. Thus, unlike vertebrate phagocytes, the direct oxidation of glucose is not stimulated by phagocytosing invertebrate blood cells. As far as we are aware, this latter conclusion has not been subsequently challenged while more recently Bertheussen & Seljelid (1982) have detected C3 receptors on coelomocyte surfaces in an echinoderm.

Cells resembling mast cells

Cells resembling vertebrate mast cells have been reported in annelids, insects, echinoderms and tunicates in which they are mainly of unknown function (Ratcliffe & Rowley 1979). These leucocytes are termed spherule cells, eleocytes, chloragogen cells, mucocytes, trephocytes, morula cells and vibratile cells, depending upon the animal group in which they occur. Typically, like vertebrate mast cells, the cytoplasm is filled with large inclusions often with a crystalline or microtubular substructure, which appears to contain various mucopolysaccharides. A recent paper by Smith & Smith (1985) emphasises just how similar invertebrate spherule cells are to vertebrate mast cells. Not only do the red spherule cells of the echinoderm, _Mellita quinquiesperforata_, closely resemble vertebrate mast cells morphologically, but it is significant that spherule cells sensitized with human serum discharge their granules. Furthermore, this degranulation occurred when animals were stressed and, even more compelling, the granules contain significant levels of histamine. The entire spherule cell response resembles that of vertebrate mast cells, which following IgE binding and subsequent degranulation develop a Type I hypersensitivity reaction.

Lymphocyte-like cells

Cells closely resembling vertebrate lymphocytes in morphology have been observed in many invertebrate groups (Ratcliffe et al. 1985). Most attention has been focussed on annelids, echinoderms and tunicates. Research using annelids was stimulated by the discovery of allograft and xenograft rejection in earthworms which exhibited both specificity and short-term memory (Valembois 1963; Duprat 1964; Cooper 1969). These processes together with adoptive transfer, graft infiltration by lymphocyte-like cells, blastogenic response towards transplantation antigens and T-cell mitogens such as PHA and Con A, and the presence of Con A membrane receptors (see refs in Cooper 1981, 1982), are similar to reactions in vertebrates. Blood cells involved in the two groups are, as mentioned previously, probably analogous due to the relative phylogenetic positions of annelids and vertebrates: any similarities may represent an example of convergent evolution. There has also been much interest in the lymphocyte-like cells of echinoderms and protochordates due to their presumed importance as possible ancestors of the vertebrates (Garstang 1928; Berrill 1955; Romer 1967). Wright (1976) postulated that a lymphocyte first makes its appearance as a distinct cell type in tunicates. The haematogenic tissue or "lymph nodules" in the pharynx and gut of tunicates and their renewing (ie rapid turnover rate) cell populations are regarded by Wright (1976) as evidence of homology for "lymphocytes" of these two groups. This conclusion has been questioned by Warr & Marchalonis (1978) and Warr (1981) due to a lack of biochemical and molecular evidence and of *in vitro* blastogenic responses of lymphocyte-like cells in the tunicate, *Pyura stolonifera*, previously exposed to a range of mitogens and allogeneic cells (Warr *et al*. 1977). Wright & Ermak (1982) also believe that tunicate "lymphocytes" can be distinguished from stem cells by their lack of a prominent nucleolus. Rowley *et al* (1984), however, have been unable to distinguish between "stem cells" and "lymphocytes" in the tunicates, *Ascidia mentula, Ciona intestinalis* and *Styela clava*, and have also failed to detect freely circulating lymphocyte-like cells in the hemichordate, *Saccoglossus horsti*, or in the cephalochordate, *Branchiostoma lanceolatum*.

The most compelling evidence for homology between the blood cells of an invertebrate and those of vertebrates is provided by recent research into the immunobiology of echinoderms. Not only can axial organ (AO) cells of the starfish, *Asterias rubens*, secrete an antibody-like substance (antibody-like functionally as detailed biochemical analyses have yet to be completed, although some progress in this direction has been made [see Brillouet & Leclerc 1986]), but exposure to pokeweed mitogen also stimulates non-adherent AO cells to release a lymphokine-like substance (Leclerc & Brillouet 1981; Leclerc *et al*. 1981). A similar factor, termed "sea star factor", has also been reported to have properties comparable with T-cell lymphokines (Prendergast *et al*. 1983). The AO cells of *A. rubens* can also be fractionated by a method used to separate mammalian B- and T-cells and the subpopulations obtained behave like these cells towards mitogens such as Con A and lipopolysaccharide (Brillouet *et al*. 1981). More recently, Leclerc *et al*. (1986) have shown that the production of the antibody-like factor by *A. rubens* not only requires the interaction of adherent and non-adherent cells which resemble B- and T-cells ultrastructurally (Anteunis *et al*. 1985), but also the participation of phagocytic cells. Finally, from experiments involving the injection of AO cells into vertebrate hosts, there is additional evidence for T-like cells and an ancestral lymphoid organ in *A. rubens* (Leclerc *et al*. 1977a,b). The inoculation of irradiated mice with AO cells induces angiogenesis, similar to that occurring after injecting allogeneic T-cells, while the injection of AO cells into chick embryos provokes splenomegaly. Neither starfish digestive organ cells, ovocytes, nor coelomocytes induce this angiogenesis. Separation experiments showed that the AO cells involved are mainly the smallest, non-adherent, lymphocyte-like cells, although co-operation with phagocytic cells may be necessary (Leclerc *et al*. 1977b). Overall, these experiments indicate, together with those of Smith (1978, 1980) and Smith & Smith

(1985) which have shown in other echinoderm species that vertebrate-like lymphoid organs and allergic reactions may occur, that the immune system of echinoderms contains cellular and humoral elements which may be homologous to components of vertebrate immunity. Confirmation awaits detailed biochemical and molecular analyses.

Erythrocyte- and thrombocyte-like cells

Cells closely resembling vertebrate erythrocytes are present in some annelids, sipunculids, molluscs, echinoderms and in several minor phyla (eg Ochi 1969; Fänge 1974; Fontaine & Lambert 1973; Valembois & Boiledieu 1980), while a few insect species also have thrombocyte-like cells (Zachary & Hoffmann 1973). Morphologically, invertebrate erythrocyte-like cells closely resemble their vertebrate counterparts. In particular, they may be similar to the nucleated erythrocytes of lower vertebrates and to early erythroblast stages of mammals (Fontaine & Lambert 1973). Points of particular similarity are their biconvex shape, uniform distribution of respiratory pigment in the cytoplasm and nucleus, the often amorphous cytoplasm with few organelles and the marginal bundles of microtubules. The respiratory pigment varies from one group to another in invertebrates and includes haemoglobin and haemerythrin. Since invertebrate "erythrocytes" often retain leucocyte features and are found in groups unrelated to vertebrate ancestors, it is most likely that these cells are analogous rather than homologous to vertebrate erythrocytes. They may have arisen by convergent evolution due to similar evolutionary/ environmental pressures. This also probably applies to thrombocyte-like cells found in dipteran insects which like their vertebrate counterparts, not only induce blood clotting but may also be phagocytic (Rowley & Ratcliffe 1976; Ferguson 1976).

CONCLUDING REMARKS

In conclusion, despite problems in comparing vertebrate with invertebrate blood cells and tracing true homologies, progress has been accelerated recently by examining leucocytes from groups assumed to be related to the vertebrate ancestor. The echinoderms, in particular, have provided compelling evidence for homology between their axial organ cells and vertebrate lymphocytes. The similarities observed, however, are based purely on functional affinities and urgently need to be confirmed from evidence provided by modern biochemical and molecular techniques. Such methods have recently been successfully applied to detecting and characterising Thy-1 and β_2-microglobulin homologues on the blood cells and in the serum of a range of invertebrate species (Shalev et al. 1981, 1983; Roch & Cooper 1983; Mansour & Cooper 1984; Mansour et al. 1985). These molecules may well form the molecular link between invertebrates and vertebrates and have arisen from a common ancestral gene which served as the progenitor for β_2-microglobulin, Thy-1, immunoglobulin domains and class I and II MHC-encoded proteins (Mansour et al. 1985), all of which, as we shall see in the remaining chapters of this book, form surface markers or receptors at the surface of vertebrate blood cells.

ACKNOWLEDGEMENTS
     Our thanks go to Professor E.L. Cooper for critically reviewing
this manuscript, and to the Science and Engineering Research Council
(Grants GR/C/30122, GR/D/21684, GR/D/54644) and the Stimulation Programme
of the European Economic Community (Grant ST2J-0139-1-UK) for financial
support.

REFERENCES

Anderson, R.S. (1981). Comparative aspects of the structure and function
          of invertebrate and vertebrate leucocytes. In Invertebrate Blood
          Cells, Vol 2, eds. N.A. Ratcliffe & A.F. Rowley, pp. 629-41.
          London: Academic Press.
Anteunis, A., Leclerc, M., Vial, M., Brillouet, C., Luquet, G., Robineaux,
          R. & Binaghi, R.A. (1985). Immunocompetent cells in the starfish
          Asterias rubens. An ultrastructural study. Cell Biol. Int.
          Rep., 9, 663-70.
Atassi, M.Z., van Oss, C.J. & Absolom, D.R. (eds.)(1984). Molecular
          Immunology. New York: Marcel Dekker.
Barnes, R.D. (1980). Invertebrate Zoology (4th edition). Philadelphia:
          W.B. Saunders Co.
Berrill, N.J. (1955). The Origin of Vertebrates. London and New York:
          Oxford University Press.
Bertheussen, K. & Seljelid, R. (1982). Receptors for complement on
          echinoid phagocytes. I. The opsonic effect of vertebrate sera on
          echinoid phagocytosis. Dev. Comp. Immunol., 6, 423-31.
Bigger, C.H., Hildemann, W.H., Jokiel, P.L. & Johnston, I.S. (1981).
          Afferent sensitization and efferent cytotoxicity in allogeneic
          tissue responses of the marine sponge Callyspongia diffusa.
          Transplantation, 31, 461-64.
Borysenko, M. (1976). Phylogeny of immunity: An overview.
          Immunogenetics, 3, 305-26.
Brillouet, C. & Leclerc, M. (1986). Production of an antibody-like factor
          in the sea star Asterias rubens: Role of the gastric haemal
          tufts. Cell Biol. Int. Rep., 10, 307-13.
Brillouet, C., Leclerc, M., Panijel, J. & Binaghi, R.A. (1981). In vitro
          effect of various mitogens on starfish (Asterias rubens) axial
          organ cells. Cell. Immunol. 57, 136-44.
Burnet, F. M. (1968). Evolution of the immune process in vertebrates.
          Nature, 218, 426-30.
Burnet, F.M. (1971). "Self-recognition" in colonial marine forms and
          flowering plants in relation to the evolution of immunity.
          Nature, 232, 230-35.
Buss, L.W. & Green, D.R. (1985). Histocompatibility in vertebrates: The
          relict hypothesis. Dev. Comp. Immunol., 9, 191-201.
Cohen, N. (1977). Phylogenetic emergence of lymphoid tissue and cells.
          In The Lymphocyte: Structure and Function, ed. J.J. Marchalonis,
          pp. 149-202. New York: Marcel Dekker.
Cooper, E.L. (1969). Specific tissue graft rejection in earthworms.
          Science, 166, 1414-15.

Cooper, E.L. (1976a). Evolution of blood cells. Ann. Immunol. (Inst. Pasteur), 127C, 817-25.

Cooper, E.L. (1976b). Cellular recognition of allografts and xenografts in invertebrates. In Comparative Immunology, ed. J.J. Marchalonis, pp. 36-79. Oxford: Blackwell.

Cooper, E.L. (1977). Evolution of cell mediated immunity. In Developmental Immunobiology, eds. J.B. Solomon & J.D. Horton, pp. 99-105. Amsterdam: Elsevier/North-Holland Biomedical Press.

Cooper, E.L. (1981). Immunity in invertebrates. CRC Crit. Rev. Immunol., 2, 1-32.

Cooper, E.L. (1982). Invertebrate defense systems. An overview. In The Reticuloendothelial System. A Comprehensive Treatise. Vol. 3, eds. N. Cohen & M.M. Sigel, pp. 1-35. New York: Plenum Press.

Cooper, E.L., Klempau, A.E., Ramirez, J.A. & Zapata, A.G. (1980). Source of stem cells in evolution. In Development and Differentiation of Vertebrate Lymphocytes ed. J.D. Horton, pp. 3-14. Amsterdam: Elsevier/North Holland Biomedical Press.

Davies, C. & Goose, J. (1981). Killing of newly excysted juveniles of Fasciola hepatica in sensitized rats. Parasite Immunology, 3, 81-96.

DuPasquier, L. (1976). Phylogenesis of the vertebrate immune system. In The Immune System, eds. F. Melchers & K. Rajewsky, pp. 101-15. Berlin: Springer-Verlag.

Duprat, P.(1964). Mise en evidence de reactions immunitaires dans les homogreffes de paroi du corps chex le Lombricien Eisenia fetida typica. C.R. Acad. Sci., 259, 4177-79.

Fänge, R. (1974). Coelomocytes and blood cells of vermiform invertebrates. In Recherches Biologiques Contemporaires, ed. L. Arvy, pp. 13-24. France: Vagner.

Farley, C.A. (1969). Probable neoplastic disease of the hematopoietic system in oysters, Crassostrea virginica and Crassostrea gigas. Nat. Cancer Inst. Monogr., No. 31, 541-55.

Ferguson, H.W. (1976). The ultrastructure of plaice (Pleuronectes platessa) leucocytes. J. Fish Biol., 8, 139-42.

Fontaine, A.R. & Lambert, P. (1973). The fine structure of the haemocyte of the holothurian, Cucumaria miniata (Brandt). Can J. Zool. 51, 323-32.

Garstang, W. (1928). The morphology of the Tunicata, and its bearings on the phylogeny of the Chordata. Quart. J. Micr. Sci., 72, 51-187.

Gateff, E. (1978). Malignant and benign neoplasms of Drosophila melanogaster. In The Genetics and Biology of Drosophila Vol 2b, eds. M. Ashburner & T.R.F. Wright, pp. 181-275. London: Academic Press.

Haeckel, E. (1874). The gastraea-theory, the phylogenetic classification of the Animal Kingdom and the homology of the germ-lamellae. Q.J. Micr. Sci., 14, 142-65; 223-47.

Hildemann, W.H. (1977). Specific immunorecognition by histocompatibility markers: The original polymorphic system of immunoreactivity characteristic of all multicellular animals. Immunogenetics 5, 193-202.

Hildemann, W.H., Clark, E.A. & Raison, R.L. (1981). Phylogeny of
    immunocompetence. In Comprehensive Immunogenetics, pp. 302-46.
    Oxford: Blackwell.
Hildemann, W.H., Johnson, I.S. & Jokiel, P.L. (1979). Immunocompetence in
    the lowest metazoan phylum: Transplantation immunity in sponges.
    Science, 204, 420-22.
Hildemann, W.H. & Thoenes, G.H. (1969). Immunological responses of
    Pacific hagfish. I. Skin transplantation immunity.
    Transplantation, 7, 506-21.
Jollie, M. (1973). The origin of the chordates. Acta Zoologica, 54,
    81-100.
Kilarski, W. & Plytycz, B. (1981). The presence of plasma cells in the
    lamprey (Agnatha). Dev. Comp. Immunol., 5, 361-66.
Lackie, A.M. (1980). Invertebrate immunity. Parasitology, 80, 393-412.
Leclerc, M. & Brillouet, C. (1981). Evidence of antibody-like substances
    secreted by axial organ cells of the starfish Asterias rubens.
    Immunology Letters, 2, 279-81.
Leclerc, M., Brillouet, C., Luquet, G., Agogue, P. & Binaghi, R.A.
    (1981). Properties of subpopulations of starfish axial organ: in
    vitro effect to pokeweed mitogen and evidence of lymphokine-like
    substances. Scand. J. Immunol., 14, 281-84.
Leclerc, M., Brillouet, C., Luquet, G. & Binaghi, R.A. (1986). Production
    of an antibody-like factor in the sea star Asterias rubens:
    involvement of at least three cellular populations. Immunology,
    57, 479-82.
Leclerc, M., Redziniak, G., Panijel, J. & El Lababidi, M. (1977a).
    Reactions induced in vertebrates by invertebrate cell suspensions
    I. Specific effects of sea star axial organ cells injection.
    Dev. Comp. Immunol., 1, 299-310.
Leclerc, M., Redziniak, G., Panijel, J. & El Lababidi, M. (1977b).
    Reactions induced in vertebrates by invertebrate cell suspensions
    II. Non adherent axial organ cells as effector cells. Dev. Comp.
    Immunol., 1, 311-20.
Manning, M.J. (1975). The phylogeny of thymic dependence. Amer. Zool.,
    15, 63-71.
Manning, M.J. (1979). Evolution of the vertebrate immune system. J. Roy.
    Soc. Med., 72, 683-88.
Manning, M.J. (1980). The evolution of vertebrate lymphoid organs. In
    Aspects of Developmental and Comparative Immunology, ed. J.B.
    Solomon, pp. 67-72. Oxford: Pergamon Press.
Manning, M.J. & Horton, J.D. (1982). RES Structure and function of the
    Amphibia. In The Reticuloendothelial System. A Comprehensive
    Treatise. Vol. 3, eds. N. Cohen & M.M. Sigel, pp. 423-59. New
    York: Plenum Press.
Manning, M.J. & Turner, R.J. (1976). Comparative Immunobiology.
    Glasgow: Blackie.
Mansour, M.H. & Cooper, E.L. (1984). Serological and partial molecular
    characterization of a Thy-1 homolog in tunicates. Eur. J.
    Immunol., 14, 1031-39.

Mansour, M.H. , De Lange, R. & Cooper, E.L. (1985). Isolation, purification, and amino acid composition of the tunicate hemocyte Thy-1 homolog. J. Biol. Chem., 260, 2681-86.

Marchalonis, J.J. (1977). Immunity in Evolution. London: Arnold.

Marchalonis, J.J. & Cone, R.E. (1973). The phylogenetic emergence of vertebrate immunity. Aust. J. Exp. Biol. Med. Sci., 51, 461-88.

Marchalonis, J.J. & Warr, G.W. (1978). Phylogenetic origins of immune recognition: Naturally occurring DNP-binding molecules in chordate sera and hemolymph. Dev. Comp. Immunol., 2, 443-60.

Miller, N.W., Sizemore, R.C. & Clem, L.W. (1985). Phylogeny of lymphocyte heterogeneity: the cellular requirements for in vitro antibody responses of channel catfish leukocytes. J. Immunol., 134, 2884-8.

Nakamura, M., Mori, K., Inooka, S. & Nomura, T. (1985). In vitro production of hydrogen peroxide by the amoebocytes of the scallop, Patinopecten yessoensis (Jay). Dev. Comp. Immunol., 9, 407-17.

Ochi, O. (1969). Blood pigments and erythrocytes found in some marine Annelida. Mem. Ehime Univ. Sci., Ser. B. (Biol.), 6, 63-131.

Prendergast, R.A., Lutty, G.A. & Scott, A.L. (1983). Directed inflammation: the phylogeny of lymphokines. Dev. Comp. Immunol., 7, 629-32.

Raison, R.L. & Hildemann, W.H. (1984). Immunoglobulin bearing blood leucocytes in the pacific hagfish. Dev. Comp. Immunol., 8, 99-108.

Ratcliffe, N.A. & Rowley, A.F. (1979). A comparative synopsis of the structure and function of the blood cells of insects and other invertebrates. Dev. Comp. Immunol., 3, 189-243.

Ratcliffe, N.A. & Rowley, A.F., eds. (1981). Invertebrate Blood Cells. Vols. 1 and 2. London: Academic Press.

Ratcliffe, N.A., Rowley, A.F., Fitzgerald, S.W. & Rhodes, C.P. (1985). Invertebrate immunity: basic concepts and recent advances. Int. Rev. Cytol., 97, 183-350.

Roch, P.G. & Cooper, E.L. (1983). A β_2-microglobulin-like molecule on earthworm (L. terrestris) leukocyte membranes. Dev. Comp. Immunol., 7, 633-36.

Romer, A.S. (1967). Major steps in vertebrate evolution. Science, 158, 1629-37.

Rowley, A.F. & Ratcliffe, N.A. (1976). An ultrastructural study of the in vitro phagocytosis of Escherichia coli by the hemocytes of Calliphora eythrocephala. J. Ultrastruct. Res., 55, 193-202.

Rowley, A.F., Rhodes, C.P. & Ratcliffe, N.A. (1984). Protochordate leucocytes: a review. Zool. J. Linn. Soc., 80, 283-95.

Schmit, A.R. & Ratcliffe, N.A. (1977). The encapsulation of foreign tissue implants in Galleria mellonella larvae. J. Insect Physiol., 23, 175-84.

Scofield, V.L., Schlumberger, J.M., West, L.A. & Weissman, I.L. (1982). Protochordate allorecognition is controlled by a MHC-like gene system. Nature, 295, 499-502.

Shalev, A., Greenberg, A.H., Logdberg, L. & Bjorck, L. (1981). β_2-microglobulin-like molecules in low vertebrates and invertebrates. J.

Immunol., <u>127</u>, 1186-91.

Shalev, A., Pla, M., Ginsburger-Vogel, T., Echalier, G., Logdberg, L., Bjorck, L., Colombani, J. & Segal, S. (1983). Evidence for β₂microglobulin-like and H-2-like antigenic determinants in <u>Drosophila</u>. J. Immunol., <u>130</u>, 297-302.

Sminia, T., van der Knaap, W.P.W. & Boerrigter-Barendsen, L.H. (1982). Peroxidase-positive blood cells in snails. J. Reticuloendothel. Soc., <u>31</u>, 399-404.

Smith, A.C. (1978). A proposed phylogenetic relationship between sea cucumber polian vesicles and the vertebrate lymphoreticular system. J. Invertebr. Pathol., <u>31</u>, 353-57.

Smith, A.C. (1980). Immunopathology in an invertebrate, the sea cucumber, <u>Holothuria cinerascens</u>. Dev. Comp. Immunol., <u>4</u>, 417-32.

Smith S.L. & Smith, A.C. (1985). Sensitization and histamine release by cells of the sand dollar, <u>Mellita quinquiesperforata</u>. Dev. Comp. Immunol., <u>9</u>, 597-603.

Stang-Voss, C. (1974). On the ultrastructure of invertebrate hemocytes: An interpretation of their role in comparative hematology. <u>In</u> Contemporary Topics in Immunobiology, Vol. IV, ed. E.L. Cooper, pp. 65-76. London: Plenum Press.

Valembois, P. (1963). Recherches sur la nature de la reaction antigreffe chez le Lombricien <u>Eisenia foetida</u> Sav. C.R. Acad. Sci. Paris, <u>257</u>, 3488-90.

Valembois, P. & Boiledieu, D. (1980). Fine structure and functions of haemerythrocytes and leucocytes of <u>Sipunculus nudus</u>. J. Morph., <u>163</u>, 69-77.

Warr, G.W. (1981). Evolution of the lymphocyte. Immunology Today, <u>2</u>, 63-68.

Warr, G.W., Decker, J.M., Mandel, T.E., DeLuca, D., Hudson, R. & Marchalonis, J.J. (1977). Lymphocyte-like cells of the tunicate, <u>Pyura stolonifera</u>: Binding of lectins, morphological and functional studies. Aust. J. Exp. Biol. Med. Sci., <u>55</u>, 151-64.

Warr, G.W., & Marchalonis, J.J. (1978). Specific immune recognition by lymphocytes: an evolutionary perspective. Q. Rev. Biol., <u>53</u>, 225-41.

Wright, R.K. (1976). Phylogenetic origin of the vertebrate lymphocyte and lymphoid tissue. <u>In</u> Phylogeny of Thymus and Bone Marrow-Bursa Cells, eds. R.K. Wright & E.L. Cooper. pp. 57-70. Amsterdam: Elsevier/North-Holland Biomedical Press.

Wright, R.K. & Ermak, T.H. (1982). Cellular defense systems of the Protochordata. <u>In</u> The Reticuloendoehtlial System. A Comprehensive Treatise. Vol.3, eds. N. Cohen & M.M. Sigel pp. 283-320. New York: Plenum Press.

Zachary, D. & Hoffmann, J.A. (1973). The haemocytes of <u>Calliphora erythrocephala</u> (Meig.)(Diptera). Z. Zellforsch. Mikrosk. Anat., <u>141</u>, 55-73.

Zapata, A., Fänge, R., Mattisson, A. & Villena, A. (1984). Plasma cells in adult Atlantic hagfish, <u>Myxine glutinosa</u>. Cell Tissue Res., <u>235</u>, 691-93.

2 Fish

A.F. Rowley, T.C. Hunt, M. Page & G. Mainwaring
Biomedical and Physiological Research Group, School of Biological
Sciences, University College of Swansea, Singleton Park, Swansea
SA2 8PP, UK.

INTRODUCTION & EVOLUTIONARY ASPECTS

Fish are by far the largest group of vertebrates with 20,000
known species (Bone & Marshall 1982). Although monophyletic, they
exhibit enormous diversity in a range of features such as life style,
morphology and, ultimately, immune responsiveness and haematology. There
are two main types of fish; the jawless, commonly called agnathans, and
the jawed or gnathostomate forms. This latter group can be divided into
the Chondrichthyes (fish with a skeleton composed almost entirely of
cartilage) and the Osteichthyes (fish with a predominantly bony
skeleton).

Fish taxonomy is in a state of flux with, for example, some authors such
as Nelson (1969) placing the agnathans in a separate superclass and all
of the jawed vertebrates in a sister group, the superclass
Gnathostomata. In turn, the gnathostomates are subdivided into the
Elasmobranchiomorpha (cartilaginous fishes) and the Teleostomi (bony
fishes, amphibians, reptiles, birds and mammals) (Nelson 1969). This
classification scheme reflects the fundamental differences between the
cartilaginous and bony vertebrates. A more traditional classification
scheme which is used in this chapter is shown in Table 1 together with
the common names of fish often used in haematological/immunological
studies.

The first fish-like vertebrates appeared in the fossil record during the
Cambrian period over 500 million years ago. These animals, collectively
termed ostracoderms, were mostly heavily armoured and jawless. The only
living descendants of the ostracoderms are the lampreys and hagfishes,
but such animals have numerous morphological and physiological
specialisations which have evolved over some 300 million years. Hence,
extant agnathans are not just a form of 'primitive' fish (Hardisty
1979).

Originally, it was widely believed that the gnathostomates evolved
during the Silurian from now extinct agnathans. Most zoologists today,
however, reject this view in favour of the gnathostomates and agnathans
as sister groups but sharing a common hypothetical ancestor (Jarvik
1980). The bony and cartilaginous fish evolved at a relatively similar
time probably from separate ancestors. During the Carboniferous period,
cartilaginous fish flourished, but since then they have been rapidly
overtaken in numbers by bony fish which now predominate in the aquatic

environment. Bony fish can be subdivided into two main groups, the ray-finned (Actinopterygii) and the fleshy-finned forms (Sarcopterygii). Within this latter group are only a few survivors of a once dominant assemblage including lungfish and coelacanths from which the first terrestrial vertebrates probably evolved.

It is widely accepted that fish, like all other vertebrates, have a common leucocyte pattern consisting of granulocytes, monocytes, lymphocytes and thrombocytes. What is still a matter of debate, however, is the identification and relative functions of these cell types. Problems in blood cell identification have largely resulted from an over-reliance on morphological techniques, and in particular the use of Romanowsky stained smears. These problems are well exemplified by the fish literature on granulocyte heterogeneity. In mammals there are three types of granulocytes, the neutrophils/heterophils, eosinophils and basophils (see Chapter 6), each with their own characteristic morphologies and functions. In fish, although most authors recognise neutrophils/heterophils, the existence of eosinophils and in particular, basophils, is often questioned (Ellis 1977a). Likewise, until recently, the existence of lymphocyte sub-populations, equivalent to mammalian B and T cells was equivocal, but as a result of hapten carrier, mitogen and receptor studies it now seems probable that fish do have such cells (Ellis 1982a; 1986; McCumber et al. 1982; Miller et al. 1985).

In summary, although fish probably have a common ancestor, their great diversity must be firmly recognised by all comparative immunologists/haematologists. In the present chapter therefore, the structure and ontogeny of the blood cell types from each group of fish will wherever possible,be considered separately. The emphasis in this review will be towards a thorough appraisal of the classification, structure and ontogeny of fish leucocytes in an attempt to provide a firm basis for future research. The functions of the different leucocyte

TABLE 1   A Classification Scheme for Living Fish

CLASS AGNATHA - lampreys and hagfishes

CLASS CHONDRICHTHYES - cartilaginous fish
    Subclass Elasmobranchii - sharks, rays, dogfish
    Subclass Holocephali - rabbit fish

CLASS OSTEICHTHYES - 'bony' fish
    Subclass Actinopterygii
        Superorder Chondrostei - sturgeons and paddlefish
        Superorder Holostei - gars and bowfins
        Superorder Teleosti - eels, salmonids, catfish, cyprinids,
                           flatfish, blennies
    Subclass Sarcopterygii - lungfish and coelacanths

types are also covered, but because of the vast literature available the reader is referred to the many excellent reviews cited in the text for more detailed accounts.

CIRCULATORY SYSTEMS

The blood vascular system of the majority of fish is described as single, i.e. blood passes through the heart only once during a single circuit of the body. As expected, the system is essentially closed and is based on the common vertebrate plan with a well-developed heart, arteries, capillary beds and veins, mostly of similar structure to their mammalian counterparts. The heart consists of four chambers; the ventricle, atrium, sinus venosus and a conus/bulbus arteriosus. This final chamber is situated between the muscular ventricle and the ventral aorta and varies in both its structure and function within the different groups. For example, in elasmobranchs and related forms, the chamber, called the conus arteriosus is muscular, while in teleosts it is non-muscular and, because of its shape, termed the bulbus arteriosus. The general anatomy and physiology of the circulatory systems of fish is described in a number of relatively recent reviews including Randall (1970), Johansen (1971) and Satchell (1971).

In hagfishes, the vascular system shares several features in common with its equivalent in the cephalochordates such as open capillary junctions, lack of capillary fenestrations and contractile ability of some blood vessels (Casley-Smith & Casley-Smith 1975; Hardisty 1979, 1982). In consequence, the vascular system is at low pressure, has a high fluid volume in relation to the body weight (17ml 100g^{-1}; Bond 1979) and is characterised by many blood sinuses between the arteries and veins. In lampreys, although these sinuses are still present, the blood pressure is higher and the relative blood volume lower (8.5ml 100g^{-1}; Bond 1979) than in hagfishes. The blood volume in elasmobranchs is again lower (6-8ml 100g^{-1} body weight; Bond 1979) and a higher pressure is maintained in the vascular system than in lampreys. Blood sinuses/lacunae are also still present.

In lungfish, the heart and associated blood vascular system are somewhat more complex. These changes are brought about by the reliance of these animals on aerial rather than wholly aquatic respiration. In consequence, the heart is partially divided to separate oxygenated blood entering via the pulmonary vein from deoxygenated blood from the main body tissues. This division of the heart consists of a spiral fold in the conus arteriosus and interatrial and interventricular septa in the atrium and ventricle respectively (Hildebrand 1982). The circulatory system of lungfish and associated forms is similar to that seen in some amphibians (see Chapter 3).

Although a detailed review of the gross anatomy of the lymphatic system of fish is beyond the scope of this chapter, it is briefly described here due to its contamination in some fish by a variety of blood cells. The anatomy of the lymphatic system has been described in several papers (e.g. Wardle 1971) and the excellent account of Kampmeier (1969), who reviewed much of the earlier literature. In the Agnatha, Chondrichthyes and some Osteichthyes, it is only partially formed and, due to its contamination with erythrocytes and other blood cells, called a haemolymphatic system (Hildebrand 1982). In teleosts, however, it is said to be totally free from erythrocytes. The main lymphatic vessels consist of four subcutaneous ducts (a dorsal, ventral and two lateral) and one or two subvertebral (cardinal) ducts together with smaller networks which connect some of these major vessels. The movement of lymph is aided by small heart-like structures called lymph propulsors. No lymph nodes or valves have been described in the lymphatic system of fish.

A recent report by Vogel & Claviez (1981) has cast doubt on the existence of lymphatic vessels in fish. They argue that the histological features used to discriminate lymph from blood vessels are erroneous and hence, what most authors have described as lymphatic vessels, are venous vessels devoid of erythrocytes. Lymphatic vessels are said to have only evolved in the amphibians as a result of the higher blood pressure associated with a terrestrial existence.

ORIGIN & FORMATION OF BLOOD CELLS

In most fishes, despite the existence of a voluminous literature on the structure and location of the myeloid and lymphoid tissues, the origin of different blood cell types is still unresolved. Much of the existing literature is unfortunately 'classical', in that it pre-dates the development of electron microscopy, radiolabelling techniques and, more importantly, modern immunobiology. There therefore remains a large gap in our understanding of the ontogenetic development of fish leucocytes.

Agnathans

Morphological studies with hagfishes have revealed two main fixed haemopoietic sites, namely scattered foci in the lamina propria of the gut wall and lymphoid and erythroid precursors in the pronephros (Holmgren 1950; Good et al. 1966; Östberg et al. 1976; Tanaka et al. 1981). Most is known of the haemopoietic foci in the gut wall which seem to be responsible for granulopoiesis, perhaps lymphopoiesis, but apparently not for erythro- or thrombo-poiesis (Tanaka et al. 1981). Despite the christening of these foci as a "primitive" or "intestinal spleen" (e.g. Good et al. 1966; Tomonaga et al. 1973a), this view has been challenged by Tanaka et al. (1981) who found no evidence for the formation of the foci during the establishment of collateral veins. Little is known of the supposed role of the pronephros in haemopoiesis, though Holmgren (1950) considered lymphoid cells, erythrocytes and spindle cells (thrombocytes or lymphocytes?) as originating from this site, at least in young specimens of the Atlantic hagfish, Myxine glutinosa. In contrast, some authors have placed more emphasis on the peripheral blood as a site for haemopoiesis (e.g. Jordan & Speidel 1930; Linna et al. 1975). Haemocytoblasts in the peripheral blood of the

Californian hagfish, Eptatretus stouti, show proliferative ability as judged by tritiated thymidine incorporation, but the haemopoietic potential of such cells is unclear (Linna et al. 1975). Mitotic "erythroblasts" in the peripheral blood have also been identified in Eptatretus burgeri (Tomonaga et al. 1973b).

Finally, of particular interest in hagfishes is the reported lack of a thymus (Riviere et al. 1975) and this, together with the poorly organized haemopoietic sites described above, suggests an evolutionary condition not far removed from some of the advanced invertebrates such as urochordates and cephalochordates (see Chapter 1).

In lampreys, the sites of haemopoiesis are far more clearly defined and well-developed than in the hagfishes, but there is still no clear differentiation between haemopoietic and lymphoid organs. Lampreys have an unusual life cycle with a protracted larval stage called an ammocoete and a relatively short-lived, usually parasitic, adult phase. As a result, the major sites of haemopoiesis undergo radical remodelling and relocation during metamorphosis. In the ammocoete, the earliest identifiable haemopoietic cells are found in the blood islands of prolarvae (Percy & Potter 1976) and such cells apparently differentiate to form erythrocytes. At the same time, a primitive spleen (often called a protospleen) develops in an infolding of the alimentary canal called the typhlosole (Figs 1-4; Potter et al. 1982). This protospleen has been studied in a number of different species of lamprey and its structure is now well researched (e.g. Finstad et al. 1964; Good et al. 1966; Tanaka et al. 1981; Fujii 1982; Page 1983; Ardavin et al. 1984). The protospleen is fed by the mesenteric artery which branches and eventually forms a complex network of sinusoidal spaces (Fig. 1). These sinusoidal spaces are mainly confined to an area dorsal of the central mesenteric artery, though a few may be seen in a ventral position. Within the sinusoids are free blood cells as well as macrophage-like cells attached to the bounding endothelium (Fig. 2). Between the sinusoidal space is the intersinusoidal parenchyma which consists of a compacted mass of lymphocytes, lymphoblasts, erythrocytes, immature and mature granulocytes, and a few macrophages (Figs 3,4). These cells are supported by a network of collagen fibres together with fibroblasts and reticular cells (Fig. 4). This intersinusoidal parenchyma appears to be the site of lymphocyte, erythrocyte and granulocyte production as judged by the number of mitotic figures and immature cells present (Figs 3,4; Page 1983; Ardavin et al. 1984). No evidence, however, for monocyte and thrombocyte differentiation could be found in Lampetra sp. (Page 1983), but in Petromyzon marinus, thrombopoiesis has been reported in the protospleen (Good et al. 1966). Mature leucocytes produced in the protospleen appear to gain entry into the sinusoidal spaces by diapedesis (Fig. 3), and then, according to Tanaka et al. (1981), pass to the general circulation via venous capillaries, situated just beneath the muscularis mucosa, which drain the protospleen. The distribution of cell types in the parenchyma of the protospleen is generally random, although a zone of lymphocytes is often found around the central mesenteric artery (Page 1983).

The second area of haemopoietic tissue in the larval lamprey is situated in the intertubular folds of the opisthonephric kidney (Fig. 5). As in the protospleen, an intersinusoidal parenchyma encloses immature blood cells, mainly lymphocytes and granulocytes, but again there is no indication of any thrombocytic or monocytic lineages (Page 1983). The relative importance of the protospleen vs. opisthonephros in haemopoiesis is also not established. One final site of haemopoiesis in larval lampreys is the blood (Percy & Potter 1976). However, this probably only accounts for the final stages in the maturation of some of the cell types as it seems doubtful if any stem cells would be present in such an environment.

Before, and during metamorphosis, the haemopoietic activity of the protospleen and opisthonephric kidney is lost and this is taken over by the supraneural body (= protovertebral arch and fat body/column of some authors; Figs 6,7) and the adult kidney (George & Beamish, 1974; Kelényi & Larsen 1976; Piavis & Hiatt 1971; Percy & Potter 1977; Page 1983; Ardavin et al. 1984). The supraneural organ is situated dorsal to the spinal cord and first becomes apparent during its invasion by immature and mature blood cells presumably from the protospleen and opisthonephric kidney. Because of its fatty nature (Fig. 6), it bears some resemblance to bone marrow (George & Beamish 1974). Piavis & Hiatt (1971) and George & Beamish (1974) have both made careful studies on the role of the supraneural organ of the sea lamprey, P. marinus in haemopoiesis. They identified myeloblasts, lymphoblasts, monoblasts, megakaryoblasts and erythroblasts. It must be emphasised, however, that their interpretations were based entirely on the morphological similarities of stained cells with mammalian counterparts and such methods are now generally believed to be unreliable. Indeed, the ontogenetic pathways for all leucocyte types proposed previously in both the blood and haemopoietic sites (e.g. Percy & Potter 1976; 1981) need careful re-examination employing techniques such as ultrastructural enzyme cytochemistry and immunocytochemistry together with radiolabelling and autoradiography.

Finally, the question of the existence of a thymus or its equivalent in lampreys needs mention. Evidence for a thymus largely comes from the observation of diffuse lymphoid accumulations in the pharynx of the ammocoete (Figs 8,9), in a position akin to the thymus of jawed fish (Manning & Turner 1976). With age, these involute but this feature, although similar to events with the thymus of other vertebrates (Manning 1981), is of little phylogenetic interest since in lampreys all other haemopoietic tissues involute/disappear during metamorphosis. Recent ultrastructural studies of these lymphoid accumulations suggest that they represent swollen lymphatic and/or venous sinuses lined by an endothelium and containing mainly lymphocytes and lymphoblasts held in place by a poorly developed network of reticular cells (Fig. 9; Page & Rowley 1982). The lymphoid cells in this area are in contact with the general circulation as evidenced by their association with colloid carbon following its intravascular injection (Page 1983). Hence, the lymphoid accumulations differ from the thymus in a number of ways. Most important is the lack of a blood barrier seen around the thymus of other vertebrates (Kendall 1981). Further functional studies are required to resolve the position on thymic homology, although the small size of the lymphoid accumulations limits the experimental design to the in vivo approach.

Chondrichthyes

Like other fish, the elasmobranchs and holocephalans (see Table 1) lack both bone marrow and lymph nodes but do, however, have a discreet lymphoid thymus, a well-differentiated spleen, which is situated outside the alimentary tract attached via mesentery, and a number of often large lymphomyeloid tissues unique to this class.

The structure and location of the thymus and the stage at which it involutes are variable in cartilaginous fish. For example, in the lesser spotted dogfish, Scyliorhinus canicula, the thymus exists as paired lobed masses dorsal to the first two gill arches which involute after only a few weeks post-hatch (Pulsford et al. 1984). In general, the cortex and medullary regions are poorly differentiated in S.canicula (Fig. 10) and most of the cells are lymphocytes(thymocytes) with a few macrophages and reticular elements (Pulsford et al. 1984). Other authors have, however, reported in some elasmobranchs that the cortex and medulla of the thymus are clearly delineated (Fig. 11) and that epithelial cysts and myoid cells are present (e.g. Good et al. 1966; Zapata 1980). Hassall's corpuscles have only been occasionally identified in the elasmobranch thymus (Beard 1902; Good et al. 1966) but their identification was based solely on light microscopy. The structural studies on the thymus of cartilaginous fish have unfortunately, rarely been extended to any functional aspects of this organ either in lymphopoiesis or immunity.

The structure of the spleen has been detailed in the elasmobranchs with only a few reports from the holocephalans (Zapata 1984). Ultrastructural studies have been made in a number of elasmobranchs, including the Atlantic nurse shark, Ginglymostoma cirratum (Fänge & Mattisson 1981), the thornback Raja clavata and marbled electric rays Torpedo marmorata (Zapata 1980), and the lesser spotted dogfish, S. canicula (Pulsford et al. 1982; Hunt & Rowley 1986a) and these have provided useful additional detail, particularly on the differentiation of blood cell types. The following description of splenic morphology is based on our own observations with S. canicula (Hunt & Rowley 1986a) but is similar in most other elasmobranchs. In adult dogfish, the spleen is large, 7-9 cm in length, attached to the latero-ventral surface of the stomach. The outermost layer, the capsule, consists of epithelial cells and underlying connective tissue to which is anchored the reticular cell/fibre network (Pulsford et al. 1982). The white and red pulp areas are fairly clearly defined (Fig. 12) but not to the same extent as in birds and mammals. Blood flows into the spleen via the lienogastric artery which divides to form numerous arterioles and eventually capillaries. These latter structures are surrounded by a sheath of macrophages and reticular cells called an ellipsoid (= periarterial macrophage sheath of Weiss et al. 1985; Figs 12,13). The function of the ellipsoids in antigen trapping will be discussed later (see section on macrophage functions). Blood passing through the ellipsoids enters the thin walled sinusoids of the red pulp with its important haemopoetic activity. Within this zone are erythrocytes and thrombocytes in various stages of maturation, lymphoid blast cells, plasma cells, fixed macrophages, a few free monocytes and mainly mature granulocytes (Pulsford et al. 1982; Fänge & Nilsson 1985; Hunt & Rowley, 1986a). There is little evidence for any granulopoiesis in the spleen of the dogfish.

Elasmobranchs and holocephalans also possess a number of unique lymphomyeloid tissues. In the elasmobranchs these are the epigonal organ and the organ of Leydig (Figs 14-16). In at least some holocephalans, lymphomyeloid tissue is present in the orbital and subcranial region, and smaller accumulations have been reported in the cardiac epithelium and the intestinal spiral fold (Fänge & Sundell 1969; Fänge 1984; Zapata 1984). The importance of these two latter tissues in haemopoiesis is undetermined, while the former seems to be equivalent to the epigonal/Leydig's organs of the elasmobranchs (Zapata 1984).

The epigonal organ is associated with either the testes or ovary and is reported to be the largest lymphomyeloid tissue in the blue shark, Prionace glauca and the nurse shark, G. cirratum (Fänge 1977; Fänge & Mattisson 1981). It is generally yellow-white in colour, which reflects its lack of erythropoiesis, and is fed by the gonadal arteries (Fänge & Pulsford 1983). The morphology of the epigonal organ has been studied in detail in G. cirratum (Fänge & Mattisson 1981), R. clavata (Zapata 1981a), S.canicula (Fänge & Pulsford 1983) and T. marmorata (Zapata 1981a). The organ is lobulate and contains numerous granulocytes in different stages of maturation, lymphoid elements and plasma cells (Figs 15,16; Fänge & Pulsford 1983). The organ of Leydig is structurally similar to the epigonal organ, except it is situated in the submucosa of the alimentary tract (Fig. 14; Fänge 1968; Fänge & Pulsford 1983; Mattisson & Fänge 1982; Honma et al. 1984). The work of Fänge & Pulsford (1983) on the epigonal and Leydig's organ of S. canicula, provides one of the most detailed description of granulopoiesis in any elasmobranch. They illustrated the existence of "blast cells" - most likely granulocytic precursors including myeloblasts, promyelocytes, myelocytes, metamyelocytes and mature granulocytes. In their view, the four different types of granulocytes (see section on blood cell structure) arise from a single population of myeloblasts/promyelocytes. This interpretation is at odds with similar studies on granulopoiesis in elasmobranchs (Zapata 1981a) and moreover mammals (see Chapter 6).

The existence of immature granulocytes and plasma cells in the epigonal and Leydig's organ indicates their role in both haemopoiesis and antibody generation. The location of Leydig's organ in the oesophagus may be important in protecting this region from parasitic assault, in a similar way to the GALT of other vertebrates. Perhaps the antibody produced by the plasma cells is released into the oesophageal mucus. The relative functioning of these two lymphomyeloid tissues vs. the spleen in antibody formation will be discussed later (see section on lymphocyte functions).

Finally, the blood may be an important haemopoietic compartment in cartilaginous fishes. Several stages in erythrocyte formation have been seen in circulation (Saunders 1966; Pederson & Gelfant 1970; Hyder et al. 1983). Furthermore, autoradiographic and cytophotometric studies of the peripheral blood of the smooth hound, Mustelus canis, showed that lymphocytes and erythroblasts, but not other cell types, synthesise DNA, RNA and protein (Pederson & Gelfant 1970).

Osteichthyes
The principal lymphomyeloid tissues of bony fish are the thymus, spleen and kidney. Most of the studies on these tissues have been carried out in teleosts although some work has been produced on chondrosteans (Clawson et al. 1966) and dipnoans (Rafn & Wingstrand 1981). Some important experimental research on the immunological and haemopoietic activity of the thymus in teleosts will be described first.

During teleost development, the thymus is the first lymphoid organ to appear (Ellis 1977b; Grace & Manning 1980; Bly 1985) although shortly after, lymphocytes make their appearance in the kidney and then a similar, though less marked, acquisition occurs in the spleen (Grace & Manning 1980). These observations suggest that the thymus seeds the kidney and spleen with lymphocytes (T cells?). Indeed, rainbow trout, which have previously been injected intrathymically with tritiated thymidine, contain radiolabelled cells (lymphocytes) in both the spleen and kidney, though the majority occur in the former organ (Tatner 1985). Furthermore, mouse anti-carp thymocyte monoclonal antibodies (from clone WCT4) which identify a population of thymocytes in the thymus of the carp, Cyprinus carpio at day 4 post-fertilisation, also label lymphocytes in the pronephric kidney some 3 days later (Secombes et al. 1983a). Antibody from clone WCT4 also binds to ectodermal cells of the thymus anlage, suggesting either an ectodermal origin or influence for the developing thymocytes (Secombes et al. 1983a). These excellent reports of Secombes et al. (1983a) and Tatner (1985) suggest a role for the thymus of teleosts as a major (primary?) lymphoid organ.

There are a number of important differences in the structure and possible functions of the thymus of teleosts and 'higher' vertebrates. The teleost thymus is rarely clearly divided into cortical and medullary zones (Ellis 1977b; Manning 1981; Chilmonczyk 1983, 1985; Cooper et al. 1983; Bly 1985; Fig. 17) although a densely stained outer region and a lightly stained inner region have been described in the thymus of Salmo gairdneri (Chilmonczyk 1983; Fig. 18) and the viviparous blenny, Zoarces viviparus (Bly 1985). The existence of Hassall's corpuscles in the teleost thymus is also very variable but such bodies, if indeed they truly exist, are never as frequent as in reptiles, birds and mammals (Manning 1981). Another difference, which probably has important functional consequences, is the position of the teleost thymus. This organ throughout its ontogeny is often closely associated with the epithelium of the pharynx (Figs 17,18). Chilmonczyk (1985) has also

recently demonstrated pores a few micrometers in diameter on the thymic
surface of young trout (Fig. 19). These pores may allow antigenic
material from the surrounding water to gain access to the thymus at
least in young fish. Other workers (e.g. Ellis 1982a) have also
questioned if the thymus of teleosts (and other fish?) forms an
immunologically closed environment protected by a complex connective
tissue capsule and a blood-thymus barrier as found in mammals (Kendall
1981). The ultrastructural identification of plasma cells/plasmablasts
in the thymus of immunologically naive fish (e.g. Zapata 1981b; Fänge &
Pulsford 1985) together with the thin surrounding capsule in teleosts
strongly points to direct contact with antigen (see also section on
lymphocyte functions).

The kidney has been found to be a major blood forming organ in bony fish
(Catton 1951; Zapata 1979a,b; Bielek 1981; Botham & Manning 1981) and
its dark red colour reflects it probable function in erythropoiesis. Two
main haemopoietic sites are present in most fish, namely the anterior,
pronephric, or head kidney, and the main, opisthonephric, or trunk
kidney (Figs 20-22), which as well as its blood forming function is also
excretory. The structure of the main haemopoietic area of the kidney
seems to differ within the teleosts. For example, Ellis and de Sousa
(1974) and Sailendri & Muthukkaruppan (1975) consider that non-lymphoid
(red pulp) and lymphoid (white pulp) zones exist in Pleuronectes
platessa and Tilapia mossambica respectively, while Zapata (1979a) found
a mainly homogeneous mass of haemopoietic tissue in Rutilus rutilus and
Gobio gobio. Perhaps these differences reflect the relative importance
of the kidney in different fish in antigen trapping and antibody
formation (see section on lymphocyte functions). Ellis et al. (1976) and
Bogner & Ellis (1977) also reported the existence of clumps of
macrophages containing melanin granules (melanomacrophage centres; Fig.
20) in the "white pulp" of the anterior kidney of salmonids - the
possible role of such structures in antibody formation will be discussed
later (see section on macrophage functions).

The haemopoietic kidney has been reported to be a site of erythrocyte,
granulocyte, lymphocyte and monocyte differentiation (Bogner & Ellis
1977; Zapata 1979a,b; Bielek 1981). The pathway of granulocyte
differentiation, which is considered to be similar to that in mammals,
has been most carefully studied (Zapata 1979b; Bielek 1981). In the two
cyprinids studied, Cyprinus carpio and Tinca tinca, heterophilic,
eosinophilic and basophilic promyelocytes, myelocytes and mature cells
were identified in the kidney (Fig. 22; Bielek 1981). This
identification was based on size, structure and staining characteristics
of the different types of granules (see also Figs 54-56 in section on
blood cell structure).

The spleen of teleost fish is basically similar to that of the
elasmobranchs, but melanomacrophage aggregates are usually present in
the white pulp (Ellis et al. 1976; Agius 1981; Zapata 1982; Fänge &
Nilsson 1985). In many fish, although the spleen is involved in
haemopoiesis, its role in this process is thought to be secondary to the
pronephros (Davies & Haynes 1975). In others, such as the perch, Perca

<u>fluviatilis</u>, the spleen is the only haemopoietic organ (Catton 1951).

Chondrosteans, holosteans and dipnoans apparently have similar haemopoietic tissue as in the teleosts i.e. spleen, kidney, thymus and anterior kidney, although in the former group, the spleen is divided into two non-connected areas within the alimentary canal (Rafn & Wingstrand 1981). Holosteans and chondrosteans also possess a granulopoietic tissue associated with the central nervous system and spinal cord (see Fänge 1984) while in sturgeons and paddlefish "lymphoid" tissue of unknown haemopoietic potential is associated with the heart (Clawson <u>et</u> <u>al</u>. 1966).

<u>Summary</u>
        Hagfishes have poorly developed haemopoietic tissues perhaps reflecting their evolutionary position near the base of the vertebrate assemblage. Their sister group, the lampreys, however, contain a protospleen in the larval stage which is probably homologous with the spleen of other vertebrates. The existence of a true thymus is unresolved.

Gnathostomate fish utilise a number of areas of the body for the location of haemopoietic cells, the most notable of which are the kidney, spleen, thymus, gonads (Leydig's organ) and oesophagus (epigonal organ). Other minor, or infrequent locations include the heart, intestine, liver and skeleton. Final stages in blood cell maturation are often also found in the blood. No fish have the bone (marrow) as a site for haemopoiesis; this apparently first became active in the amphibians with the switch from an aquatic to a terrestrial existence (see Chapters 1 & 3). Cooper <u>et</u> <u>al</u>. (1980) have suggested that this change is associated with the protection of haemopoietic stem cells to exposure from ionising radiations. In aquatic animals such as fish, water acts as a good barrier against radiation, while with the switch to a terrestrial existence the previously used haemopoietic sites would be vunerable and so these areas were relocated within the radiopaque bone.

Although the sites at which haemopoiesis takes place in fish are now well established, our knowledge of which cell types are produced in the different organs is very incomplete. For example, few researchers have reported on the location and stages in the differentiation of monocytes and thrombocytes. Also,even when the lineage of different leucocyte types is reported, it is often at best tentative. Finally, the embryological origin of blood cells needs careful study to identify possible haemopoietic stem cells. In this respect, work with fish has lagged very much behind that with amphibians, birds and mammals (see Chapters 3, 5 & 6).

<u>BLOOD CELL STRUCTURE & CLASSIFICATION</u>
        Fish blood contains numerous nucleated erythrocytes, thrombocytes, lymphocytes, granulocytes and monocytes. The fluid phase, or plasma, consists of a variety of inorganic ions, sugars, together

with clotting factors and alpha, beta and gamma globulins. In general, the concentration of proteins in the blood of fish is lower than in mammals. The distribution of albumin in the plasma of fish is variable; for example, it is absent in elasmobranchs but present in most bony fish (Gunter et al. 1961). These authors suggest that analbuminaemia among fish seems to be a primitive characteristic.

In most fish the osmotic balance of blood is maintained by inorganic ions such as Na^+ and K^+, while in elasmobranchs and holocephalans, which are isosmotic with sea water, urea is stored for this purpose.

Erythrocytes (Figs 23-25)

Erythrocytes are the dominant cell type in the blood of the vast majority of fish. Some, however, such as the icefish contain few erythrocytes and lack haemoglobin (Ruud 1954; Barber et al. 1981). In the icefish, Chaenocephalus aceratus, although a few erythrocytes are found in the blood they are invariably senescent and fragile (Barber et al. 1981).

Numerous studies provide basic information on haematological parameters such as haematocrit (packed cell volume), erythrocyte sizes, numbers, and haemoglobin content (Table 2) and the reader is referred to detailed accounts elsewhere (see for example, Hesser 1960; Blaxhall 1972; Blaxhall & Daisley 1973). Table 2 illustrates some of the major trends and differences in these parameters among the main groups of fish. Haematocrit values are very variable both within one and between different species. There are no distinct phylogenetic trends and instead the differences observed in haematocrit as well as haemoglobin content correlate with the life style of the fish. Fast active swimmers tend to have high haematocrits/haemoglobin levels while slower, less active forms, have low values for both of these parameters (Hall & Gray 1929; Eisler 1965).

Several researchers have suggested that the haematocrit might be a useful routine indicator of general fish health (e.g. Snieszko 1960; Blaxhall 1972; Wedemeyer et al. 1983), however, the haematocrit can vary depending on experimental procedures, such as the use of anaesthesia prior to bleeding, and whether or not an anticoagulant is used (Blaxhall 1973; Smit et al. 1977; Lowe-Jinde & Niimi 1983). Haematocrit values sometimes vary dramatically too during the life cycle and throughout the seasons. This variation has been particularly well researched in lampreys (Macey & Potter 1981; Potter et al. 1982) and Atlantic salmon (Conroy 1972). During the early stages of metamorphosis in the lamprey, Geotria australis, the haematocrit and erythrocyte numbers fall, but later these rise with the formation of new adult erythrocytes (Macey & Potter 1981). In the Atlantic salmon, Salmo salar, markedly high haematocrits are found in pre-spawning adults while at sea, but levels fall again during spawning (Conroy 1972).

The size and shape of mature erythrocytes is also variable. Erythrocytes of agnathans, Chondrichthyes and Dipnoi are generally larger than their

TABLE 2 Erythrocyte and General Haematological Parameters

Genus/species	Haematocrit (%)	Erythrocyte dimensions (μm)	Erythrocyte no. (x 10^6mm^{-3})	References
Agnathans:				
Myxine glutinosa	19	15 - 35[a]	ND	Mattisson & Fänge (1977)
Geotria australis	41.5 (L)[b] 46.2 (A)[b]	12.4 x 3.3 (L) 12.8 x 4.1 (A)	1.8 (L) 1.2 (A)	Macey & Potter (1981)
Lampetra fluviatilis	36.5	ND	1.7	Ivanova-Berg & Sokolova (1959)
Chondrichthyes:				
Heterodontus portjacksoni	14-25	17.5 x 15	1.5 - 2.6	Stokes & Firkin (1971)
Prionace glauca	22.3	22.8 x 15	ND	Johansson-Sjöbeck & Stevens (1976)
Raja erinacea	ND	20.3 x 12.9	ND	Hartman & Lessler (1964)
14 species of elasmobranchs	ND	15.1 - 25.5 x 12.0 - 17.9	0.4 - 0.7	Saunders (1966)
Osteichthyes:				
Anguilla rostrata	ND	11.5 x 7.6	ND	Hartman & Lessler (1964)
Carassius auratus	ND	13.4 x 8.8	2.0	Watson et al. (1963)
C. auratus	ND	5.8	1.6	Javaid & Akhtar (1977)
Cyprinus carpio	61	ND	ND	Field et al. (1943)
Neoceratodus forsteri (lungfish)	ND	41 x 17	2.6	Ward (1969)
Pleuronectes platessa	24 - 62	ND	1.1 - 1.5	Preston (1960)
Salmo salar	24 - 67[c]	14 x 8	0.8 - 1.2	Conroy (1972)
Salmo trutta	20 - 43	15 x 9	0.6 - 1.3	Blaxhall & Daisley (1973)

a) Diameter given is of major axis
b) A = adult; L = larval stage (ammocoete)
c) For details see text

equivalents in teleosts (Table 2; for further information on size differences in red blood cells within fish and other vertebrates see Wintrobe 1933). Lamprey erythrocytes are flattened and biconcave (Potter et al. 1974) and, with the exception of the nucleus, bear a strong resemblance to the erythrocytes of mammals (see Chapter 6). In other fish, mature erythrocytes are biconvex with a central 'hump' corresponding to the position of the nucleus and surrounding cytoplasm.

As discussed previously (see "Blood Cell Formation"), the blood is an important site for erythropoiesis in fishes, particularly in agnathous and cartilaginous fish (Pederson & Gelfant 1970; Mattisson & Fänge 1977). Although many researchers have mentioned the existence of immature erythrocytes, few have attempted to categorise the different maturation stages found in the blood. In the plaice, P. platessa, Ellis (1976) distinguished proerythroblasts, erythroblasts, late erythroblasts, proerythrocytes and young erythrocytes but he failed to note which types were only found in the haemopoietic tissues. In the agnathans and elasmobranchs in particular, cells which share certain features with mammalian erythroblasts can be found in the peripheral blood (Saunders 1966; Pederson & Gelfant 1970; Mattisson & Fänge 1977). In smears, they are round to spherical with a centrally placed nucleus enclosing reticulated chromatin. The cytoplasm appears as a thin but distinct rim and unlike mature erythrocytes is strongly basophilic after Romanowsky staining (Fig. 23). Few ultrastructural studies on fish blood have identified and illustrated this cell type. Those which purport to do so (e.g. Pederson & Gelfant 1970) may be looking at a form most closely related to the late erythroblast of mammals. We have seen such cells in the peripheral blood of dogfish, S. canicula, especially following the intravenous injection of antigenic material when immature erythrocytes are released from the splenic store as part of a general stress reaction (Figs 23, 24). Ultrastructurally, such cells have a densely stained cytoplasm containing many free ribosomes, which account for the basophilia seen in smears, and a central round to reniform nucleus (Fig. 24). These cells cannot be confused with 'small' lymphocytes as they are larger and only rarely observed.

Proerythrocytes and "young erythrocytes" (=reticulocytes of some authors e.g. Conroy 1972) are readily seen in blood smears of many species of normal fish. Characteristic changes between these two maturation stages include a gradual decrease in basophilia and an increase in cytoplasmic haemoglobin which stains pinkish with Giemsa (Sekhon & Beams 1969; Sekhon & Maxwell 1970; Ellis 1976). "Young erythrocytes" in plaice are only distinguished from mature forms by the residual patchy basophilia of the cytoplasm (Ellis 1976). Young erythrocytes in dogfish, S. canicula, occasionally have some basophilia but more characteristic is a ring of eosinophilic material around the nucleus (Fig. 23). Andrew (1965) showed that mature erythrocytes of lampreys still retain small patches of basophilia. Identification of "proerythrocytes" and "young erythrocytes" from ultrastructural studies is hampered by the terminology used. For example, Sekhon and colleagues (Sekhon & Beams 1969; Sekhon & Maxwell 1970) refer to all of these cells as "immature erythrocytes", even though they found several different morphologies

within this category. "Immature erythrocytes" have a number of
organelles which are lost in the mature form; these include a prominent
Golgi complex, mitochondria, mono- and poly-ribosomes, centrioles and
rough and smooth endoplasmic reticulum (Sekhon & Beams 1969; Sekhon &
Maxwell 1970; Hyder et al. 1983). During maturation, these organelles
undergo lysosomal digestion and the remnants of this process, the myelin
whorls, are either stored or expelled from the erythrocyte.
Concomitantly, there is a progressive increase in the amount of
haemoglobin seen in the cytoplasm.

Mature erythrocytes either have a few or no surviving cytoplasmic
organelles apart from the marginal bands (microtubular bundles) which
are responsible for the maintenance of cell shape. The nucleus is
retained in the mature erythrocyte but rarely contains a nucleolus (Fig.
25).

Information on the erythrocytes of fish is not just morphological. For
example, there is an extensive literature on the structure of
haemoglobin and on the antigenic variation in the red cell membrane.
Fish, like other vertebrates, have been found to have blood groups, the
biological basis of which is linked to antigenic variation of the
erythrocyte membranes (Cushing 1964). These blood groups are genetically
controlled and have been found to be useful markers in the analysis of
population structure in many species of fish. Interest in the structure
of fish haemoglobins stems from the use of these molecules as probes for
biochemical evolution. The haemoglobin molecule is particularly useful
as it is found in the blood of many invertebrates and vertebrates and
has a relatively linear rate of amino acid changes making it a reliable
'molecular clock' (Koler 1979). The primary structure of haemoglobin is
now known for a number of fish species but most interest has focussed on
the agnathans due to their evolutionary position near the base of the
vertebrate phylogenetic tree (see reviews of Coates 1975; Perutz 1976;
Hardisty 1979; Koler 1979). In hagfishes, the primary structure of the
haemoglobin molecule is variable from species to species and most
closely resembles the beta-globin chain of 'higher' vertebrates, whereas
in lampreys the structure is highly conserved and most like the
alpha-chain (Hardisty 1979). Unlike other vertebrates in which a four
chain structure ($\alpha_2\beta_2$) is present, the agnathan haemoglobin exists
primarily as a monomer with only one haem group to each molecule. In
lampreys, the deoxygenated form of haemoglobin is dimeric but with
identical sub-units. Hardisty (1979) has used the likeness of the
lamprey and hagfish haemoglobins to the alpha and beta-chains to argue
that gene duplication to give rise to these chains might have occurred
at a much earlier stage in vertebrate evolution than most researchers
believe and at, or before, the common ancestor of the agnathan and
gnathostomate fishes arose. More traditional views are supported in the
majority of textbooks and reviews on the subject due to the acquisition
of the typical $\alpha_2\beta_2$ quarternary structure of haemoglobin only following
the evolution of cartilaginous and bony fishes.

<u>Lymphocytes</u> (Figs 26-30)

Lymphocytes are usually the most commonly observed leucocyte type present in the blood of fishes accounting for as much as 85% of the total leucocyte population. A perusal of the literature illustrates that the figures available for lymphocyte frequency are highly variable. This is due to two major factors. First, not all authors include thrombocytes in the differential white cell count with the result that the proportion of leucocytes identified as lymphocytes in such studies is increased. Second, lymphocyte identification in smears is difficult as they can be confused with thrombocytes and, occasionally, erythroblasts. Much of the older literature on fish haematology describes the existence of a number of lymphoid populations such as "lymphoid haemoblasts" (e.g. Jordan & Speidel 1924; Catton 1951) which these authors considered to be the precursors of as well as lymphocytes, also erythrocytes, thrombocytes and granulocytes. Studies with mainly mammalian lymphocytes have now shown this view to be erroneous but the term lymphoid haemoblast has been a surprising survivor in the literature on fish haematology (e.g. Weinreb & Weinreb 1969).

Based on morphological criteria, the lymphoid elements in the blood of fishes can be subdivided into three categories namely, lymphocytes (=small lymphocytes), lymphoblasts and plasma cells. Lymphocytes are small cells ca. 5-8 μm in diameter with a high nuclear:cytoplasmic ratio and a basophilic cytoplasm (Fig. 26). Ultrastructurally, they are easily identifiable from other leucocyte types such as thrombocytes, due to the central nucleus with dense patches of heterochromatin and an occasional nucleolus, and a poorly developed thin rim of cytoplasm containing a few vesicles, vacuoles, mitochondria, RER and an occasional Golgi complex (Figs 27, 28; Ferguson 1976a; Cannon et al. 1980a; Morrow & Pulsford 1980; Page & Rowley 1983). Cells which have morphological but not necessarily functional equivalence to lymphoblasts, have been identified in fish (Bogner & Ellis 1977; Page 1983; Page & Rowley 1983). Page (1983) found lymphoblast-like cells in the blood of lampreys which are characterised by a more voluminous cytoplasm than lymphocytes and contain many polyribosomes, a few profiles of RER and a nucleus with little heterochromatin and a nucleolus (Fig. 29). Such an interpretation of these cells as lymphoblasts is open to criticism as they bear some resemblance to erythroblasts which also occur in the general circulation (compare Fig. 24 with Fig. 29). If such cells are lymphoblasts, then how they fit in the lineage of normal or activated lymphocytes remains unclear. Plasma cells have been found in the peripheral blood of many fish albeit in very small numbers (Barber et al. 1981). They are easily identified due to the presence of swollen rough endoplasmic reticulum and an active Golgi complex associated with immunoglobulin synthesis and storage (Fig. 30). The scarcity of plasma cells in the circulation reflects their limited release from lymphoid and lymphomyeloid tissues. The distribution of fixed plasma cells will be described later (see section on lymphocyte functions).

An intriguing subject to comparative immunologists is the evolution of the lymphocyte and the stages in the differentiation to form subpopulations i.e. T and B cells (Warr & Marchalonis 1978; Warr 1981;

see also Chapter 1). Fish are important in these studies because they
are the first animals with <u>clearly</u> defined lymphoid organs and
functional lymphocytes (Warr 1981). The question still remains, however,
do fish have T and B cells? The available data from functional studies,
aspects of which will be reviewed later, show that both cellular (T
cell-mediated) and humoral (B cell-mediated) acquired immunity is
present in all fishes so far studied (Manning & Turner 1976; Ellis
1982a). This suggests the existence of a lymphoid system operating in
these animals with some heterogeneity. The main approaches utilised in
studies evaluating the presence of T and B cell equivalents are (i)
mitogen responsiveness, (ii) identification of cell surface determinants
such as immunoglobulin, and (iii) the behaviour of lymphocytes during
cell separation procedures such as density gradient centrifugation and
adherence to glass/nylon wool. Invariably, most investigators use a
combination of these methods (e.g. Sizemore <u>et al.</u> 1984; Miller <u>et al.</u>
1985).

<u>Mitogenicity studies</u>. The blastogenic effect of mitogens on different
populations of lymphocytes is well established in both birds and mammals
(see Chapters 5 & 6). In essence, lipopolysaccharide (LPS) is considered
a B cell mitogen while concanavalin A (Con A) and phytohaemagglutinin
(PHA) predominantly stimulate only T cells.

Studies on mitogenicity responses in agnathans are few. Cooper (1971)
showed that lymphocytes from larval lampreys respond to PHA <u>in vitro</u> by
the production of blast cells suggesting the existence of T-like
lymphocytes. Research with cartilaginous fish, especially sharks, is
more extensive (Lopez <u>et al.</u> 1974; Sigel <u>et al.</u> 1978; Pettey & McKinney
1981). In experiments on density gradient centrifugation separated
lymphocytes from the nurse shark, Sigel <u>et al.</u> (1978) found that
following centrifugation three layers containing lymphocytes were
produced and these cells from all bands responded to high concentrations
of Con A. With PHA, only lymphocytes from the bottom layer responded by
blastogenesis. This study, although showing some difference in the
response to mitogens, only indicates the existence of T-like cells. More
comprehensive studies, again with nurse sharks have been performed by
Pettey and McKinney (1981). They separated the leucocytes according to
their adherence to glass into an adherent and a non-adherent population.
Only the adherent cells showed cytotoxic reactions following stimulation
with Con A, PHA and LPS and formed E-rosettes, while the non-adherent
cells had surface associated immunoglobulin. Interpretation of these
results in terms of B and T cell dichotomy is difficult as the
experiments used mixed populations of cells and both adherent and
non-adherent cultures consisted of lymphocytes, monocytes, thrombocytes,
granulocytes and blast cells, although more lymphocytes were found
within the non-adherent population.

Mitogenicity studies with teleost lymphocytes are far more numerous and
meaningful. Lymphocytes from carp, <u>Cyprinus carpio</u> (Caspi <u>et al.</u> 1982,
1984; Rosenberg-Wiser & Avtalion 1982; Cenini & Turner 1983), channel
catfish, <u>Ictalurus punctatus</u> (Clem <u>et al.</u> 1981; Clem <u>et al.</u> 1984;

Sizemore et al. 1984; Miller et al. 1985), rainbow trout, Salmo
gairdneri (Etlinger et al. 1976; Bogner & Ellis 1977; DeLuca et al.
1983), brown trout, Salmo trutta (Blaxhall & Sheard 1985) and bluegill,
Lepomis macrochirus (Cuchens & Clem 1977) and from other species, have
been used in mitogenicity studies. Some of the earliest work was carried
out by Etlinger et al. (1976) with Con A and LPS. In these experiments,
rainbow trout thymocytes responded well to the T cell mitogen Con A but
not to the B cell mitogen LPS, while lymphocytes from the anterior
kidney responded to LPS but not Con A. These results could be explained
by the existence of T-like cells in the thymus and mainly B-like cells
in the anterior kidney. In C. carpio, Caspi et al. (1984) found that
peripheral blood leucocytes in culture would respond to PHA, Con A and
LPS. More importantly, PHA and Con A exposed cells ceased to synthesise
DNA when the mitogen was removed while LPS treated cells continued to
proliferate in the absence of the mitogen. Furthermore, PHA and Con A
incubated cells when transferred into a medium containing only LPS
failed to maintain their proliferation showing the specificity of these
lymphocytes for a particular mitogen. Scanning and transmission electron
microscopy of LPS and PHA induced cells in soft agar cultures revealed
that there were morphological differences between these two populations
of blast cells. PHA induced cells were elongated, contained many
mitochondria and were smooth surfaced, whereas LPS incubated cells were
more rounded, packed with RER and ribosomes and were often covered in
microvilli.

Lymphocytes from the bluegill, L. macrochirus also showed interesting
results with mitogens (Cuchens & Clem 1977). These cells were isolated
from the anterior kidney, thymus and spleen. Cells derived from the
thymus, unlike those in the rainbow trout, responded to both classical B
(LPS) and T (Con A and PHA) cell mitogens. Some heterogeneity was
observed, however, using cells from the anterior kidney, with one
population responding to PHA and Con A but not LPS (T-like) and another
to only LPS (B-like). These sub-populations were further investigated by
incubating lymphocytes from the anterior kidney with rabbit antiserum to
bluegill brain tissue. The rationale for this approach is that murine T
lymphocytes and brain cells exhibit the surface Thy-1 antigen and by
raising antiserum to these brain cells and incubating this with
complement and lymphocytes the T cell population is destroyed. In the
bluegill, lymphocytes from the anterior kidney treated with antiserum to
brain tissue failed to respond to the T cell mitogen PHA, whereas the
LPS response was enhanced compared with untreated lymphocyte cultures
from the anterior kidney. Hence, the PHA sensitive lymphocytes from the
bluegill share a common antigenic determinant with brain tissue in a way
similar to murine T cells. Further evidence for lymphocyte heterogeneity
in bluegills was that the PHA and Con A sensitive cells did not, while
the LPS reactive lymphocytes, did form stable rosettes with rabbit
erythrocytes. Approximately 75% of the lymphocyte population from the
bluegill anterior kidney are T cell-like as determined by PHA/Con A
responsiveness, while the remaining 25% are B cell like in their
reaction to LPS.

Recent studies by Clem and colleagues with the channel catfish, I.

punctatus have provided the most compelling evidence for the existence
of T and B cell analogues in fish (Clem et al. 1981; Clem et al. 1984;
Sizemore et al. 1984; Miller et al. 1985). They used a number of
techniques including cell separation and mitogenicity assays to
characterise the lymphocytes. Their significant results will be
discussed later in this section.

Surface immunoglobulin. In mammals, B lymphocytes readily display
surface immunoglobulin (sIg) where it acts as a receptor. The nature of
this receptor has been subject to sophisticated biochemical and
molecular studies which have shown it to be an IgM-like monomeric
immunoglobulin with 2 heavy (H) and 2 light (L) chains. On the other
hand, the existence of surface immunoglobulin as a receptor on T
lymphocytes has been a matter for conjecture for a number of years.
Current thought is that the receptor is composed of two chains with
variable and constant regions which have structural homologies with Ig
domains (Roitt et al. 1985). This receptor is in close association with
the T3 glycoprotein and these co-operate during antigen recognition.

Early studies using polyclonal antisera to free Ig to demonstrate the
localisation of sIg on the lymphocytes of fish produced some surprising
results. Surface immunoglobulin positive cells were found not only in
anterior kidney and spleen, as expected, (i.e. B cells) but also in the
thymus (Ellis & Parkhouse 1975; Clem et al. 1977; Etlinger et al. 1977;
Warr et al. 1977, 1979; DeLuca et al. 1978). These results might suggest
that only one population of lymphocytes exists in fish, but important
differences have been demonstrated in the receptors on thymus-derived
and kidney/spleen-derived lymphocytes. Fiebig et al. (1977) studying the
molecular weights and polypeptide chain composition of
membrane-associated Ig from carp lymphocytes found that the Ig monomer
of anterior kidney lymphocytes has a molecular weight of 220,000 while
the molecular weight of that from the thymocytes is 260,000. Blood
lymphocytes (a mixture of T and B cells?) displayed both molecular
weight varieties. They also reported that the thymocyte Ig possessed two
unique components with molecular weights of 35-40,000 and 110,000. This
work was later extended in a series of studies (Fiebig et al. 1980,
1983) to show that the thymocyte sIg has H chains but no L chains and
does not have all the determinants specific for the μ chain of humoral
(free) IgM. Other studies on sIg have also been undertaken by Warr and
colleagues with rainbow trout (Marchalonis et al. 1978; Warr et al.
1979, 1980, 1983) and goldfish (Warr et al. 1976, 1977; Ruben et al.
1977; Warr & Marchalonis 1977) as model animals. For a detailed review
of these studies and the available methodology see Warr & Marchalonis
(1982) and Warr et al. (1984). They too showed that the sIg molecules on
T-like and B-like lymphocytes of these fish differed. Warr et al. (1976)
found that the sIg of goldfish thymic lymphocytes was unlike that of
splenic lymphocytes in both solubility properties and in mobilities of
the H chains on electrophoresis in SDS-containing buffers. Unlike in
carp though, both L and H chains were found associated with the membrane
of thymic lymphocytes in goldfish (Warr et al. 1976).

More recently, studies on sIg on fish lymphocytes have employed
monoclonal antibodies as more specific probes to these molecules. The
immunohistochemical studies of Secombes et al. (1983b) with carp
lymphocytes used a series of mouse monoclonal antibodies raised to carp
thymocytes and carp serum Ig. Selected clones of mouse monoclonal
antibody producing cells showed differences between reactions to
thymocytes, peripheral lymphocytes and reticular cells consistent with
T/B cell diversity. Fiebig et al. (1983) have also examined carp
lymphocytes from various tissues with a panel of 10 monoclonals raised
to carp IgM. Not all the monoclonal antibodies reacted with the carp
thymocytes, whereas all 10 monoclonals bound to the peripheral blood
lymphocytes, albeit with varying avidities. Lobb & Clem (1982) from
their studies with monoclonal antibodies to channel catfish Ig noted
that one population of lymphocytes contained abundant sIg (B cells?)
while another did not (T cells?). One other important finding was that
polyclonal antisera to Ig demonstrated almost 100% sIg positive cells
while with the monoclonals, only ca. 40% of the lymphocytes showed
binding. This result raises the question of just what does the
polyclonal antisera react with on the lymphocyte surface? One
possibility is cross-reactivity of some anti Ig antisera with
carbohydrate moieties on the cell membranes (Yamaga et al. 1978). This
former work has since been further extended using separated populations
of sIg positive and sIg negative lymphocytes, and macrophages (Sizemore
et al. 1984; Miller et al. 1985). Cells which are sIg positive responded
to LPS regardless of the presence of macrophages, while the sIg negative
cells undergo blastogenesis following exposure to either Con A or LPS
but only with the addition of macrophages. These results reinforce the
view that cells with B and T cell-like properties exist in fish and also
show a role for accessory cells such as macrophages in these processes.
The possible functioning of macrophages in specific immunity will be
discussed later in this review.

Lymphocyte separation methods. Leucocyte separation is usually only the
first stage in any study on lymphocyte heterogeneity. B-like cells from
the goldfish have been found to bind to nylon wool in a manner akin to
some mammalian B lymphocytes (Ruben et al. 1977). Some separation of
different populations of lymphocytes was also achieved by
Percoll/Ficoll-Hypaque density gradient centrifugation (Sigel et al.
1978; Blaxhall & Sheard 1985). The 'panning' method of Sizemore et al.
(1984) used coated dishes to bind with the sIg positive cells leaving
the non-adherent sIg negative cells free in the supernatant. One final
study is also included, for it provides much needed evidence for
lymphocyte heterogeneity in the Pacific hagfish (Gilbertson et al.
1986), a descendant of some of the first vertebrates. Flow cytometric
analysis demonstrated two populations of cells from the peripheral
blood; one of large leucocytes consisting of monocytes and granulocytes
and one of small leucocytes consisting of lymphocytes. Subsequent
density gradient centrifugation provided a partially pure population of
small leucocytes over 70% of which were sIg positive. These were further
studied by flow cytometric analysis with monoclonal antibodies raised to
hagfish leucocytes and demonstrated some heterogeneity of the small
leucocytes. These cells (lymphocytes only?) need further functional

studies similar to those described above to see if T and B cell analogues are present.

In summary, while we still probably cannot give fish lymphocytes the label "T" and "B", there is ample evidence for subpopulations in both agnathans and gnathostomates. This is without doubt one of the most rewarding fields in fish immunobiology and should provide not only vital comparative information but also some insight into the nature and functions of the T cell receptor.

Thrombocytes (Figs 31-35)

Thrombocytes are commonly found in the peripheral blood of fish where following smearing and Romanowsky staining they appear in four morphological forms, namely, spiked, spindle, oval or fragmented (Ellis 1977a) (Figs 31, 32). If smearing is carried out too slowly, or the blood allowed to partially clot before smearing, then the thrombocytes are often seen in aggregates displaying a ragged appearance (Fig. 32). When blood is collected in a suitable anticoagulant such as sodium citrate, however, the thrombocytes are oval - spindle-shaped (Shepro et al. 1966) and this is presumably their normal in vivo appearance. If placed on clean glass slides, thrombocytes will attach and spread out to reveal a number of small cytoplasmic granules (Fig. 33) but these are lost during subsequent incubation and the cells eventually become degenerate (Mainwaring & Rowley 1985a). Under similar conditions, granulocytes and monocytes spread out and locomote but unlike thrombocytes maintain their viability for long periods. Due to the different morphologies of the thrombocyte in blood smears, their identification has proven difficult. Indeed, the oval form can be easily mis-identified as a lymphocyte (Saunders 1968). Hence, differential cell counts for thrombocytes from Romanowsky stained blood smears are likely to be inaccurate with investigators categorising lymphocytes as thrombocytes and vice versa. Suitable cytochemical methods can improve matters. For example, in studies with plaice, Pleuronectes platessa, Ellis (1976) found in contrast to lymphocytes, that acid phosphatase activity in thrombocytes mainly occurred as a single mass close to the nucleus, while PAS stained a small number of coarse granules. Although such staining patterns are likely to be variable from species to species similar results were reported in the dogfish, S. canicula (Mainwaring unpublished) and in the channel catfish, Ictalurus punctatus (Cannon et al. 1980a). In this latter fish, thrombocytes stained supravitally with Janus green and neutral red with a densely stained mass close to the nucleus. Such methods may therefore prove useful for distinguishing thrombocytes from lymphocytes in some species of fish.

Such cytochemical studies on fish thrombocytes are relatively few but even so similar results are seen for many of the tests employed (Table 3). For example, thrombocytes usually lack peroxidase but stain for acid phosphatase and PAS. Results for Sudan black B and alkaline phosphatase are somewhat more variable.

Ultrastructural studies have shown that the cytoplasm of fish

TABLE 3 Cytochemical Features of Fish Thrombocytes

Genus/species	Common name	Acid phosphatase	Alkaline phosphatase	Peroxidase	PAS	Sudan black B	References
				Cytochemical Test			
Lampetra planeri	Brook lamprey	-	-	-	+	-	Fey (1965)
Lampetra fluviatilis	River lamprey	+	ND	-	ND	ND	Page & Rowley (1983)
Scyliorhinus canicula	Dogfish	-	-	-	+	-	Parish et al.(1986)
Raja clavata	Thornback ray	-	-	-	+	-	Mainwaring (unpub.)
Blennius pholis	Blenny	+	+	-	+	-	Mainwaring (unpub.)
Carassius auratus	Goldfish	+/-	-	-	+	ND	Fey (1965)
Cyprinus carpio	Carp	+	-	-	+	ND	Fey (1965)
Pleuronectes platessa	Plaice	+	-	-	+	-	Ellis (1976)
Salmo trutta	Brown trout	ND	ND	-	-	-	Blaxhall & Daisley (1973)
Scardinius erythrophthalmus	Rudd	+	-	-	+	+	Mainwaring (unpub.)

KEY: + = positive; - = negative; +/- = inconsistent or weak; ND = not determined

thrombocytes resembles mammalian platelets with a surface-connected
canalicular system, peripheral bands of microtubules and small granules
(Figs 34, 35; compare with Figs 48-51 in Chapter 6) (Shepro et al. 1966;
Tomonaga et al. 1973b; Ferguson 1976a; Cannon et al. 1980a, Zapata &
Carrato 1980; Barber et al. 1981; Daimon & Uchida 1985). In mammals, two
types of granules are found in platelets, the alpha granules containing
hydrolytic enzymes and cationic substances which increase vascular
permeability, and the dense bodies containing ADP, ATP, calcium and
5-hydroxytryptamine (Zucker-Franklin & Grusky 1979; Taussig 1984). In
fish, only a single population of granules has been reported (Cannon et
al. 1980a) and these have a clear space between the granule contents and
the bounding membrane which is characteristic of lysosomes (Daems et al.
1972). These organelles, therefore, are more like the alpha granules
than the dense bodies of platelets. The degree of granulation and
vacuolation of fish thrombocytes is variable and this has led some
workers to suggest the existence of sub-populations (e.g. Zapata &
Carrato 1980; Pica et al. 1983). In our view, this variation probably
represents either different maturation stages of one cell type or some
functional heterogeneity within the thrombocytes. The nuclear shape and
chromatin pattern of thrombocytes is also very characteristic in thin
sections. The nucleus is usually strongly indented with prominent
heterochromatin which transverses the nucleus (Shepro et al. 1966;
Cannon et al. 1980a; Barber et al. 1981; Parish et al. 1986). These
features further enable thrombocytes to be distinguished from
lymphocytes and monocytes in thin sections.

Finally, some mention of the so-called "spindle cells" of hagfishes
should be made due to their similarity to thrombocytes. Hagfishes have
been found to have both thrombocytes, with the characteristic
ultrastructural features seen in other fishes (Mattisson & Fänge 1977;
Tomonaga et al. 1973b), and spindle cells, which in smears look similar
to the spindle-shaped thrombocytes of some jawed fish. In vitro, the
spindle cells round off to form "lymphocyte-like cells" (Smith 1970)
particularly in the presence of vinblastine and colchicine (Fänge et al.
1974; Mattisson & Fänge 1977). Ultrastructurally such features as a
clefted nucleus and vesicles/granules in the cytoplasm are reminiscent
of thrombocytes but these cells do not contain a canalicular system and
the cytoplasm is filled with free ribosomes (Mattisson & Fänge 1977). It
is likely that such cells are either some form of immature stem cells or
lymphocytes.

Granulocytes (Figs 36-60)
No other leucocyte type has caused so much confusion in the
fish literature or been a subject for such dispute. Fish haematologists
have spent the last few decades debating the structure, heterogeneity
and functions of these cells. As in other vertebrates, the
classification of fish granulocytes has been based firstly, on their
appearance following Romanowsky staining and secondly, on their
ultrastructural features. Only rarely have functional studies been
utilised to aid in the identification of the different granulocyte
types. It is important for the reader to realise that although the

staining reaction of a particular granulocyte is for example
eosinophilic, this does not necessarily imply any structural or
functional homology with mammalian eosinophilic granulocytes.

Agnathans. Hagfishes have been reported to possess only a single
population of granulocytes which most authors have categorised as
neutrophilic/heterophilic (Table 4). Östberg et al. (1976) found that
the granulocytes of Myxine glutinosa were slightly eosinophilic but even
so believed that they had functional homology (though largely without
any evidence) with the neutrophilic granulocyte of mammals. In the
granulocytes of M. glutinosa, Mattisson & Fänge (1977) recognised only a
single population of granules in the cytoplasm, while Linthicum (1975)
in a study on the haematology of Eptatretus stouti, identified both
"primary" and "secondary" granules. The "primary" granules were large
and rounded and predominated in the metamyelocyte stage, whereas the
"secondary" granules were more electron-dense and cylindrical or
rod-shaped and increased in number with granulocyte maturation. Whether
the primary and secondary granules are separate populations or part of
the same maturation series is unclear. Cytochemical studies are limited
to a report on the absence of peroxidase in these cells (Johansson
1973).

The appearance of different granulocytes types seems to have taken place
relatively early in vertebrate evolution as lampreys have been found to
possess neutrophilic/heterophilic, eosinophilic, and occasionally,
basophilic granulocytes (Table 4). What is unusual, at least in Lampetra
spp., is that both the neutrophils/heterophils and eosinophils are
present in ammocoete larvae but only the former cell type is found in
the adults (Potter et al. 1982; Rowley & Page 1985a).
Neutrophils/heterophils have a bilobed or trilobed nucleus with patches
of pale blue-purple (azurophilic) granules, whereas the eosinophils are
filled with coarse bright orange-red granules and the nucleus is
eccentric and occasionally bilobate (Figs 36, 37). The fine structure of
lamprey neutrophils/heterophils and eosinophils has been described by
Kelényi & Larsen (1976), Potter et al. (1982), Page & Rowley (1983) and
Rowley & Page (1985a). Both immature and mature neutrophils/heterophils
are present in the blood of Lampetra fluviatilis and Petromyzon marinus.
Immature forms are characterised by an often non-lobate nucleus and a
cytoplasm with a few profiles of RER, granules and prominent Golgi
complexes (Fig. 38). Mature forms have a lobed nucleus and the cytoplasm
contains many granules but little RER and no Golgi complex (Figs 36,
39). The cytoplasmic granules in both immature and mature
neutrophilic/heterophilic granulocytes are morphologically heterogeneous
(Figs 38-40). The smallest (Type I) are round-oval, 0.1-0.3µm in
diameter and contain a fine flocculent material. Type II granules are
generally larger, 0.2-0.4µm in diameter, more electron-dense and round
to sausage-shaped. The final type (Type III), are sausage-shaped,
0.25-0.6µm in length, contain a central electron-dense or
electron-lucent crystalline rod, and the surrounding matrix is usually
homogeneous or has a striated pattern (Fig. 40; Page & Rowley 1983).
Type I and II granules predominate in the cytoplasm of immature cells

TABLE 4   Granulocyte Heterogeneity in Agnathans

Genus/species	Neutrophilic/heterophilic granulocytes	Eosinophilic granulocytes	Basophilic granulocytes	References
Hagfishes:				
Eptatretus burgeri	+	-	-	Tomonaga et al. (1973b)
Eptatretus stouti	+	-	-	Linthicum (1975)
Myxine glutinosa	+	-	-	Östberg et al. (1976)
Myxine glutinosa	+ (50)[a]	-	-	Mattisson & Fänge (1977)
Lampreys:				
Lampetra sp. (larva)	+ (8.0)	+ (1.2)	-	Rowley & Page (1985a)
Lampetra fluviatilis and L. planeri (larva)	+	+	-	Percy & Potter (1976)
Lampetra planeri (larva & adult)	+	+	+	Fey (1966a)
Lampetra fluviatilis (adult)	+ (49.6)	-	-	Page & Rowley (1983)
Lampetra fluviatilis (adult)	+	-	-	Potter et al. (1982)
Lampetra japonica (adult)	+	-	-	Fujii (1981)
Petromyzon marinus (adult)	+	+	+ (very rare)	Piavis & Hiatt (1971)

a) Values in parentheses are the % of the total leucocyte count

while Type II and III are common in the mature neutrophils/heterophils
(compare Figs 38, 39). This feature, together with an increase in acid
phosphatase activity between Type I - III granules, suggests that they
are part of a single maturation series (Kelényi & Larsen 1976; Page &
Rowley 1983). Cytochemical studies with lamprey neutrophils/heterophils
are limited to a few reports (Kelényi & Larsen 1976; Page 1983). The
cells are acid phosphatase, PAS and esterase positive but peroxidase and
Sudan black B negative.

Immature and mature eosinophilic granulocytes are found in the blood of
larval Lampetra sp. They are readily distinguished from the
heterophils/neutrophils in smears due to obvious staining differences,
and in thin sections because of their large, ca. 0.6μm in diameter,
homogeneous, electron-dense granules (Figs 37, 41). Cells equivalent to
eosinophilic promyelocytes and myelocytes have been found in the
protospleen and kidney of larval Lampetra sp. (Page 1983), demonstrating
their separate lineage from the neutrophilic/heterophilic forms.
Cytochemical and histochemical data on these cells is sparse, although
in Lampetra they are peroxidase negative at pH 7.6 and 11, and acid
phosphatase positive (Rowley & Page 1985a). Fey (1966a) has reported a
variable reaction for peroxidase in the eosinophils of adult brook
lampreys, L. planeri, but an absence of this enzyme in the larvae.

Except for Fey (1966a) and Piavis & Hiatt (1971), who have described
basophilic granulocytes in Lampetra and Petromyzon, respectively, all
other authors have failed to identify this cell type (Table 4). The
existence of true basophils in lampreys is questionable. Piavis & Hiatt
(1971) stated that they are rarely seen in peripheral blood and are
absent from the supraneural body, which is the main haemopoietic organ.
Ultrastructural studies (e.g. Barber, unpublished, reported in Potter et
al. 1982; Page & Rowley 1983) have failed to find any granulocytes which
do not resemble either the neutrophilic/heterophilic or eosinophilic
forms.

Chondrichthyes. Studies on elasmobranch haematology reveal a complex
picture of granulocyte heterogeneity in this group (Table 5). Many
authors have noted the existence of eosinophilic, heterophilic and
neutrophilic granulocytes but few have reported the presence of
basophilic granulocytes.

Confusion exists in much of the elasmobranch literature in the use of
the term heterophil. For example, Mattisson & Fänge (1982) classified
one type of granulocyte from the shark, Etmopterus spinax, as
heterophilic even though the granules stained with eosin. Furthermore,
Saunders (1966) regarded cells with small rod-shaped eosinophilic
granules as heterophils. This could account for the apparent failure to
mention eosinophilic granulocytes in other reports (e.g. Sherburne
1974). Clearly, if a particular granulocyte stains strongly with eosin
then it is better classified as eosinophilic even if it may resemble
mammalian neutrophils/heterophils in other ways (e.g. granule structure,
nuclear lobation etc.). If, however, this cell type shares many

TABLE 5   Granulocyte Heterogeneity in Elasmobranchs

Genus/species	Common name	Heterophils/neutrophils	Eosinophils	Basophils	References
Sharks and related forms:					
Etmopterus spinax	blue velvet shark	+[a]	+[b] (5)	–	Mattisson & Fänge (1982)
Ginglymostoma cirratum	nurse shark	+[c]	+	–	Hyder et al. (1983)
Mustelus canis		+ (5)	+ (17.5)	–	Reznikoff & Reznikoff (1934)
Mustelus asterias	stellate smooth hound	+[1 type] (4)	+[3 types] (31)	–	Mainwaring (unpublished)
Prionace glauca	blue shark	+	+	–	Johansson-Sjöbeck & Stevens (1976)
Scyliorhinus canicula	lesser spotted dogfish	+[1 type] (2)[d]	+[3 types] (36)[d]	–	Morrow & Pulsford (1980); Mainwaring & Rowley (1985a,b)
S. canicula	lesser spotted dogfish	+[1 type]	+[2 types]	–	Parish et al. (1985, 1986)
Squalus acanthias	spurdog (spiny dogfish)	+[2 types][e] (9.3)	–	–	Sherburne (1974)
Squalus acanthias	spurdog (spiny dogfish)	+[1 type] (1)	+[3 types] (41)	–	Mainwaring (unpublished)
13 species of sharks	–	+[a]	+	–	Saunders (1966)
Rays:					
Raja clavata	thornback ray	–	+[2 types] (16)	–	Mainwaring & Rowley (1985a)
Raja microcellata	painted (small-eyed)ray	–	+[2 types] (38)	–	Mainwaring & Rowley (1985a)
Torpedo marmorata	marbled electric ray	+[f] (21.3)	+ (19.8)	–	Pica et al. (1983)

a) Granulocytes classified as heterophils but have eosinophilic granules (see text)
b) Values in parentheses are the % of the total leucocyte count
c) EM study only and so classification as a heterophil/heterophil is tentative
d) Figures given from Morrow & Pulsford (1980) though highly variable
e) One type described as heterophilic and one as neutrophilic
f) Cell type is described as neutrophilic by authors

functional properties with the heterophils/neutrophils of higher
vertebrates, then there are grounds for terming it heterophilic. Table 5
also emphasises that two, or occasionally three types of eosinophilic
granulocytes are found in some species, and in these fish elucidating
the functional implications in this heterogeneity is particularly
intriguing (see "Granulocyte Functions").

The structure and cytochemistry of elasmobranch granulocytes is well
illustrated by cells of the lesser spotted dogfish, Scyliorhinus
canicula, which has been extensively investigated (Morrow & Pulsford
1980; Fänge & Pulsford 1983; Mainwaring & Rowley 1985a b; Parish et al.
1985 1986; Hunt & Rowley 1986a-d). No 'universally' accepted
classification scheme exists for the granulocytes of S. canicula (Table
6). Morrow & Pulsford (1980) and Mainwaring & Rowley (1985a,b) recognise
three morphologically distinct types of eosinophilic granulocytes and
one population of neutrophils, while Parish et al. (1985 1986) have
distinguished only two types of eosinophils and one of neutrophils. We
(Mainwaring & Rowley 1985a b) have termed these granulocytes G1 - G4 (=
Type I - Type IV granulocytes of Morrow & Pulsford 1980). G1
granulocytes are often the most common and have an eccentric,
occasionally bi-lobed, nucleus with a cytoplasm filled with
eosinophilic, round-ovoid granules which in thin sections are
unstructured and electron-dense (Figs 42-44; Table 6). Other
characteristic ultrastructural features include a few profiles of RER
and numerous small vesicles. The second most common granulocyte type is
the G4 (Type IV). This cell type is moderately eosinophilic and because
of its shape can be difficult to differentiate from thrombocytes (Figs
45, 46) but in thin sections the granules in the G4 are very distinctive
with their square or angular shapes (Fig. 47). G4 granulocytes have been
likened to the mammalian mast cell (Morrow & Pulsford 1980) while Parish
et al. (1986) failed to classify this cell type as a granulocyte;
instead they believe it to be a type of thrombocyte. Although this cell
type can attach to glass surfaces by fine pseudopodia (Fig. 46), unlike
the other granulocytes, it does not spread out or locomote (Mainwaring &
Rowley 1985a). G3 granulocytes are strongly eosinophilic and are easily
identified in smears and thin sections due to the position and shape of
the nucleus (Figs 48-50; Table 6). The granules are unstructured and
electron-dense, like those of the G1 and G4 granulocytes, but
characteristically are rod-shaped (Figs 49, 50). Finally, the G2
neutrophilic granulocyte is easily identified in thin section with its
small heterogeneous granules and clefted nucleus, but extremely
difficult to find in smears (Figs 51, 52). Consequently, cells
identified and classified as G2 granulocytes in smears may not be the
same cell type as the G2 granulocyte of electron microscopical studies!
Cell separation studies using Percoll density gradient centrifugation
have failed to resolve this inconsistency because this cell type is not
recovered from the bands following separation (Mainwaring & Rowley
1985a). Morrow & Pulsford (1980) have suggested that the G2 may be an
immature form of one of the other granulocyte types.

Cytochemical and histochemical studies on the G1-G4 granulocytes have
shown the absence of peroxidase, beta glucuronidase and Sudan black B

TABLE 6   Structural and Functional Features of Dogfish (S. canicula) Granulocyte Sub-populations[a]

Feature	Granulocyte type			
	G1	G2	G3	G4
% of total leucocyte count	27.5[b]	1.5	3.0	9.5
Maximum diameter (μm)	20	12	13	18
Nucleus	irregular-non lobate	horseshoe shaped	lobate (2-3 lobes)	elongate, often clefted
Granules	1 population, round-ovoid	2 populations, variable morphology	1 population, elongate	1 population, square (cuboid?)
Cytochemistry:				
Acid phosphatase	++[c]	++	++	+/-
Alkaline phosphatase	-	-	++	+
Peroxidase	-	-	-	-
Naphthyl AS-D chloroacetate esterase	+	+	++	+
PAS	++	+/-	+	-
Sudan black B	-	-	-	-
Phagocytic activity	+	?	-	-
Localisation at sites of inflammation	+	?	-	-
Amoeboid activity	+	?	+	-
Chemokinesis/chemotaxis to:				
1. zymosan activated serum/plasma	-	?	-	-
2. synthetic peptides	-	?	-	-
3. bacterial culture filtrates	-	?	-	-
4. activated immune serum	-	?	-	-
5. leukotriene B_4	+	?	+	-

a) Data from Mainwaring & Rowley (1985a,b); Hunt & Rowley (1986 b-d)
b) Values are very variable
c) Key: ++ strong activity; + positive reaction product; +/- variable reaction product; - no reaction product

staining, but the presence of a range of enzymes including acid
phosphatase, aryl sulphatase, and acid naphthyl AS-D chloroacetate
esterase (Mainwaring & Rowley 1985b; Table 6). None of these
cytochemical methods allows for a precise identification of the
granulocyte types using a particular enzyme or group of enzymes.
Information on the possible interrelationships of the four granulocyte
types is still awaited, but the distinct granule morphologies, the
variation in their functional properties (see section on granulocyte
functions) and the existence of all of these cell types in the
haemopoietic tissues, suggests that they are probably separate cell
types.

Other members of the shark and dogfish assemblage show similar
subpopulations of eosinophilic granulocytes (Table 5). Most readily
identified are those probably morphologically and functionally(?)
equivalent to the G1 and G4 granulocytes. For example, in the stellate
smooth hound, Mustelus asterias, and the spurdog, Squalus acanthias,
four granulocyte types can be distinguished with some morphological and
cytochemical similarities to the G1-G4 granulocytes already described in
the dogfish, S.canicula (Mainwaring & Rowley unpublished observations).
Identification of the neutrophilic granulocyte (G2) in thin sections is
again a problem in these two species. In the nurse shark, Gingylmostoma
cirratum, in contrast to the above reports, Hyder et al. (1983) only
found one population of eosinophils. These cells have granules with a
substructure reminiscent of murine eosinophils with a central dense
crystalline core surrounded by a matrix with a paracrystalline
substructure (Fig. 53a-c). Two populations of eosinophils have also been
found in the rays, Raja clavata and R. microcellata (Mainwaring & Rowley
1985b). These are probably most closely related to the G1 and G3
granulocyte types of dogfish, though how they compare functionally is
unknown. What is particularly interesting, however, is the lack of
heterophils/neutrophils and cells morphologically similar to the G4 type
of sharks and dogfish (Table 5).

Finally, it is pertinent at this stage to compare some of the
morphological features of elasmobranch eosinophilic granulocytes with
their supposed counterparts in mammals. In this context, Kelényi &
Nemeth (1969) produced an extremely useful list of the main
morphological characteristics of mammalian eosinophils. As described by
Mainwaring & Rowley (1985b), the eosinophilic granulocytes from
elasmobranchs share few of the morphological/cytochemical
characteristics of mammalian eosinophils. This, together with the
results from limited functional studies (see section on granulocyte
functions), makes the existence of an homology between elasmobranch and
mammalian eosinophils very unlikely.

Osteichthyes. There is great variation within the bony fishes in both
the abundance and staining reactions of the granulocytes (Tables 7-10).
This suggests that different populations of granulocytes have arisen
several times during the divergent evolution of this group. Most
morphological and functional information on the granulocytes of bony

TABLE 7   Granulocyte Heterogeneity in Teleosts – 1. Cyprinids

Genus/species	Common name	Neutrophilic/heterophilic granulocytes	Eosinophilic granulocytes	Basophilic granulocytes	References
Carassius auratus	Goldfish	+ (5.1)	+ (2.2)	+ (0.2)	Watson et al. (1963)
"	"	+	+	+ (rare)	Weinreb (1963)
"	"	+ (37.3)	+ (36.8)	– (0)	Davies & Haynes (1975)
"	"	+ (5.8)	+ (1)	–	Javaid & Akhtar (1977)
"	"	+	+	–	Barber & Westermann (1978)
"	"	+	+	+	Garavini & Martelli (1981)
Cyprinus carpio	Carp	+	+	+	Bielek (1981)
"	"	+ (1)	+ (8)	+ (1)	Cenini (1984)
Cyprinodon variegatus	Stripped killifish	–	+ (5.3)	–	Gardner & Yevich (1969)
Fundulus heteroclitus	Common mummichog	–	+	–	Gardner & Yevich (1969)
Tinca tinca	Tench	+	+	+	Bielek (1981)

fishes comes from research with the cyprinids (carp, goldfish etc.) and
salmonids (salmon, trout), and so this will be considered first. A brief
description of granulocyte structure and heterogeneity in some other
fish of economic and/or evolutionary importance will then be considered.

Table 7 shows the distribution, staining characteristics and relative
abundance of granulocytes in the cyprinids. The most commonly studied
fish is the goldfish, Carassius auratus, and all studies report the
presence of neutrophilic/heterophilic and eosinophilic granulocytes,
though the existence of basophils is again doubtful. Both Watson et al.
(1963) and Weinreb (1963), however, have commented on the presence of
basophils but had difficulty finding such cells in the peripheral blood.
Garavini & Martelli (1981) also gave no details of the structure or
relative abundance of basophils, only that they are alkaline phosphatase
and peroxidase negative, and no micrographs have been presented for this
cell type. Weinreb (1963) reported that the neutrophil was the most
commonly encountered granulocyte in goldfish, but gave no quantitative
data. Such cells were described as ca. 10μm in diameter with an
eccentric nucleus, and as in other fish, only a partially lobate
appearance (Weinreb 1963). Details of the granule substructure cannot be
commented on due to the poor fixation seen in the micrographs
accompanying this latter paper. Garavini & Martelli (1981) stated that
two types of granules are formed in immature and mature goldfish
neutrophils. Regarding eosinophils, according to Weinreb (1963), they
are smaller cells, ca. 7.5μm in diameter with an eccentric nucleus which
may be indented, sausage-shaped or bilobed. The granules are larger than
those of neutrophils, averaging 0.4μm in diameter, and have a few
centrally located crystalline rods not unlike those seen in the
neutrophilic granulocytes of lampreys (Fig. 40) and some other bony
fish. Similar substructures were reported in the study of Garavini &
Martelli (1981). Finally, the basophilic granulocytes according to
Weinreb (1963) are some 10-20μm in diameter with a cytoplasm enclosing
basophilic granules. Ultrastructurally, these granules are homogeneous
and average 0.8μm in diameter.

Another commonly researched cyprinid is the carp, Cyprinus carpio with
which two major ultrastructural studies have been undertaken (Bielek
1981; Cenini 1984). Both report the presence of all three types of
granulocyte and there is general agreement about the identification of
neutrophils, eosinophils and basophils and interpretation of granule
substructure (Table 8). Neutrophils/heterophils are characterised by
granules, some with a central electron-dense or electron-lucent
crystalline rod (Fig. 54) which may have a fibrillar appearance (Cenini
1984). The substructure of these granules is similar to that in the
eosinophils of C. auratus and the heterophils/neutrophils of lampreys
(Page & Rowley 1983) and some bony fishes (see Figs 40, 60). Bielek
(1981) found that immature granules of carp heterophils/neutrophils are
markedly peroxidase positive at pH 7.6 but only weakly so at pH 9.0
(Table 8). Mature granules are also peroxidase negative at both pH
values. The heterogeneity in staining, size and peroxidase content of
the granules in the heterophils/neutrophils could indicate either the
presence of several subpopulations or of different developmental stages

TABLE 8   Major Characteristics of the Granulocytes in the Carp, *Cyprinus carpio*

(data from Bielek 1981 and Cenini 1984)

Feature	Heterophilic/neutrophilic granulocytes	Eosinophilic granulocytes	Basophilic granulocytes
% of total leucocyte count	1	8	1
Cell diameter (μm)	5–8	6–8	?
Nucleus	kidney-shaped, eccentric	oval-shaped, eccentric	eccentric
Granules: size (μm)	0.1–0.4 (small) 0.5–0.6 (large)	0.15–0.8	0.15 – 1.4
shape	round-ovoid	round–sausage shaped	round
substructure	central crystalline rod with 3.5mm periodicity substructure in some large granules	variable, electron-dense and electron-lucent zones	variable, electron-dense and electron-lucent zones
peroxidase	in large granules, positive at pH 7.6 weak at pH 9	weak at pH 7.6, strong at pH 9.0	negative

of a single population. Bielek (1981) apparently favours a single
population but does mention a small separate, peroxidase-negative type
of granule formed at the myelocyte stage. Further cytochemical and
biochemical studies are necessary to resolve this problem.

The eosinophilic granulocytes have granules which can be difficult to
differentiate from those in the basophils, although according to Cenini
(1984) they differ in size and structure (Table 8; Figs 55, 56). Cenini
(1984) also believes that intermediates between eosinophils and
basophils are present in carp but this was not substantiated by Bielek
(1981). The relative ease with which the granulocyte types can be
identified in C. carpio should make this an ideal animal for further
functional and structural studies. Cell separation methods have also
been developed for carp leucocytes and these should facilitate this work
(Bayne 1986).

In salmonids, heterophilic/neutrophilic granulocytes predominate with
eosinophils and basophils either absent or present in very low numbers
(Table 9). To our knowledge, no detailed ultrastructural and
cytochemical studies have been performed with salmonid granulocytes.

The channel catfish, a siluriform fish closely related to the cyprinids,
has been widely used in immunological studies and so it is important to
review their haematology with particular reference to the granulocytes.
According to Cannon et al. (1980a,b) only "neutrophils" are present in
the blood of this species, though Williams & Warner (1976), in a light
microscope study, also identified eosinophils and basophils (Table 9).
"Neutrophils" of the channel catfish are characterised by a cytoplasm
filled with rod-shaped granules, the mature form of which have a
crystalline or striated appearance in the centre (Cannon et al. 1980a).
These granules are also strongly peroxidase positive with a reaction
product spread evenly throughout the granule contents (Cannon et al.
1980b). Similar substructures have been found in the granules from
heterophils/neutrophils of a number of teleosts including the blenny,
Blennius pholis (Mainwaring, unpublished; Figs 57, 58), rockling,
Ciliata mustela (Mainwaring unpublished), striped bass, Morone saxatilis
(Bodammer 1986) and in plaice, Pleuronectes platessa (Ferguson 1976a).
In these fish, with the exception of M. saxatilis, only one granulocyte
type has been found, the heterophil/neutrophil (Table 9), and only a
single population of granules is present in the cytoplasm. Studies with
other teleosts are relatively few but include the rudd, Scardinius
erythrophthalmus, in which the granules of the heterophils contain
crystalline cores remarkably similar to those in their counterparts in
lampreys (compare Figs 59, 60 with Fig. 40). They also have a similar
maturation series to the Type I-III granules of the latter fish.

The meagre information available on the dipnoans is given in Table 10.
It is interesting that all three granulocyte types are apparently
present but that eosinophils predominate (Ward 1969). To our knowledge,
ultrastructural and cytochemical studies of these cells are lacking.

TABLE 9   Granulocyte Heterogeneity in Teleosts - 2. Salmonids and Others

Genus/species	Common name	Neutrophilic/heterophilic granulocytes	Eosinophilic granulocytes	Basophilic granulocytes	References
Salmonids:					
Onchorhynchus nerka	Sockeye salmon	+	+/-[a]	+/-[a]	Watson et al. (1956)
Salmo gairdneri	Rainbow trout	+	-	-	Barber & Westermann (1978)
"	"	+	+/-[b]	-	Bielek (1981)
"	"	+	-	-	Klontz (1972)
Salmo salar	Atlantic salmon	+ (5-55)[c,d]	+ (0.5)[d]	-	Conroy (1972)
Salmo trutta	Brown trout	+	-	+[e]	Catton (1951)
"	"	+	-	-	Blaxhall & Daisley (1973)
Others:					
Blennius pholis	Blenny	+ (21.8)	-	-	Mainwaring (unpublished)
Chaenocephalus aceratus	Antarctic icefish	-?	+[2 types]	-	Barber et al. (1981)
Ciliata mustela	Rockling	+ (21.2)	-	-	Mainwaring (unpublished)
Ictalurus punctatus	Channel catfish	+	+/-	+	Williams & Warner (1976)
"	"	+ (15)	-	-	Cannon et al. (1980a,b)
Pleuronectes platessa	Plaice	+ (7)	-	-	Ellis (1976), Ferguson (1976a)
Scardinius erythrophthalmus	Rudd	+ (19.7)	-	-	Mainwaring (unpublished)

a) Eosinophils and basophils only present in diseased/starved fish
b) Author reports that this cell type is very scarce
c) Values in parentheses are the % of the total leucocyte count
d) Variation seen during the life cycle. Values do not take into account the thrombocytes
e) Cells rarely seen and termed 'coarse granulocytes' by author

TABLE 10   Granulocyte Heterogeneity in Lungfish (Dipnoi)

Genus/species	Common name	Heterophilic/neutrophilic granulocytes	Eosinophilic granulocytes	Basophilic granulocytes	References
Neoceratodus forsteri	Australian lungfish	+ (4)[a]	+[2 types] (70)	+ (1)	Ward (1969)
Protopterus	African lungfish	+	+	+	Barber & Westermann (1978)
Protopterus ethiopicus	African lungfish	-	+[3 types]	+	Jordan & Speidel (1931)

a) Values in parentheses are the % of the total leucocyte count

Monocytes (Figs 61, 62)

Monocytes account for a small proportion of the total
leucocyte count in fish blood. For example, Ellis (1976) reported a
value of only 0.1-0.2% in plaice, P. platessa, although other accounts
present values of between 1-8% (e.g. Cannon et al. 1980a) which agrees
with the normal figure for mammals (see Chapter 6). Some authors have
been unable, however, to find monocytes (Blaxhall & Daisley 1973), while
others fail to mention their existence (e.g. Stokes & Firkin 1971).
Whether this is due to an oversight or a real absence of such cells is
unclear.

The nomenclature used to describe monocytes in fishes is variable. For
example, they have been termed "haemoblasts" (Jakowska 1956) and
"macrophages" (e.g. Klontz 1972; Barber et al. 1981). The reader is
directed to Ellis (1977a) for a full account of the problems in monocyte
terminology. Like Ellis (1977a), and the more recent papers/reviews, the
term macrophage for fixed cells is only used here to remain in step with
the current mammalian usage for this term.

Monocytes are characterised by a prominent eccentric, kidney-shaped to
bilobed nucleus, which stains blue-purple with Romanowsky stains and an
agranular grey-blue cytoplasm (Fig. 61). The peripheral region of these
cells may be indistinct or ragged in smears perhaps due to the formation
of protoplasmic extensions. Ultrastructural reports on monocytes of
agnathans (Linthicum 1975; Potter et al. 1982; Page & Rowley 1983),
elasmobranchs (Morrow & Pulsford 1980; Hyder et al. 1983; Hunt & Rowley
1986b) and Osteichthyes (Ferguson 1976a; Cannon et al. 1980a,b; Cenini
1984) show that these cells are structurally similar from one species to
another. The cytoplasm contains numerous profiles of RER, an active
Golgi complex and centrioles usually located in the nuclear cleft, and
Golgi-derived vesicles/granules (Fig. 62). Also present are
heterophagosomes, attesting to the phagocytic activity of such cells,
and many peripheral pinocytotic vesicles. Protoplasmic extensions
commonly occur, particularly in actively phagocytic cells. All of these
features are found in the monocytes of other vertebrates (see Chapters
3-6 in this volume and Ackerman & Douglas 1980). There are relatively
few cytochemical studies on fish monocytes, and with the exception of
Cannon et al. (1980b), Bielek (1981) and Page & Rowley (1983), limited to
light microscopy (Table 11). Few trends can be seen from the results
given in Table 11, except that monocytes are acid phosphatase and PAS
positive but variably active for peroxidase. In this latter respect,
Bielek (1981) and Fey (1966b) differ in their work on the monocytes of
carp, with only the former reporting a strong reaction for peroxidase.

Macrophages (Figs 2, 63)

Although macrophages are by definition fixed cells and
rarely found in the peripheral blood, they are included here because
they are most likely derived from circulating monocytes. This latter
assumption has yet to be confirmed in any species of fish. Macrophages
are found in connective tissue (Ellis 1977a), most lymphomyeloid tissues
(see section on macrophage functions and Figs 2 & 63) and

TABLE 11   Cytochemical Features of Fish Monocytes

Genus/species	Common name	Cytochemical Test							References
		Acid phosphatase	Alkaline phosphatase	Peroxidase	α NAE[a]	NASDCE[b]	PAS	Sudan black B	
Lampetra fluviatilis	River lamprey	++[c]	ND	-	ND	ND	ND	ND	Page & Rowley (1983)
Lampetra planeri	Brook lamprey	+/-	+	+/-	ND	ND	+	ND	Fey (1966b)
Raja clavata	Thornback ray	++	-	-	-	+	++	+/-	Mainwaring (unpublished)
Scyliorhinus canicula	Lesser spotted dogfish	++	+	-	++	+	+	-	Mainwaring (unpublished)
Blennius pholis	Blenny	++	+	+	+	+	+	+	Mainwaring (unpublished)
Carassius auratus	Goldfish	+	-	++	ND	ND	+	-	Fey (1966b)
Ciliata mustela	Rockling	++	+/-	-	+/-	-	+	++	Mainwaring (unpublished)
Cyprinus carpio	Carp	+	+/-	++	ND	ND	++	-	Fey (1966b)
Cyprinus carpio	Carp	ND	ND	-	ND	ND	ND	ND	Bielek (1981)
Ictalurus punctatus	Channel catfish	ND	ND	-	ND	ND	ND	ND	Cannon et al. (1980b)

a) α NAE = α naphthyl acetate esterase
b) NASDCE = naphthyl AS-D chloroacetate esterase
c) ++ = strongly positive; + = positive; +/- = inconsistent or weak; - = negative; ND = not determined

non-lymphomyeloid organs/tissues such as the liver (Awaya et al. 1977;
Ferri & Sesso 1981; Hunt & Rowley 1986a), and the peritoneal cavity
(MacArthur et al. 1984; Bodammer 1986). In many species, a proportion of
the macrophages contain melanin and other pigments including lipofuscin
and haemosiderin (Agius 1985); such cells are commonly called
melano-macrophages (Fig. 63). These melanin granules are invariably
found in vacuoles together with other debris (Rowley & Page 1985b; Hunt
& Rowley 1986a) and this reinforces the view that pigment is
incorporated into macrophages by the ingestion of melanocytes and
melanophores (Agius 1985). Ultrastructurally, macrophages have many of
the features seen in monocytes but also have an increased number of
primary and secondary lysosomes (Fig. 2).

Mast Cells
The existence of fixed cells of similar structure and
function to mast cells has until recently been a matter of conjecture.
Earlier studies (reviewed in Ellis 1977a) were at best unconvincing,
though Roberts et al. (1972) found cells in the dermis of plaice that
stained metachromatically and contained histamine but not
5-hydroxytryptamine. Progress in this field has been largely made by
Ellis (1982b; 1985). Injection of a crude extract of exotoxin from the
pathogenic bacterium, Aeromonas salmonicida, produces a marked increase
in the levels of histamine in the blood and a degranulation and
disappearance of eosinophilic granular cells in the stratum compactum of
the gut wall. Such cells are either directly or indirectly involved in
the histamine release and have therefore been likened to mast cells. A
morphological study on these cells in Salmo gairdneri by Ezeasor &
Stokoe (1980) has shown that they differ markedly from mast cells in
structure and cytochemistry. What seems to be important is that the
morphological criteria normally used to detect mast cells in tissues,
i.e. metachromasia with toluidine blue and similar stains, is not of any
assistance for studies with fish. Eosinophilic granular cells have also
been reported in the dermis and the gills; both sites where mast cells
would be expected to occur.

FUNCTIONS OF BLOOD CELLS

Erythrocytes
The role of erythrocytes in the transport of oxygen, and to
a far lesser extent carbon dioxide, is of course well known. Mention
should be made, however, of the physiological consequences of the
molecular evolution of haemoglobin within the fishes which has already
been described (see section on erythrocyte structure). Coates (1975)
reviews the possible stages in the evolution of haemoglobin within the
vertebrates. In the agnathans, although multiple haemoglobins are
present, they are formed from a single basic subunit unlike the alpha
and beta chains of gnathostomate haemoglobin (see, however, Hardisty
1979). Hagfishes represent the basic vertebrate pattern where both the
deoxygenated and oxygenated form of the molecule exists as a monomer.
Due to the existence of only monomeric haemoglobin, there is little sign

of a Bohr effect nor co-operativity between the haem binding sites and this results in a hyperbolic rather than sigmoidal haemoglobin oxygen dissociation curve (Maclean 1978). It must be borne in mind that the Bohr effect is an important modulator of haemoglobin function and allows for effective release of oxygen in tissues with a high carbon dioxide (and hence low pH) concentration (Maclean 1978). In lampreys, the haemoglobin molecule forms dimers or tetramers on deoxygenation, mimicking the situation in gnathostomates, but these dissociate to monomers with oxygenation. Hence, these molecules show some co-operativity in oxygen binding and a significant Bohr effect which generally improves the ability of the haemoglobin molecule to load and unload oxygen under different environmental conditions. According to Coates (1975) and Perutz (1976), all vertebrates above the lampreys and hagfishes have haemoglobins which form tetramers with two alpha and two beta chains. The next stage in the functional evolution of haemoglobin in fishes is its sensitivity to low molecular weight metabolites of red cells such as ATP and later on 2, 3-diphosphoglycerate. This phenomenon first occurs in the elasmobranchs though is more marked in the Osteichthyes (Coates 1975).

Lymphocytes

Lymphocytes play a central role in both the cellular and humoral arms of the specific immune system of fish. To fully review the information available on all of the functional aspects of lymphocytes would be impossible in the space available. Instead, a basic general summary will follow together with more detailed accounts on some recent findings of particular interest, such as non-specific cytotoxicity and cell co-operation reactions in antibody formation.

As in other vertebrates, contact with antigen leads to the proliferation of lymphocytes with the resultant production of humoral immunoglobulin (Ig) and/or activated T-like cells involved in graft rejection-type responses. The development of fish farming with its associated problem of disease, has resulted in an impetus for research into immunisation as a method of controlling outbreaks of diseases. Hence, a tremendous wealth of information now exists on the humoral immunopotential of these animals. Areas particularly well researched include the dynamics of antibody production (Rijkers 1982), the ontogeny of such responses (Manning et al. 1982), the practical methods available for immunization and their efficacy (Michel 1982; Ward 1982) and the structure and properties of the resultant Ig (Marchalonis 1977). In cell mediated immunity, the nature and dynamics of graft rejection have been studied in agnathans (Perey et al. 1968; Hildemann & Thoenes 1969), elasmobranchs (Perey et al. 1968) and bony fish (Hildemann & Haas 1960; Manning et al. 1982). Chronic rejection of initial allografts is the characteristic response of agnathans and Chondrichthyes, while Osteichthyes show acute reactions. All fish, however, exhibit accelerated second set graft rejection. Valuable information is now becoming available on the possible role of products such as lymphokines (e.g. Secombes 1986) and monokines (see section on macrophage functions) in the regulation of the immune responses of fish.

The site of immunoglobulin synthesis. Table 12 summarises the location of Ig synthesis in the major groups of fish. Tomonaga et al. (1984) showed the importance of both the spleen and the lamina propria of the valvular intestine of elasmobranchs in antibody formation. Some plasma cells were also found in the liver, epigonal and Leydig's organ, but the thymus enclosed no cells containing intracellular Ig. From these and other results, the elasmobranch spleen appears to be of major importance in Ig synthesis, though splenectomy in two species of sharks has been shown to make no difference to antibody levels produced following challenge (Ferren 1967).

Particularly interesting is the recent work on secretory humoral immunity. Intraperitoneal injection of sheep erythrocytes in the rainbow trout, S. gairdneri, results in the production of Ig in both the mucus and serum (St Louis-Cormier et al. 1984). The existence of plasma cells in the cutaneous dermis and mucus synthesising anti-sheep erythrocyte Ig could also be demonstrated by immunofluorescent antibody techniques. Similar results, showing secretory Ig in mucus from either the skin and/or the alimentary canal have also been reported in channel catfish (Ourth 1980) and plaice (Fletcher & Grant 1969). The GALT-like tissue identified in many different fishes (e.g. Tomonaga et al. 1984; Hart et al. 1986) is probably responsible for the Ig in the intestinal mucus. Although the Ig from mucus and serum are antigenically identical (Fletcher & Grant 1969; Lobb & Clem 1981a), molecular differences are found in the Ig of serum, bile and cutaneous mucus in the sheepshead, Archosurgus probatocephalus (Lobb & Clem 1981a,b). The presence of Ig, agglutinins (e.g. Kamiya & Shimizu 1980) and lysozyme (e.g. Fletcher & White 1973; Ourth 1980) in the mucus of fish is obviously very important in protection from macrobial and microbial agents using the skin as a site of invasion/attachment.

TABLE 12   Location of Antibody-forming Cells in Different Types of Fish

Fish	Location[a]	References
Lampreys larval stage	protospleen	Zapata et al. (1981) Fujii (1982)
adult	supraneural body	Hagen et al. (1983)
Elasmobranchs	spleen > intestine > liver, Leydig organ, epigonal organ	Tomonaga et al. (1984)
Chondrosteans	spleen, pericardial haemopoietic tissue	Clawson et al. (1966)
Teleosts	pronephric and mesonephric kidney > spleen[b]	Rijkers et al. (1980)

a) A small number of plasma cells is also found in the blood of many fish
b) Ninety per cent of the total plaque forming cells following injection of sheep erythrocytes are in the kidney, 5% in the spleen, < 1% in thymus, heart and peripheral blood.

Finally, antibody production has been shown to occur in the thymus of some fish as demonstrated by the presence of plaque forming cells (Ortiz-Muniz & Sigel 1971) and plasma cells (Zapata 1981b) in this organ. The presence of an incomplete or thin capsule around the thymus and the existence of pores, perhaps allowing communication between the thymocytes and the surrounding water (Fig. 19), indicate that at least in some fish, and in early stages of ontogeny, antigen may make direct contact with the lymphoid cells of the thymus. Lymphocytes and macrophages with sequestered antigen could also migrate into the thymus if the capsule and blood barrier are incomplete, but studies to date on lymphocyte circulation have failed to demonstrate such a process (Ellis & de Sousa 1974). Although there are such reports of the direct participation of the thymus in antibody formation, other researchers have failed to find such an event (e.g. Tomonaga et al. 1984).

Evidence for cell co-operation in antibody production. The possible existence of T- and B-like lymphocytes in fish, based on evidence from differential mitogenicity responses and the presence of surface immunoglobulin, has already been described (see section on lymphocyte structure). An alternative approach to show that lymphocyte populations analogous to T and B cells exist, and that these co-operate in the production of humoral antibody, is to use hapten-carrier complexes. Haptens are small molecules such as trinitrophenyl (TNP) which on their own are unable to evoke an immune response, but if linked to a carrier molecule such as bovine serum albumin (BSA), human gamma globulin (HGG) etc. these complexes evoke antibody production to occur to the hapten. The reaction has been shown to be carrier specific i.e. if the primary immunisation is with TNP-HGG and rechallenge occurs with the same hapten-carrier complex then a normal elevated secondary response takes place to TNP. If, however, for the rechallenge a different carrier is linked to TNP then the elevated response is lacking. This "carrier effect" is explained by the cells involved in the antibody response recognising two parts of the antigenic complex with T cells reacting with the carrier while B cells recognise the hapten. The carrier primed T cells involved in this response are T-helper cells. Using similar experimental approaches, a "carrier effect" has been demonstrated in the winter flounder, Pseudopleuronectes americanus (Stolen & Mäkelä 1980), the goldfish, C. auratus (Ruben et al. 1977; Warr et al. 1977) and in the channel catfish, I. punctatus (Miller & Clem 1984; Miller et al. 1985). The studies with catfish investigated the primary in vitro anti-hapten responses of sIg positive lymphocytes (B cells), sIg negative lymphocytes (T cells) and macrophages (monocytes) isolated from peripheral blood to thymus-dependent (TNP-KLH) and thymus-independent (TNP-BSA) hapten-carrier complexes. As expected, sIg positive cells together with macrophages produce a strong response to the thymus-independent complex, whereas both lymphocyte sub-populations and macrophages are required for a response to TNP-KLH. The nature of the hapten and carrier specificities were also tested by the depletion of antigen-reactive sIg positive and/or sIg negative lymphocytes. The results clearly demonstrated that carrier specificity is linked to sIg negative cells which probably act in a similar way to T helper cells.

One other important finding from these studies indicates how low
temperatures suppress antibody formation (Miller & Clem 1984). In fish,
it is a well-established principle that low temperatures suppress the
production of an immune response and that elevation to normal
environmental levels allows optimal humoral and/or cellular
responsiveness (see review by Avtalion 1981). Miller & Clem (1984) found
that low temperatures suppress antibody responses to thymus-dependent
but not thymus-independent antigens and therefore, the effect is
probably mediated via the T-helper cells. How temperature
inhibits/suppresses the T-helper response is unclear but it may involve
differences between T and B cells in cell membrane fluidity at low
temperatures (Abruzzini et al. 1982; Miller & Clem 1984).

The structure of immunoglobulins. The structure of fish Ig has been the
subject of a number of excellent reviews (e.g. Litman 1976; Marchalonis
1977; Warr & Marchalonis 1982) to which the reader is referred. The
following account gives a brief synopsis of the major findings and
incorporates some of the more recent results.

The majority of fish produce a single type of Ig most closely related to
IgM of other vertebrates, but there is growing evidence for the presence
of other Ig classes and subclasses.

Lampreys and hagfishes have been shown to form Ig to a range of
antigens. In hagfishes, earlier investigations failed to induce Ig
formation (e.g. Papermaster et al. 1964), and it was only in 1970 that
Thoenes & Hildemann first reported success in raising antibody to
erythrocytes in these animals. The antibody produced is of high
molecular weight and considered to be similar to IgM. Further detailed
descriptions of the molecule are wanting. In lampreys, the Ig is
characterised by a tendency to dissociate due to the lack of disulphide
bonds between the heavy(H) and light(L) chains. There is still some
debate as to its exact structure (see Marchalonis 1977; Hardisty 1979)
as studies are hampered by the low levels of Ig found in the body fluids
(Marchalonis 1977).

The Ig of the Chondrichthyes (Elasmobranchii and Holocephali) has been
characterised in a number of species. Most information is available on
the elasmobranchs where an IgM-like molecule is found which typically
exists in two forms: a 17-19S pentamer and a 7S monomer (Marchalonis
1977). In most species, these two forms have the same antigenic
determinants and therefore do not constitute different classes. Some
species, particularly in the Rajoidei, do however, have two classes of
Ig (Kobayashi et al. 1984; Tomonaga & Kobayashi 1985; Tomonaga et al.
1985). In the skate, Raja kenojei, two Ig types are found in the serum:
a high molecular weight 18S form (mw 840,000) with a pentameric
structure and a J chain component, and a low molecular weight 9S form
(mw 320,000) which is a dimer associated by non-covalent forces. The L
chains of both the low and high molecular weight forms are similar (mw
23,000) but the H chains are distinct in both molecular weights and in
antigenic determinants. Furthermore, separate populations of plasma
cells are involved in the synthesis of these two types of Ig (Tomonaga

et al. 1984). Tomonaga & Kobayashi (1985) have suggested that this Ig diversity in the elasmobranchs has only evolved within the rays, as sharks and dogfish apparently only produce one class of antibody.

Particular emphasis has been placed on the nature of the antigen combining site of Ig from sharks (Shankey & Clem 1980; Clem & Leslie 1982a,b). The main point in question is does the Ig display heterogeneity in the combining site to a single antigen as do antibodies in birds and mammals? Analysis of amino acid composition of the L chain has been interpreted as indicating the presence of both constant and variable regions. Furthermore, nurse sharks possess a relatively large number of different L chain amino acid sequences which are compatible with antibody combining sites to a specified antigen. Hence, the antibody combining site repertoire of nurse sharks, at least, is probably quite extensive.

There are few studies on the structure of Ig in the holocephalans. A naturally-occurring haemagglutinin with properties similar to Ig has been described in the ratfish, Callorhynchus callorhynchus (Sanchez et al. 1980). The molecule has a molecular weight of 960,000 with H and L chains but no J chain.

In the Osteichthyes, Ig often exists in two forms - a high molecular weight (600,000-800,000 daltons) tetrameric molecule and a monomeric form, and these are antigenically related (Litman 1976; Warr & Marchalonis 1982). There are also some reports of the presence of a J chain associated with the tetrameric Ig (see Marchalonis 1977). Recently, some heterogeneity in the high molecular weight form of the Ig molecule in the sheepshead, A. probatocephalus, has been found (Lobb & Clem 1981a,b) with one population of tetramers and another of dimers. Heterogeneity has also been found in the Ig of catfish with different populations of molecules which vary in their covalent structure as well as in their component L chains (Lobb 1986). As pointed out by Lobb (1986), we are now in a position to determine if the different forms of Ig have specific functions in the immunological repertoire of fish.

Non-specific cytotoxicity. The leucocytes from immunologically naive and mitogen-stimulated fish show cytotoxic reactions towards a range of normal and transformed cell lines of fish and mammalian origin (Hinuma et al. 1980; Graves et al. 1984; Evans et al. 1984a,b,c; Hayden & Laux 1985; Moody et al. 1985). The nature of this cytotoxicity has been carefully evaluated in the channel catfish, I. punctatus (Evans et al. 1984a,b,c; Graves et al. 1984; Carlson et al. 1985). In smears, the cytotoxic cells are morphologically monocyte-like, but non-adherent to plastic dishes and also fail to ingest colloidal iron. This non-adherence does not necessarily rule out the possibility that these cells are mononuclear phagocytes as Clem et al. (1985) found that monocytes of the channel catfish do not attach to plastic unless pretreated with fibronectin. Furthermore, in other fish, the cytoxicity reactions have been accredited to monocytes/macrophages rather than lymphocytes (Hinuma et al. 1980; see section on macrophage functions)

Catfish cytotoxic cells are also a relatively homogeneous population as they band together following Percoll density gradient centrifugation. Scanning and transmission electron microscopy of these cells demonstrates a similar morphological appearance to the natural killer (NK) cells of mammals, though in catfish, the cytotoxic cells do not contain the small granules seen in NK cells. Cytotoxicity requires physical contact with the target cell and large numbers of cytotoxic cells are required to kill a single target cell. Further analysis of the interaction between cytotoxic cells and the target cells has shown its dependence on divalent cations, intact microtubules and microfilaments, and a requirement for an intact secretory apparatus (Carlson et al. 1985). Hence, these reactions involve cell locomotion and are possibly mediated by secretory factors produced by the cytotoxic cells. It is possible that the cytotoxic cells of catfish are an evolutionary primitive form of the NK cells of mammals.

Although the studies with catfish described above indicate a single population of cytotoxic cells closely related to mammalian NK cells, in other fish some killing may be due to monocytes/macrophages and/or granulocytes.

Thrombocytes (Fig. 64)

These cells have an important role in blood coagulation and in general haemostasis. They may also have limited phagocytic ability.

Blood coagulation. Although fish blood clots in response to injury, the speed and effectiveness of this process is variable. For example, in elasmobranchs most authors, with the exception of Saunders (1966), have noted a protracted or "haemorrhagic" clotting pattern (Doolittle & Surgenor 1962). This protracted clotting can be improved if extrinsic factors such as skin or high levels of calcium are added (Doolittle & Surgenor 1962). The observation that addition of sea water to whole blood causes rapid clotting (Stokes & Firkin 1971) explains how external wounds on elasmobranchs do not result in massive haemorrhaging. Clotting is a much more rapid process in bony fishes with a firm clot usually formed within a few minutes (Doolittle & Surgenor 1962; Ward 1969; Mainwaring & Rowley 1985c). Similar results have been obtained for lampreys (Page, unpublished), although the clot formed, particularly in adult forms, is less solid.

In fish, the involvement of thrombocytes in coagulation is thought to be fundamentally similar to the process in other vertebrates, i.e. factors from the thrombocytes mediate the conversion of prothrombin to thrombin, which in turn results in the production of fibrin from soluble plasma fibrinogen (Doolittle & Surgenor 1962). The precise role of fish thrombocytes in prothrombin activation at the molecular level is, however, still unclear. In contrast, Doolittle and colleagues have made a careful study of the conversion of fibrinogen to fibrin in lampreys and compared the molecular events with those occurring in mammals (Doolittle 1965a,b; Cottrell & Doolittle 1976; Laudano & Doolittle

1980). In lampreys, during the conversion of fibrinogen to fibrin, two
fibrinopeptides termed A and B are formed as a result of the cleavage of
fibrinogen under the control of thrombin. Fibrinopeptide A is a six
stranded length of amino acids, while fibrinopeptide B is a single
stranded 36 amino acid chain. Both of these fibrinopeptides are very
much different from their mammalian counterparts. In experiments where
mammalian thrombin is substituted for lamprey thrombin, only the
fibrinopeptide B is cleaved off the fibrinogen molecule (Cottrell &
Doolittle 1976). Conversely, lamprey thrombin releases only
fibrinopeptide A from mammalian fibrinogen (Doolittle & Wooding 1974).
Although there is this species specificity for fibrinogen cleavage, the
amino acid sequences on the fibrin side of both junctions split by
thrombin are highly conserved and virtually identical with those of the
alpha and beta chains of mammalian fibrin. Finally, the fibrin molecule
formed in lampreys is larger than its mammalian counterpart but has the
same basic structure with paired alpha, beta and gamma chains (Doolittle
& Wooding 1974).

The involvement of thrombocytes in coagulation has been observed in
vitro (Gardner & Yevich 1969; Stokes & Firkin 1971; Wardle 1971). Wardle
(1971) found that in the absence of anticoagulants, plaice (Pleuronectes
platessa) thrombocytes became disrupted and the fibrin fibres radiated
from these cells. Subsequently, erythrocytes were entrapped in this
sticky mass and then leucocytes, identified as lymphocytes, were
mobilised and migrated towards the clot (chemotaxis?). Similar events
also occur with thrombocytes of the dogfish, S. canicula (Fig. 64).

In vivo thrombus formation has not been observed in fishes but, the
aggregation of thrombocytes in blood smears in which the cytoplasm is
ragged (Fig. 32), suggests that such a process occurs following
bleeding. In mammals, platelet aggregation to form a thrombus is
initiated by collagen/thrombin which causes these cells to change shape
and release the contents of the alpha and dense granules. The conversion
of arachidonic acid in the platelets to thromboxane A_2, and lower levels
of prostaglandins D_2, E_2 and $F_{2\alpha}$ is of prime importance because the
thromboxane is a potent inducer of platelet aggregation and also plays
some role in the control of this process (Taussig 1984; Gerrard 1985).
Thromboxane A_2 is unstable and is rapidly converted to the stable form
of the molecule, thromboxane B_2. The part played by prostaglandins in
platelet aggregation in mammals is thought to be insignificant compared
with the effects of thromboxane (Taussig 1984). In contrast, recent
experiments by Kayama et al. (1985) have revealed that incubation of
carp thrombocytes with radiolabelled arachidonic acid results in its
conversion to mainly prostaglandins with little, if any, thromboxane.
The implications of these findings in thrombus formation by carp
thrombocytes is unclear but this system would make an excellent model
for determining the role of endogenous prostaglandins in thrombosis in
the absence of thromboxanes. Other differences between fish thrombocyte
and mammalian platelet aggregation have been elucidated by Stokes &
Firkin (1971) with the Port Jackson shark (Heterodontus portusjacksoni).
They found that shark thrombocytes readily aggregate in vitro but in
contrast to platelets, this process is temperature reversible and is

also independant of thrombin and ADP. Hence, the control and outcome of
thrombocyte aggregation in fishes may not necessarily be the same as in
mammals and warrants further biochemical and pharmacological studies
with purified populations of such cells.

Phagocytosis. Many authors have reported that the injection of foreign
particulate material into the vascular system of fishes results in its
association with the cytoplasm of thrombocytes (Ferguson 1976a; Morrow &
Pulsford 1980; Hunt & Rowley 1986b). They have, however, mostly
expressed some doubt as to whether this association represents true
phagocytosis because the injected material seems to be within the
canalicular system which is in direct contact with the external medium
surrounding the thrombocytes (Daimon & Uchida 1985). If this is indeed
the case, then the material is still extracellular and this process is
therefore not true phagocytosis. The finding that while small
particulates such as colloidal carbon and latex spheres ($0.8\mu m$ in
diameter) are 'ingested' by the thrombocytes, while larger particles
such as erythrocytes and yeast are not (Stokes & Firkin 1971; Parish et
al. 1985; Hunt & Rowley 1986b) could be then explained by the small
particles dropping into the canalicular system. Furthermore, the choice
of colloidal carbon as a test particle in phagocytosis experiments is
unwise, as single carbon particles are small enough to be pinocytosed,
as evidenced by the bristle coat around the vacuoles containing this
material in other cells (e.g. Page & Rowley 1984), and only aggregated
complexes are phagocytosed. The perplexing lack of published micrographs
showing actual stages of the ingestion process also indicates that, if
present, then phagocytosis by fish thrombocytes cannot be a widespread
phenomenon.

Some support for true phagocytosis by fish thrombocytes is provided by
the presence of large vacuoles (ca. $5\mu m$ in diameter) surrounding test
particles (Ferguson 1976a). The size of these vacuoles would necessitate
a 10-20 fold dilation of the canalicular system for their formation and
this seems unlikely. On balance, evidence is marginally in favour of
thrombocytes not having true phagocytic powers. Mammalian platelets also
become associated with latex spheres and carbon following incubation and
the nature of this process has also been questioned (White 1972). By
using lanthanum nitrate, which stains the external membrane of platelets
and other cells, White (1972) was able to prove that the 'vacuoles'
enclosing latex were in continuity with the outer membrane and hence had
not pinched off into the cytoplasm to form phagosomes. Even if fish
thrombocytes are phagocytic, their activity is probably insignificant in
comparison with that of monocytes and macrophages (see sections on
monocyte and macrophage functions).

Granulocytes
Granulocytes function in the non-specific defences of fish by
migrating to sites of inflammation and there destroying the invaders by
phagocytosis or by a cytotoxic-like response. What is unclear, however,
is the functional heterogeneity in those fish where several

morphologically distinct granulocyte types are present.

Participation in inflammatory responses. Although there are relatively
few experimental studies on the inflammatory responses in fish, most
reports indicate an active role for the granulocytes in this process.
Finn & Nielsen (1971), for example, observed that both "PMN leucocytes",
macrophages and lymphocytes accumulate at sites of inflammation in
rainbow trout following injection of staphylococci, adjuvant or
cauterisation of the integument. Invariably, the earliest response
involves the accumulation of granulocytes which peaks at ca. 12-24h and
is followed later by macrophages and lymphocytes. This biphasic response
of granulocytes and macrophages has also been found in plaice (P.
platessa) in which intra-peritoneal injection of glycogen followed by
endotoxin elicits the immigration of neutrophils with a peak number at
ca. 48h, after which macrophages also start to appear (MacArthur et al.
1984). Although a biphasic response is evident in tilapia following the
intraperitoneal injection of liquid paraffin, peak numbers of
macrophages/monocytes occur at 4 days while maximal neutrophil/eosinophil
numbers are present after 10 days (Suzuki 1986). Both "neutrophilic" and
"eosinophilic" granulocytes together with macrophages migrate into the
peritoneal cavity of the striped bass (Morone saxatilis) following
injection of Bacillus cereus (Figs 65, 66), but no indication of the
relative abundance of the different cells or the dynamics of this
response have been presented (Bodammer 1986).

Inflammatory reactions to macrobial parasites are also of interest,
particularly to determine if a response occurs similar to that in mammals
in which eosinophils attach to and destroy the parasite (see Chapter 6).
Whiting (Merlangius merlangus) naturally infected with cestode
(Contracaecum and Anisakis) larvae are encapsulated by "neutrophils",
macrophages and proliferating fibroblasts (Elarifi 1982). A similar
reaction occurs to the cestode Ligula intestinalis in the roach, Rutilus
rutilus (Hoole & Arme 1982). It is not known if these two species of fish
possess eosinophils.

Chemotactic/chemokinetic responses. The active migration of granulocytes
described in the above in vivo studies (e.g. Finn & Nielsen 1971;
MacArthur et al. 1984) is indicative that a chemotactic/chemokinetic
response might be involved. There are a number of factors - both
endogenous and exogenous - that can act as chemoattractants. The
endogenous factors include complement fragments such as C5a and
leukotrienes, in particular leukotriene B_4. Exogenous substances,
including bacterial culture filtrates and synthetic peptides such as
F-Met-Phe and F-Met-Leu-Phe, which are thought to be present in these
filtrates, are also very potent chemoattractants for most mammalian
granulocytes (Wilkinson 1982). Enhanced migration (chemokinesis) of
plaice neutrophils has been shown in response to F-Met-Leu-Phe at levels
as low as 10^{-9}M, endotoxin-activated plaice serum, and 5-100 fold
dilutions of culture filtrates from Vibrio alginolyticus (MacArthur et
al. 1985; Nash et al. 1986). In dogfish, however, the granulocytes fail

to respond to bacterial culture filtrates, synthetic peptides such as
F-Met-Phe or to fish and mammalian complement fragments (Hunt & Rowley
1986c; & unpublished observations). G1 and G3 granulocytes do show
enhanced migration to synthetic leukotriene B$_4$ (Table 6) but it is not
known if this is chemotaxis or simple chemokinesis (Hunt & Rowley 1986d).

Phagocytosis. Despite the obvious involvement of granulocytes in
inflammation, such cells are, surprisingly, not always phagocytic. For
example, intravascular injection of bacteria, latex or sheep erythrocytes
into the dogfish (S. canicula), although resulting in the aggregation and
accumulation of G1 granulocytes in the gills (Fig. 67), does not lead to
any phagocytosis by this cell type (Hunt & Rowley 1986b). If, however,
the G1 granulocytes are challenged in vitro with the same range of
foreign particles, they are actively phagocytic (Hunt & Rowley 1986b). In
the gar, Lepisosteus platyrhinius, the granulocytes also fail to ingest
yeasts and Staphylococcus epidermis both in vivo and in vitro, while
under the same conditions between 13-59% of the mononuclear phagocytes
are phagocytic (McKinney et al. 1977).

Phagocytosis by fish granulocytes does, however, occur both in vivo
(MacArthur et al. 1984; MacArthur & Fletcher 1985; Rowley & Page 1985a;
Bodammer 1986) and in vitro (Fujii 1981; Parish et al. 1985) in a number
of species and with various test particles. The rate of phagocytosis is
often affected by serum factors such as Ig and complement, though it is
not always clear whether one or both of these molecules are involved.
Neutrophilic/heterophilic granulocytes from the lamprey, Lampetra
japonica, show enhanced phagocytosis of sheep erythocytes opsonised with
specific lamprey antiserum to sheep erythrocytes but not with antisera
raised in bullfrogs (Fujii 1981). Fujii (1981) believed that this opsonic
activity resided with Ig in the serum although no controls were employed
to rule out the participation of complement. Indeed, complement receptors
have been found to be important in the uptake of Aeromonas salmonicida by
"neutrophil-like" cells from the peritoneal cavity of various salmonids
(Sakai 1984). In the presence of specific antiserum and an intact
complement system, enhanced phagocytosis was apparent and substitution
with mammalian complement plus antiserum or with fish complement alone,
failed to elicit a similar response. The opsonic activity is probably
generated via the classical complement pathway and shows some species
specificity. Wrathmell & Parish (1980) were unable to find both Ig and
complement receptors on dogfish (S. canicula) leucocytes, although in a
closely related elasmobranch, Mustelus canis, the granulocytes bear
receptors for heat aggregated homologous Ig (Weissmann et al. 1975).

The biochemical events involved in the intracellular killing and
digestion of material within fish granulocytes are little known.
Granulocytes do produce a range of microbicidal enzymes such as lysozyme
(Murray & Fletcher 1976) and myeloperoxidase (Kanner & Kinsella 1983) but
the absence of peroxidase in the granulocytes of many species of fish
(see section on granulocyte structure) rules out the omni-presence of the
myeloperoxidase system. Dogfish (Mustelus canis) granulocytes also show
cytotoxic reactions against sea urchin eggs when employed as target cells
(Weissmann et al. 1978). The eggs are most readily attacked if first

coated with aggregated dogfish Ig and some of the killing observed may be
brought about as a result of degranulation (i.e. enzyme release) by the
attached granulocytes.

Functional heterogeneity in fish granulocytes. Although the morphological
heterogeneity of fish granulocytes is now well established (see section
on granulocyte structure), we still have little idea how these different
populations function in the non-specific defence reactions. The
eosinophilic and heterophilic granulocytes of larval lampreys have
varying phagocytic abilities. The neutrophils phagocytose both bacteria
and colloidal carbon, while the eosinophils only ingest the former
particle and then not as avidly as the neutrophil (Rowley & Page 1985a).
Whether this trend is present with other test particles is unknown. The
eosinophils may react to metazoan parasites in a similar way to their
morphological counterparts in amphibians and mammals (see Chapters 3 & 6)
but this has yet to be investigated. There are several species of fish
which would be good models for studies into the functional heterogeneity
of granulocytes. In teleosts, carp are well-suited for such studies, as
they possess neutrophilic, eosinophilic and basophilic granulocytes, and
separation methods exist for these cells (e.g. Bayne 1986). In
elasmobranchs, the properties and functions of Percoll-separated G1-G4
granulocytes in dogfish, S. canicula, have been analysed (Table 6). Only
the G1 granulocytes are apparently involved in inflammation as judged by
their migration into the peritoneal cavity following injection of
endotoxin (Mainwaring, unpublished). Both G1 and G3 granulocytes respond
to the presence of leukotrienes indicating that they display receptors
for these molecules, but the G3s fail to phagocytose foreign particles
either in vivo or in vitro (Hunt & Rowley 1986b). The functions of the
G3, and also for that matter the G2 and G4 granulocytes, are therefore
unknown.

Monocytes
           Monocytes and macrophages as mentioned previously (see
section on monocyte functions), are probably part of the same cell
lineage and therefore have similar properties and functions. In this
review we have made a distinction between these two cell types with the
term "monocyte" retained for cells in circulation and "macrophage" for
fixed or wandering cells. The functions of these two cell types will
thus be discussed separately, although much of what can be said for
monocytes is also probably true for macrophages.

Inflammation. Circulating monocytes in mammals respond to
microbial/macrobial invasion by leaving the circulation, migrating to
the area of inflammation and then phagocytosing or walling off the
foreign invader. This involvement can be seen in both acute and chronic
inflammation.

Monocytes are participants in acute inflammatory responses of fish in
which they actively phagocytose invaders (Finn & Nielsen 1971; MacArthur

et al. 1984; MacArthur & Fletcher 1985). As discussed with granulocytes, do monocytes exhibit chemotactic/chemokinetic responses and is there evidence for such cells leaving the circulation to migrate to the site of inflammation? Unfortunately, no chemotaxis experiments have been undertaken employing purified populations of peripheral blood monocytes probably due to the difficulty in obtaining such cells (Mainwaring & Rowley 1985a). Griffin (1984) found that in chemotaxis assays run with a mixed leucocyte population, consisting of mainly granulocytes and monocytes, the granulocytes were the most actively migrating cell type. These experiments, however, may not have been run for long enough periods to demonstrate enhanced migration of monocytes in the presence of chemotactic agents. Similar findings are apparent with dogfish leucocytes in which the G1 granulocytes migrate quickly and are followed at a later stage by monocytes (Hunt & Rowley 1986d). Monocyte migration to focal areas of inflammation can be inferred from the experiments of Finn and Nielsen (1971) where such cells were apparently seen migrating out of blood capillaries. MacArthur et al. (1985) also described plaice monocytes entering the circulation from the kidney in response to the intravascular injection of turbot erythrocytes. Clearly, further studies are required with purified populations of monocytes and macrophages to clarify which agents, if any, enhance migration and also if this is a chemotactic or chemokinetic response.

Phagocytosis. Monocytes are actively phagocytic cells as revealed by both in vivo (Ellis 1976; Morrow & Pulsford 1980; Hunt & Rowley 1986a,b) and in vitro (McKinney et al. 1977; Parish et al. 1985; Hunt & Rowley 1986b) studies. Normally, although there are fewer monocytes than granulocytes in the blood (ca. 1-2% as compared with 20-30% of the total leucocyte count, respectively), they are much more avidly phagocytic. Intravascular injection of a range of particulate and non-particulate material into the dogfish, S.canicula, results in its rapid clearance by fixed phagocytes (mainly macrophages) and monocytes. These monocytes not only phagocytose the materials but also drop out of circulation and aggregate in the gills. Here, they often form large clumps in both the secondary lamellae and cavernous body region at the base of each filament (Hunt & Rowley 1986b). In fish, factors mediating the ingestion and killing phases of phagocytosis by the monocytes may be present (Scott & Klesius 1981; Scott et al. 1985). For example, channel catfish peripheral blood phagocytes (including many monocytes) phagocytose zymogen more actively in the presence of normal serum as judged by an increase in chemiluminescence (Scott & Klesius 1981). This is partial proof for the presence of a complement-mediated phagocytic process in this species. Further experimental results, again using chemiluminescence as an indicator of phagocytosis, showed the importance of both specific antibody and complement in the phagocytosis of the enteric pathogenic bacterium, Edwardsiella ictaluri (Scott et al. 1985). Monocytes therefore probably have receptors on their surfaces for Ig and complement fragments (C3b?). Contrasting results have been found in the carp Cyprinus carpio where immune serum to Staphyloccus aureus had no apparent effect on the uptake of this particle by blood leucocytes (cell type unspecified but probably monocytes) (Avtalion & Shahrabani 1975).

By comparison, the intracellular killing rates of carp leucocytes from immunised animals showed a marked increase over those of naive fish. This might be an example of either specific or non-specific activation of monocytes (and granulocytes?) but these experiments need repeating with purified cells and without serum components such as natural antibody and complement.

Macrophages (Figs 6,13,63)

The last twenty years of intensive study with mammalian macrophages has shown that these cells are more than mere phagocytes concerned with the sequestration of effete cells. For example, they produce a number of secretory products including complement factors, some types of interferon, interleukin 1 (IL-1), prostaglandins and leukotrienes and can stimulate other cells, such as T lymphocytes, to synthesise a range of products (e.g. IL-2 and B cell growth factor). They also have important functions in humoral immunity acting as antigen presenting cells, and in cellular immunity in which they become specifically activated or "armed" to show increased powers of phagocytosis and intracellular killing as well as limited cytotoxicity against tumour cells. In this section, we review some of these activities in fish and determine if they can be attributed to macrophages.

Interleukin production. There is now mounting evidence for the interaction beween lymphocytes and mononuclear phagocytes in fish (e.g. Smith & Braun-Nesje 1981; Miller et al. 1985), and so it is possible that factors such as IL-1 and IL-2 are produced during this process. There are a few studies which have considered IL-1 production in fish, the most notable of which is a recent report by Sigel and colleagues (Sigel et al. 1986). Their valuable work shows that catfish peripheral blood leucocytes recognise and respond to human IL-1 and that an epithelial cell line from hypertrophic skin lesions of carp produces a factor with many of the characteristic properties of this molecule. It is still unknown if the cell type involved in the production of IL-1 in fish is the mononuclear phagocyte and whether these cells can stimulate lymphocytes to produce IL-2. Fish may well produce IL-2 since supernatants from mitogen stimulated cultures of carp leucocytes cause proliferation of homologous "T-like" lymphoblasts (Caspi & Avtalion 1984). Similarly, cultures of carp leucocytes also respond to mammalian IL-2. It has yet to be elucidated if this IL-2 production in fish is brought about by T cell and macrophage interaction.

Interferon synthesis. Fish synthesise interferon-like molecules in response to viral infection (e.g. Beasley & Sigel 1967; de Kinkelin et al. 1981) but only one type of interferon is produced (Galabov 1973) and no studies have identified macrophages as responsible. Alexander (1985) believes that fish interferon is most likely of the alpha and/or beta classes which in mammals are produced by leucocytes and fibroblasts, respectively.

Prostaglandin & leukotriene release. In mammals, prostaglandins are produced by a multitude of cells including macrophages, and have many immunological and non-immunological functions (Gerrard 1985). Prostaglandins have been implicated in some immunological functions in fish such as immediate cutaneous hypersensitivity reactions to fungal extracts (Anderson et al. 1979) but the cell types responsible for their production in this study were not determined.

Leukotrienes, unlike prostaglandins, are probably produced by a smaller range of mammalian cell types including granulocytes, monocytes/macrophages, platelets, lymphocytes, mast cells and keratinocytes (Gerrard 1985; Grabbe et al. 1985). The principal leukotriene, as far as immunological functions are concerned, is leukotriene B_4 (LTB_4) which promotes the chemotaxis and aggregation of granulocytes and mononuclear cells (e.g. Smith et al. 1980) and the enhancement of IL-1 production (Rola-Pleszczynski & Lemaire 1985). Evidence for LTB_4 production by fish comes from in vivo work with perfused gills of Anguilla (Piomelli 1985) and in vitro experiments with dogfish leucocytes stimulated by the calcium ionophore A23187 (Rowley & Barrow unpublished results). In this latter study, a mixed population of G1 granulocytes and monocytes was used and the culture supernatants analysed by GC mass spectrometry for leukotrienes and prostaglandins. These cells release both prostaglandins and leukotrienes when exposed to the ionophore but it has yet to be determined which of these cells types is responsible for their synthesis. As described in the section on granulocyte functions, dogfish G1 and G3 granulocytes respond with enhanced migration to synthetic LTB_4 (Table 6), but mononuclear phagocytes have yet to be thoroughly tested.

Role in antigen trapping/presentation. Macrophages are an important component of the reticulo-endothelial system (RES) of all vertebrates. In fish, macrophages are found in both lymphomyeloid organs such as the spleen (Fig. 13), thymus, anterior kidney and also in non-lymphoid organs such as the heart, liver and gills. Presumably, for these latter macrophages to play a role in antigen presentation they would need to migrate to lymphoid tissues. It therefore seems more likely that such cells function in some quasi-immunological role associated more with maintaining the circulation free from effete material rather than antigen presentation. A number of studies exist on the RES of various fish species. In lampreys, colloid carbon and bacteria are internalised by macrophages in the supraneural body, liver, kidney (Fig. 63) and specialised macrophage-like cells, usually termed "cavernous body cells", in the gills. There is also some minor uptake by connective tissue macrophages and cells in the heart (Page & Rowley 1984; Rowley & Page 1985b). Of particular interest are the macrophages (which contain melanin) in the supraneural body because this is the main site of antibody formation in lampreys (Hagen et al. 1983). These cells are situated in sinusoids (Fig. 6) where material in circulation can be easily trapped and perhaps presented to adjacent lymphocytes.

Morphological aspects of the RES in elasmobranchs have also been studied

(Pulsford et al. 1982; Hunt & Rowley 1986a,b). Macrophages are found in
the ellipsoids (Fig. 13) and red pulp of the spleen, liver sinusoids,
thymus, epigonal/Leydig's organ and also in the general connective
tissue. Injection of a range of materials, including dextran, carbon,
bacteria, latex spheres and erythrocytes, revealed that little of this
is localised in the epigonal and Leydig's organ, but significant amounts
were found in the spleen and liver (Hunt & Rowley 1986a). Some uptake of
smaller materials such as colloidal carbon and latex is achieved by the
cavernous body cells but the cells involved do not appear to be
macrophages (Hunt & Rowley 1986b).

Studies on the RES of bony fish are more extensive. Soluble antigens
such as BSA and HGG are trapped by both the spleen and anterior kidney
but little is found in other sites such as the liver (Ellis 1980;
Secombes & Manning 1980). In the spleen of both carp and plaice, soluble
antigens soon after their injection were found associated with the
reticular (collagen) fibres of the ellipsoid as antigen/antibody
complexes. Later on in the plaice only, some intracellular antigen was
present in the macrophages within the ellipsoids and the
melano-macrophage centres of the spleen (Ellis 1980). Retention of
soluble antigens in the kidney was of a different nature in both plaice
and carp as it was present within macrophage-like cells in the red pulp
and, in the plaice at least, in close association with the
melanomacrophage centres. As Ellis (1980) points out, both the
ellipsoids and melano-macrophage centres are regions where metabolically
active small lymphocytes pass through and hence these could become
influenced by the trapped immune complexes. Two major questions remain
unanswered from these studies. Firstly, are the immune complexes trapped
on the reticular cell fibres or on the reticular cells themselves? Ellis
(1980) is of the opinion that they are on the fibres, but the reticular
cells have so many fine cytoplasmic processes that in the light
microscope it is impossible to differentiate between fibres and cells.
Secondly, the function of the melano-macrophage centres of the spleen
and kidney, and ellipsoids of the spleen in antigen retention has yet to
be fully assessed. Ferguson (1976b) proposed that macrophages laden with
carbon migrated to the melano-macrophage centres and these latter
structures were likened to primitive germinal centres. Circumstantial
evidence also points to migration of cells to these centres as diseased
fish often show an increase in the size of their melano-macrophage
centres and this may have resulted from the addition of
melano-macrophages from other sites such as ellipsoids. There is no
evidence of macrophage migration from the ellipsoids of the dogfish, S.
canicula, but in these fish, melano-macrophage centres are lacking (Hunt
& Rowley 1986a). The melano-macrophage centres perhaps function by
accumulating antigen which is then released to immunocompetent cells in
a controlled manner so as to avoid immunological paralysis.

There is good experimental evidence for the requirement of macrophages
in humoral immunity in fish (Miller et al. 1985). Separated populations
of T-like and B-like lymphocytes and macrophages from the channel
catfish were employed to investigate the cellular requirement for
antibody production as measured by plaque forming assays. The results

showed that macrophages are needed for both responses to 'classical' thymus-dependent and thymus-independent antigens in vitro. Of particular importance to our discussion of the role of macrophages as antigen presenters are some preliminary results reported by Clem et al. (1985). They found that culture filtrates of macrophages could substitute for intact cells in these assays. Hence, the requirement for macrophages in these experiments is not as antigen processers/presenters, but in the production of soluble factors such as monokines (IL-1?). These experiments leave in doubt, for the moment, the position of fish macrophages as antigen presenting cells.

Macrophage activation. Rainbow trout macrophages isolated from the anterior kidney can be activated by muramyl peptide and by lymphokine-like substances (Secombes 1986). This activation takes the form of increased spreading but such treatments had little effect on phagocytosis or migration inhibition assays. Atlantic salmon macrophages are stimulated in a similar manner to mammalian macrophages, by the addition of insoluble glycans such as carrageenin in a similar manner to mammalian macrophages (Smith & Braun-Nesje 1981), but the phagocytic or killing potential of such cells were not examined.

Cytotoxicity reactions by fish macrophages. Fish leucocytes show cytotoxic reactions without prior stimulation (see lymphocyte functions) but the nature of the cell type involved is unresolved. Hinuma et al. (1980) suggested that the cytotoxic cells in various cyprinids were macrophages, while Moody et al. (1985) observed cytotoxic activity in non-adherent cells (lymphoid cells?) but not macrophages from the Atlantic salmon, Salmo salar.

Iron storage. In teleosts, iron is accumulated in the melano-macrophages of the spleen, but not liver and kidney, probably in the form of haemosiderin (Agius 1979 1985). This compound is a breakdown product of haemoglobin and is presumably found in the splenic macrophages due to the phagocytosis of effete red blood cells in this organ. Although haemosiderin could not be found in the melano-macrophages of elasmobranchs (Agius 1985), Hunt and Rowley (1986a) have described erythrophagocytosis by splenic macrophages in the dogfish, S. canicula.

Phagocytosis. Fish macrophages are avidly phagocytic both in vitro (Braun-Nesje et al. 1981; Olivier et al. 1986; Secombes 1986) and in vivo (Ellis et al. 1976; Fishman et al. 1979; Bodammer 1986; Hunt & Rowley 1986a,b; Fig. 63). No studies have, however, been carried out to see if they, like mammalian macrophages, are more active in this process than monocytes. One major aspect of this macrophage function will be assessed in the following discussion, namely, are there receptors on fish macrophages similar to those in the equivalent cells in 'higher' vertebrates? Mammalian macrophages display a number of receptors on their surface which are involved in the recognition and subsequent

ingestion of foreign material. Three main types of receptor are found,
namely, those which bind Ig and the C3b component of complement, and
lectin receptors capable of reacting with sugar moieties on the surface
of micro-organisms and effete erythrocytes either directly or via
humoral lectins (Sharon 1984). This final type of receptor is
potentially of greatest interest because it is present on the phagocytes
of invertebrates and vertebrates (Ratcliffe et al. 1985). Evidence for
the existence of lectin receptors on fish macrophages comes from the
work of Ozaki et al. (1983) and Smedsrud et al. (1984). The amago
(Oncorhyncus rhodurus), a Japanese trout, contains appreciable levels of
a lectin distinct from Ig and with specificity for α -L-ramnose- and
α -D-galactose - like molecules (Ozaki et al. 1983). Receptors for this
lectin can be found on both inflammatory peritoneal macrophages and head
kidney macrophages but these are only expressed after 4 days incubation.
Hence, the evolutionary primitive mechanism of opsonisation of foreign
material via lectin receptors has probably been retained in the fishes.

Regarding Ig and C3b receptors, peritoneal macrophages from various
salmonids show some increase in phagocytic activity following the
addition of test particles and heat-inactivated serum but a markedly
increased response is produced when normal fish serum is added (Sakai
1984). This is interpreted as strong evidence for the existence of
opsonic activity and receptors for complement and Ig with complement
activation via the classical pathway. The importance of complement in
opsonisation has also been demonstrated by Olivier et al. (1986) with
trout peritoneal macrophages and Johnson & Smith (1984) who showed that
agarose beads coated with human C3b and C3bi are more rapidly
phagocytosed than uncoated beads. Furthermore, agents which
destroy/inactivate the complement pathway e.g. EDTA and heat, also
abolish this opsonisation reaction.

CONCLUDING REMARKS

Over a century of research into various aspects of fish
haematology has shown that these animals contain most, if not all, of
the types of leucocytes found in the blood of other vertebrates. Hence,
the haematological pattern first established in fish has been conserved
during the evolution of the other vertebrate classes. There are,
however, some differences in this basic haematological pattern. For
example, although many fish have two or sometimes more morphologically
distinct granulocyte types it remains unresolved if these are homologous
to the heterophilic/neutrophilic, eosinophilic and basophilic
granulocytes of "higher" vertebrates. Therefore, it is a matter of some
importance to determine the roles of these granulocyte types in a range
of fish. Progress towards this goal has been hampered by an inability to
clearly identify different granulocyte types and the difficulties faced
in separating these using techniques such as density gradient
centrifugation. Clearly, it is not possible to elucidate the functional
diversity of granulocytes in a large range of fish and it would be
prudent to concentrate on species such as carp, tilapia, catfish and
salmon/trout which have economic importance and have been the subject of
many immunological studies. Attention should also be paid to fish of

phylogenetic interest including lampreys, hagfish, sharks and lungfish.

Great advances have been made in our understanding of lymphocyte diversity in fish within the last few years mainly as a result of the adaptation of techniques originally developed for unravelling the complexities of the immune system of mammals. Most notable is the use of mitogenicity experiments with separated sub-populations of lymphocytes and accessory cells such as macrophages, because these have clearly demonstrated that T and B like cells exist in fish and that cell co-operation between lymphoid cells and macrophages and/or their products plays an important role in the development of an effective immune response.

Finally, before we can have further in-depth appreciation of the diversity and function of fish leucocytes, several basic questions need resolving. For example, what are the stages in the ontogenetic development of the different leucocyte types, where do these take place, and can the haemopoietic stem cells be identified and isolated? The availability of free-living embryonic stages and panels of monoclonal antibodies against leucocyte antigens should be vital in designing suitable experiments to answer these questions.

Acknowledgements
We are grateful to Professor A. Zapata, Dr E. Bielek, Dr J Bodammer, Dr S. Chilmonczyk and Dr S.L. Smith for their generosity in providing unpublished and published material for this chapter. Our original work quoted in this review was supported by the Natural Environment Research Council, the University of Wales, the SmithKline (1982) Foundation and the Royal Society.

REFERENCES

Abruzzini, A.F., Ingram, L.O. & Clem, L.W. (1982).
    Temperature-mediated processes in teleost immunity:
    homeoviscous adaptation in teleost lymphocytes. Proc. Soc.
    Exp. Biol. Med., 169, 12-8.
Ackerman, S.K. & Douglas, S.D. (1980). Monocytes. In The
    Reticuloendothelial System a Comprehensive Treatise. Vol. 1,
    Morphology, eds. I. Carr & W.T. Daems. pp. 297-327. New York:
    Plenum Press.
Agius, C. (1979). The role of melano-macrophage centres in iron
    storage in normal and diseased fish. J. Fish Disease, 2,
    337-43.
Agius, C. (1981). Preliminary studies on the ontogeny of the
    melano-macrophages of teleost haemopoietic tissues and age
    related changes. Dev. Comp. Immunol., 5, 597-606.
Agius, C. (1985). The melano-macrophage centres of fish: a review. In
    Fish Immunology, eds. M.J. Manning & M.F.Tatner. pp. 85-105.
    Orlando: Academic Press.
Alexander, J.B. (1985). Non-immunoglobulin humoral defense mechanisms
    in fish. In Fish Immunology, eds. M.J. Manning & M.F. Tatner.
    pp. 133-40. Orlando: Academic Press.
Anderson, A.A., Fletcher, T.C. & Smith, G.M. (1979). The release of
    prostaglandin E_2 from the skin of the plaice, Pleuronectes
    platessa L. Br. J. Pharmac., 66, 547-52.
Andrew, W. (1965). Comparative Hematology. New York: Grune & Stratton.
Ardavin, C.F., Gomariz, R.P., Barrutia, M.G., Fonfria, J. & Zapata,A.
    (1984). The lympho-hemopoietic organs of the anadromous sea
    lamprey Petromyzon marinus: a comparative study throughout
    its lifespan. Acta. Zool., 65, 1-15.
Avtalion, R.R. (1981). Environmental control of the immune response
    in fish. CRC Crit. Rev. Environ.Control, 11, 163-92.
Avtalion, R.R. & Shahrabani, R. (1975). Studies on phagocytosis in
    fish I. In vitro uptake and killing of living Staphylococcus
    aureus by peripheral leucocytes of carp (Cyprinus carpio).
    Immunology, 29, 1181-7.
Awaya, K., Tomonaga, S. & Yamaguchi, K. (1977). Mononuclear
    phagocytes (Kupffer cells) in liver sinusoids of hagfish. J.
    Electron. Microsc., 26, 228.
Barber, D.L. & Westermann, J.E.M. (1978). Occurrence of the periodic
    acid-schiff positive granular leucocytes (PAS-GL) in some
    fishes and its significance. J. Fish. Biol., 12, 35-43.
Barber, D.L., Westermann, J.E.M. & White, M.G. (1981). The blood
    cells of the Antarctic icefish Chaenocephalus aceratus
    Lönnberg: light and electron microscopic observations. J.
    Fish Biol., 19, 11-28.
Bayne, C.J. (1986). Pronephric leucocytes of Cyprinus carpio:
    isolation, separation and characterization. Vet. Immunol.
    Immunopathol., 12, 141-51.
Beard, J. (1902). The origin and histogenesis of the thymus in Raja
    batis. Zool. Jahrb., 17, 403-19.

Beasley, A.R. & Sigel, M.M. (1967). Interferon production in
    cold-blooded vertebrates. In Vitro, 3, 154-65.
Bielek, E. (1981). Developmental stages and localization of
    peroxidatic activity in the leucocytes of three teleost
    species (Cyprinus carpio L.; Tinca tinca L.; Salmo gairdneri
    Richardson). Cell & Tissue Res., 220, 163-80.
Blaxhall, P.C. (1972). The haematological assessment of the health of
    freshwater fish. J. Fish Biol., 4, 593-604.
Blaxhall, P.C. (1973). Error in haematocrit value produced by
    inadequate concentration of ethylenediamine tetra-acetate.
    J. Fish Biol., 5, 767-9.
Blaxhall, P.C. & Daisley, K.W. (1973). Routine haematological methods
    for use with fish blood. J. Fish Biol., 5, 771-82.
Blaxhall, P.C. & Sheard, P.R. (1985). Preliminary investigation of
    the characteristics of fish lymphocytes separated on a
    Percoll discontinuous gradient. J. Fish Biol., 26, 209-216.
Bly, J.E. (1985). The ontogeny of the immune system in the viviparous
    teleost Zoarces viviparus L. In Fish Immunology, eds. M.J.
    Manning & M.F. Tatner. pp. 327-41. London: Academic Press.
Bodammer, J.E. (1986). Ultrastructural observations on peritoneal
    exudate cells from the striped bass. Vet. Immunol.
    Immunopathol., 12, 127-40.
Bogner, K.-H & Ellis, A.E. (1977). Properties and functions of
    lymphocytes and lymphoid tissues in teleost fish. Fish &
    Environment, 4, 59-72.
Bond, C.E. (1979). Biology of Fishes. Philadelphia, Saunders College
    Publishing.
Bone, Q. & Marshall, N.B. (1982). Biology of Fishes. Glasgow &
    London: Blackie.
Botham, J.W. & Manning, M.J. (1981). The histogenesis of the lymphoid
    organs in the carp Cyprinus carpio L. and the ontogenetic
    development of allograft reactivity. J. Fish Biol., 19,
    403-14.
Braun-Nesje, R., Bertheussen, K., Kaplan, K. & Seljelid, R. (1981).
    Salmonid macrophages: separation, in vitro culture and
    characterisation. J. Fish Disease, 4, 141-51.
Cannon, M.S., Mollenhauer, M.H., Eurell, T.E., Lewis, D.H., Cannon,
    A.M. & Tompkins, C. (1980a). An ultrastructural study of the
    leukocytes of the channel catfish, Ictalurus punctatus. J.
    Morphol., 164, 1-23.
Cannon, M.S., Mollenhauer, H.H., Cannon, A.M., Eurell, T.E. & Lewis
    D.E. (1980b). Ultrastructural localization of peroxidase
    activity in neutrophil leukocytes of Ictalurus punctatus.
    Can. J. Zool., 58, 1139-43.
Carlson, R.L., Evans, D.L. & Graves, S.S. (1985). Nonspecific
    cytotoxic cells in fish (Ictalurus punctatus) V. Metabolic
    requirements of lysis. Dev. Comp. Immunol., 9, 271-80.
Casley-Smith, J.R. & Casley-Smith, J.H. (1975). The fine structure of
    the blood capillaries of some endocrine glands of the
    hagfish, Eptatretus stouti: implications for the evolution of
    blood and lymph vessels. Rev. Suisse Zool., 82, 35-40.

Caspi, R.R. & Avtalion, R.R. (1984). Evidence for the existence of an
    IL-2 like lymphocyte growth promoting factor in a bony fish
    (Cyprinus carpio).Dev. Comp. Immunol., 8, 51-60.
Caspi, R.R. Rozenszaln, L.A., Gothelf, Y., Pergamenikov-Litvak, T. &
    Avtalion, R.R. (1982). The cells involved in the immune
    response of fish. II. PHA-induced clonal proliferation of
    carp lymphocytes in soft agar culture. Dev. Comp. Immunol.,
    6, 683-92.
Caspi, R.R., Shahrabani, R., Kehati-Dan, T. & Avtalion, R.R. (1984).
    Heterogeneity of mitogen-responsive lymphocytes in carp
    (Cyprinus carpio). Dev. Comp. Immunol.,8, 61-70.
Catton, W.T. (1951). Blood cell formation in certain teleost fishes.
    Blood, 6, 39-60.
Cenini, P. (1984). The ultrastructure of leucocytes in carp (Cyprinus
    carpio). J. Zool. (Lond.), 204, 509-20.
Cenini, P. & Turner, R.J. (1983). In vitro effects of zinc on
    lymphoid cells of the carp, Cyprinus carpio L. J.Fish Biol.,
    23, 579-583.
Chilmonczyk, S. (1983). The thymus of the rainbow trout (Salmo
    gairdneri) light and electron microscopic study. Dev. Comp.
    Immunol., 7, 59-68.
Chilmonczyk, S. (1985). Evolution of the thymus in rainbow trout. In
    Fish Immunology, eds. M.J. Manning & M.F.Tatner. pp. 285-92.
    London: Academic Press.
Clawson, C.C., Finstad, J. & Good, R.A. (1966). Evolution of the
    immune response. V. Electron microscopy of plasma cells and
    lymphoid tissue of the paddlefish. Lab. Invest., 15, 1830-47.
Clem, L.W. & Leslie, G.A. (1982a). Phylogeny of immunoglobulin
    structure and function. XIV. Peptide map and amino acid
    composition studies of shark antibody light chains. Dev.
    Comp. Immunol., 6, 263-70.
Clem, L.W. & Leslie, G.A. (1982b). Phylogeny of immunoglobulin
    structure and function. XV. Idiotypic analysis of shark
    antibodies. Dev. Comp. Immunol., 6, 463-72.
Clem, L.W., Faulmann, E., Miller, N.W., Ellsaesser, C., Lobb, C.J. &
    Cuchens, M.A. (1984). Temperature-mediated processes in
    teleost immunity: differential effects of in vitro
    temperatures on mitogenic responses of channel catfish
    lymphocytes. Dev. Comp. Immunol., 8, 313-22.
Clem, L.W., Lobb, C.J., Faulmann, E. & Cuchens, M.A. (1981).
    Lymphocyte heterogeneity in fish: differential environmental
    effects on cellular functions. In Develop. Biol. Standard
    49, pp. 279-84. Basel: S. Karger.
Clem, L.W., McLean, W.E., Shankey, V.T. & Cuchens, M.A. (1977).
    Phylogeny of lymphocyte heterogeneity. I. Membrane
    immunoglobulins of teleost lymphocytes. Dev. Comp. Immunol.,
    1, 105-18.
Clem, L.W., Sizemore, R.C., Ellsaesser, C.F. & Miller, N.W. (1985).
    Monocytes as accessory cells in fish immune responses. Dev.
    Comp. Immunol., 9, 803-9.

Coates, M.L. (1975). Hemoglobin function in the vertebrates: an
     evolutionary model. J. Mol. Evol., 6, 285-307.
Conroy, D.A. (1972). Studies on the haematology of the Atlantic
     salmon (Salmo salar) L.). Symp. Zool. Soc. London., 30,
     101-27.
Cooper, A.J. (1971). Ammocoete lymphoid populations in vitro. In
     Fourth Annual Leucocyte Culture Conference, ed. O.R.
     McIntyre, pp. 137-47. New York: Appleton-Century-Crofts.
Cooper, E.L., Klempau, A.E., Ramirez, J.A. & Zapata, A.G. (1980).
     Source of stem cells in evolution. In Development and
     Differentiation of Vertebrate Lymphocytes, ed. J.D. Horton,
     pp. 3-15. Amsterdam: Elsevier/North-Holland Biomedical Press.
Cooper, E.L., Zapata, A., Garcia Barrutia, M. & Ramirez, J.A. (1983).
     Aging changes in lymphopoietic and myelopoietic organs of the
     annual cyprinodont fish, Notobranchus guentheri. Exp.
     Gerontol., 18, 29-38.
Cottrell, B.A. & Doolittle, R.F. (1976). Amino acid sequences of
     lamprey to fibrinopeptides A and B and characterization of
     the junctions split by lamprey and mammalian thrombins.
     Biochim. Biophys. Acta, 453, 426-38.
Cuchens, M.A. & Clem, L.W. (1977). Phylogeny of lymphocyte
     heterogeneity. II. Differential effects of temperature on
     fish T-like and B-like cells. Cell Immunol., 34, 219-30.
Cushing, J.E. (1964). The blood groups of marine animals. Adv. Mar.
     Biol., 2, 85-131.
Daems, W.Th., Wisse, E. & Brederoo, P. (1972). Electron microscopy of
     the vacuolar apparatus In Lysosomes A Laboratory Handbook,
     ed. J.T. Dingle. pp. 150-99. Amsterdam: North-Holland Pubs.
Daimon, T. & Uchida, K. (1985). Ultrastructural evidence of the
     existence of the surface connected canalicular system in the
     thrombocytes of the shark (Triakis scyllia). J. Anat., 141,
     193-200.
Davies, H.G. & Haynes, M.E. (1975). Light and electron microscope
     observations on certain leukocytes in a teleost fish and a
     comparison of the envelope-limited monolayers of chromatin
     structural units in different species. J. Cell Sci., 17,
     287-306.
de Kinkelin, P., Dorson, M. & Hattenberger-Baudouy, A.M. (1981).
     Interferon synthesis in trout and carp after viral
     infection. Dev. Comp. Immunol., Suppl. 2, 167-74.
DeLuca, D., Warr, G.W. & Marchalonis, J.J. (1978). Phylogenetic
     origins of immune recognition: lymphocyte surface
     immunoglobulins and antigen binding in the genus Carassius
     (Teleostii). Eur. J. Immunol., 8, 525-30.
DeLuca, D., Wilson, M.R. & Warr, G.W. (1983). Lymphocyte
     heterogeneity in the trout, Salmo gairdneri, defined with
     monoclonal antibodies to IgM. Eur. J. Immunol., 13, 546-51.
Doolittle, R.F. (1965a). Differences in blood clotting of lamprey
     fibrinogen by lamprey and bovine thrombins. Biochem. J., 94,
     735-41.

Doolittle, R.F. (1965b). Characterisation of lamprey fibrinopeptides.
    Biochem. J., <u>94</u>, 742-50.
Doolittle, R.F. & Surgenor, D.M. (1962). Blood coagulation in fish.
    Amer. J. Physiol., <u>203</u>, 964-70.
Doolittle, R.F. & Wooding, G.L. (1974). The subunit structure of
    lamprey fibrinogen and fibrin. Biochim. Biophys. Acta, <u>271</u>,
    277-82.
Eisler, R. (1965). Erythrocyte counts and hemoglobin content in nine
    species of marine teleosts. Chesapeake Sci., <u>6</u>, 119-20.
Elarifi, A.E. (1982). The histopathology of larval anisakid nematode
    infections in the liver of whiting, <u>Merlangius merlangus</u>
    (L.), with some observations on blood leucocytes of the
    fish. J. Fish Disease, <u>5</u>, 411-9.
Ellis, A.E. (1976). Leucocytes and related cells in the plaice
    <u>Pleuronectes</u> <u>platessa</u>. J. Fish Biol., <u>8</u> 143-56.
Ellis, A.E. (1977a). The leucocytes of fish: a review. J. Fish
    Biol., <u>11</u>, 453-91.
Ellis, A.E. (1977b). Ontogeny of the immune response in <u>Salmo salar</u>.
    Histogenesis of the lymphoid organs and appearance of
    membrane immunoglobulin and mixed leucocyte reactivity. <u>In</u>
    Developmental Immunobiology, eds. J.B. Solomon & J.D. Horton.
    pp. 225-31. Amsterdam: Elsevier/North-Holland Biomedical
    Press.
Ellis, A.E. (1980). Antigen-trapping in the spleen and kidney of the
    plaice <u>Pleuronectes</u> <u>platessa</u> L. J. Fish Disease, <u>3</u>, 413-26.
Ellis, A.E. (1982a). Differences between the immune mechanisms of
    fish and higher vertebrates <u>In</u> Microbial Diseases of Fish,
    ed. R.J.Roberts. pp. 1-29. London: Academic Press.
Ellis, A.E. (1982b). Histamine, mast cells and hypersensitivity
    responses in fish. Dev. Comp. Immunol., <u>Suppl. 2</u>, 147-55.
Ellis, A.E. (1985). Eosinophilic granular cells (EGC) and histamine
    responses to <u>Aeromonas salmonicida</u> toxins in rainbow trout.
    Dev. Comp. Immunol., <u>9</u>, 251-60.
Ellis, A.E. (1986). The function of teleost fish lymphocytes in
    relation to inflammation. Int. J. Tiss. Reac., <u>8</u>, 263-70.
Ellis, A.E., & de Sousa, M. (1974). Phylogeny of the lymphoid system.
    I. A study of the fate of circulating lymphocytes in plaice.
    Eur. J. Immunol., <u>4</u>, 338-43.
Ellis, A.E. & Parkhouse, R.M.E. (1975). Surface immunoglobulin on the
    lymphocytes of the skate <u>Raja naevus</u>. Eur. J. Immunol., <u>5</u>,
    726-8.
Ellis, A.E., Munroe, A.L.S. & Roberts, R.J. (1976). Defence
    mechanisms of fish. I. A study of the phagocytic system and
    the fate of intraperitoneally injected particulate material
    in the plaice <u>(Pleuronectes</u> <u>platessa</u> L.). J. Fish Biol., <u>8</u>,
    67-78.
Etlinger, H.M., Hodgins, H.O. & Chiller, J.M. (1976). Evolution of the
    lymphoid system. I. Evidence for lymphocyte heterogeneity in
    rainbow trout revealed by the organ distribution of mitogenic
    responses. J. Immunol., <u>116</u>, 1547-53.

Etlinger, H.M., Hodgins, H.O. & Chiller, J.M. (1977). Evolution of the
    lymphoid system. II. Evidence for immunoglobulin determinants
    on cell rainbow trout lymphocytes and demonstration of mixed
    leukocyte reaction. Eur. J. Immunol., 7, 881-7.
Evans, D.L., Grave, S.S., Cobb, D. & Dawe, D.L. (1984a). Nonspecific
    cytotoxicity cells in fish (Ictalurus punctatus). II.
    Parameters of target cell lysis and specificity. Dev. Comp.
    Immunol., 8, 303-12.
Evans, D.L., Hogan, K.T., Graves, S.S., Carlson, R.L., Floyd, E. &
    Dawe, D.L. (1984b). Nonspecific cytotoxic cells in fish
    (Ictalurus punctatus). III. Biophysical and biochemical
    properties affecting cytolysis. Dev. Comp. Immunol., 8,
    599-610.
Evans, D.L., Carlson, R.L., Graves, S.S. & Hogan, K.T. (1984c).
    Nonspecific cytotoxic cells in fish (Ictalurus punctatus).
    IV. Target cell binding and recycling capacity. Dev. Comp.
    Immunol., 8, 823-33.
Ezeasor, D.N. & Stokoe, W.M. (1980). A cytochemical, light and
    electron microscopic study of the eosinophilic granule cells
    in the gut of the rainbow trout, Salmo gairdneri Richardson.
    J. Fish Biol., 17, 619-634.
Fänge, R. (1968). White blood cells and lymphomyeloid tissues in
    fish. Bull. Off. int. Epiz., 69, 1357-63.
Fänge, R. (1977). Size relations of lymphomyeloid organs in some
    cartilaginous fish. Acta. Zool., 58, 125-28.
Fänge, R. (1984). Lymphomyeloid tissues in fishes. Vidensk. Meddr.
    dansk. naturh. Foren., 145, 143-62.
Fänge, R., Johansson-Sjöbeck, M.-L. & Kanje, M. (1974).
    Transformation of spindle cells into lymphocyte-like cells in
    the blood from Myxine glutinosa. Acta. Physiol. Scand., 91,
    13A-4A.
Fänge, R. & Mattisson, A. (1981). The lymphomyeloid (hemopoietic)
    system of the Atlantic nurse shark, Ginglymostoma cirratum.
    Biol. Bull. Woods Hole, 160, 240-9.
Fänge, R. & Nilsson, S. (1985). The fish spleen: structure and
    function. Experientia, 41, 152-8.
Fänge, R. & Pulsford, A. (1983). Structural studies on lymphomyeloid
    tissues of the dogfish, Scyliorhinus canicula L. Cell &
    Tissue Res., 230, 337-51.
Fänge, R. & Pulsford, A. (1985). The thymus of the angler fish,
    Lophius piscatorius (Pisces:Teleosti) a light and electron
    microscopic study. In Fish Immunology, eds. M.J. Manning &
    M.F. Tatner. pp. 293-311. London: Academic Press.
Fänge, R. & Sundell, G. (1969). Lymphomyeloid tissues, blood cells
    and plasma proteins in Chimaera monstrosa (Pisces,
    Holocephali). Acta Zool. 50, 155-68.
Ferguson, H.W. (1976a). The ultrastructure of plaice (Pleuronectes
    platessa) leucocytes. J. Fish Biol., 8, 139-142.
Ferguson, H.W. (1976b). The relationship between ellipsoids and
    melano-macrophage centres in the spleen of turbot
    (Scophthalmus maximus). J. Comp. Path., 86, 377-80.

Ferren, F.A. (1967). Role of the spleen in the immune response of teleosts and elasmobranchs. J. Florida Med. Assoc., 54, 434-7.

Ferri, S. & Sesso, A. (1981). Ultrastructural study of Kupffer cells in teleost liver under normal and experimental conditions. Cell Tissue Res., 220, 387-391.

Fey,F. (1965). Vergleichende hämozytologie niederer Vertebraten II. Thrombozyten. Folia Haematol., 85, 205-17.

Fey, F. (1966a). Vergleichende hämozytologie niederer Vertebraten III. Granulozyten. Folia Haematol., 86, 1-20.

Fey, F. (1966b). Vergleichende hämozytologie niederer Vertebraten. IV. Monozyten-plasmozyten-lymphozyten. Folia Haematol., 86, 133-47.

Fiebig, H., Scherbaum, I. & Ambrosius, H. (1977). Zelloberflä chenimmunglobulin von lymphozytin des karpfens (Cyprinus carpio L.). Acta. Biol. Med. Germ., 36, 1167-78.

Fiebig, H., Scherbaum, I. & Ambrosius, H. (1980). Evolutionary origin of the T lymphocyte receptor. 2. Production and partial characterization of antisera to the immunoglobulin like cell membrane protein from thymocytes of the carp. Acta. Biol. Med. Germ., 39, 845-54.

Fiebig, H., Scherbaum, I., Kraus, G. & Kupper, H. (1983). Characterization of carp lymphocyte receptors by monoclonal antibodies and conventional antisera. Dev. Comp. Immunol., 7, 755-6.

Field, J.B., Elvehjem, C.A. & Juday, C. (1943). A study of the blood constituents of carp and trout. J. Biol. Chem., 148, 261-9.

Finn, J.P. & Nielsen, N.O. (1971). The inflammatory response of rainbow trout. J. Fish Biol., 3, 463-78.

Finstad, J., Papermaster, B. W. & Good, R.A. (1964). Evolution of the immune response II. Morpholigic studies on the origin of the thymus and organised lymphoid tissue. Lab. Invest., 13, 490-512.

Fishman, J.A., Daniele, R.P. & Pietra, G.G. (1979). Lung defenses in the African lungfish (Protopterus): cellular responses to irritant stimuli. J. Reticuloedothelial Soc., 25, 179-95.

Fletcher, T.C. & Grant, P.T. (1969). Immunoglobulins in the serum and mucus of the plaice (Pleuronectes platessa). Biochem. J., 115, 65-9.

Fletcher, T.C. & White, A. (1973). Lysozyme activity in the plaice (Pleuronectes platessa L.). Experientia, 29, 1283-5.

Fujii, T. (1981). Antibody-enhanced phagocytosis of lamprey polymorphonuclear leucocytes against sheep erthrocytes. Cell Tissue Res., 219, 41-51.

Fujii, T. (1982). Electron microscopy of the leucocytes of the typhlosole in ammocoetes, with special attention to the antibody producing cells. J. Morphol., 173, 87-100.

Galabov, A.S. (1973). Interferonogenesis and phylogenesis. Bull. Inst. Pasteur, 71, 233-47.

Garavini, C. & Martelli, P. (1981). Alkaline phosphatase and peroxidase in goldfish (<u>Carassius</u> <u>auratus</u>) leukocytes. Basic Appl. Histochem., <u>25</u>, 133-9.

Gardner, G.R. & Yevich, P.P. (1969). Studies on the blood morphology of three estuarine Cyprinodontiform fishes. J. Fish Res. Bd. Can., <u>26</u>, 433-47.

George, J.C. & Beamish, F.W.H. (1974). Haemocytology of the supraneural myeloid body in the sea lamprey during several phases of life cycle. Can. J. Zool., <u>52</u>, 1585-9.

Gerrard, J.M. (1985). Prostaglandins and Leukotrienes. New York: Marcel Dekker, Inc.

Gilbertson, P., Wotherspoon, J. & Raison, R.L. (1986). Evolutionary development of lymphocyte heterogeneity: leucocyte subpopulations in the Pacific hagfish. Dev. Comp. Immunol., <u>10</u>, 1-10.

Good, R.A., Finstad, J., Pollara, B. & Gabrielsen, A.E. (1966). Morphologic studies on the evolution of the lymphoid tissues among the lower vertebrates. <u>In</u> Phylogeny of Immunity eds. R.T. Smith, P.A. Miescher & R.A. Good, pp. 149-70. Gainesville: University of Florida Press.

Grabbe, J., Rosenbach, T. & Czarnetzki, B.M. (1985). Production of LTB_4-like chemotactic arachidonate metabolites from human keratinocytes. J. Immunol., <u>85</u>, 527-30.

Grace, M.F. & Manning, M.J. (1980). Histogenesis of the lymphoid organs in rainbow trout, <u>Salmo</u> <u>gairdneri</u>: Rich. 1836. Dev. Comp. Immunol., <u>4</u>, 255-64.

Graves, S.S., Evans, D.L., Cobb, D. & Dawe, D.L. (1984). Nonspecific cytotoxic cells in fish (<u>Ictalurus</u> <u>punctatus</u>). I. Optimum requirements for target cell lysis. Dev. Comp. Immunol., <u>8</u>, 293-302.

Griffin, B.R. (1983). Opsonic effect of rainbow trout (<u>Salmo</u> <u>gairdneri</u>) antibody on phagocytosis of <u>Yersinia</u> <u>ruckeri</u> by trout leukocytes. Dev. Comp. Immunol., <u>7</u>, 253-60.

Griffin, B.R. (1984). Random and directed migration of trout (<u>Salmo</u> <u>gairdneri</u>) leukocytes: activation by antibody, complement, and normal serum components. Dev. Comp. Immunol., <u>8</u>, 589-98.

Gunter, G., Sulya, L.L. & Box, B.E. (1961). Some evolutionary patterns in fishes' blood. Biol. Bull. Woods Hole, <u>121</u>, 302-6.

Hagen, M., Filosa, M. & Youson, J.H. (1983). Immunocytochemical localization of antibody producing cells in adult lamprey. Immunol. Lett., <u>6</u>, 87-92.

Hall, F.G. & Gray, I.E. (1929). The haemoglobin concentration of the blood of marine fishes. J. Biol. Chem., <u>81</u>, 589-94.

Hardisty, M.W. (1979). Biology of the Cyclostomes. London: Chapman & Hall.

Hardisty, M.W. (1982). Lampreys and hagfishes: analysis of cyclostome relationships. <u>In</u> The Biology of Lampreys Vol. 4B, eds. M.W. Hardisty & I.C. Potter. London: Academic Press.

Hart, S., Wrathmell, A.B. & Harris, J.E. (1986). Ontogeny of
gut-associated lymphoid tissue (GALT) in the dogfish
Scyliorhinus canicula L. Vet. Immunol. Immunopathol., 12,
107-16.

Hartman, F.A. & Lessler, M.A. (1964). Erythrocyte measurements in
fishes, amphibia and reptiles. Biol. Bull. Woods Hole, 126,
83-8.

Hayden, B.J. & Laux, D.C. (1985). Cell-mediated lysis of murine
target cells by nonimmune salmonid lymphoid preparations.
Dev. Comp. Immunol., 9, 627-39.

Hesser, E.F. (1960). Methods for routine fish hematology. Prog. Fish
Cult., 22, 164-71.

Hildebrand, M. (1982). Analysis of Vertebrate Structure 2nd edn. New
York: J. Wiley & Sons.

Hildemann, W.H. & Thoenes, G.H. (1969). Immunological responses of
Pacific hagfish. I. Skin transplantation immunity.
Transplantation, 7, 506-21.

Hildemann, W.H. & Haas, R. (1960). Comparative studies of
homotransplantation in fishes. J. Cell. Comp. Physiol., 55,
227-31.

Hinuma, S., Abo, T., Kumagai, K. & Hata, M. (1980). The potent
activity of freshwater fish kidney cells in cell-killing. I.
Characterization and species distribution of cytotoxicity.
Dev. Comp. Immunol., 4, 653-66.

Holmgren, N. (1950). On the pronephros and the blood in Myxine
glutinosa. Acta Zool., 31, 233-348.

Honma, Y., Okabe, K. & Chiba, A. (1984). Comparative histology of the
Leydig and epigonal organs in some elasmobranchs. Jap. J.
Ichthyol., 31, 47-54.

Hoole, D. & Arme, C. (1982). Ultrastructural studies on the cellular
response of roach, Rutilus rutilus L., to the plerocercoid
larva of the pseudophyllidean cestode, Ligula intestinalis.
J. Fish Disease, 5, 131-44.

Hunt, T.C. & Rowley, A.F. (1986a). Studies on the reticuloendothelial
system of the dogfish, Scyliorhinus canicula. Endocytic
activity of fixed cells in the spleen and liver. Can. J.
Zool. (in press)

Hunt, T.C. & Rowley, A.F. (1986b). Studies on the
reticulo-endothelial system of the dogfish, Scyliorhinus
canicula. Endocytic activity of fixed cells in the gills and
peripheral blood leucocytes. Cell Tissue Res., 244, 215-26.

Hunt, T.C. & Rowley, A.F. (1986c). Preliminary studies on the
chemotactic potential of dogfish (Scyliorhinus canicula)
leucocytes using the bipolar shape formation assay. Vet.
Immunol. Immunopathol., 12, 75-82.

Hunt, T.C. & Rowley, A.F. (1986d). Leukotriene B$_4$ induces enhanced
migration of fish leucocytes in vitro. Immunology 59, 563-8.

Hyder, S.L., Cayer, M.L. & Pettey, C.L. (1983). Cell types in
peripheral blood of the nurse shark: an approach to structure
and function. Tissue & Cell, 15, 437-55.

Ivanova-Berg, M.M. & Sokolova, M.M. (1959). Seasonal changes in the blood of the river lamprey (Lampetra fluviatilis L.). Vop. Ikhtiol., 13, 156-62.

Jakowska, P. (1956). Morphologie et nomenclature des cellules du sang des Téléostéens. Revue Hemat., 11, 519-39.

Jarvik, E. (1980). Basic Structure and Evolution of Vertebrates Vol. I & II. London: Academic Press.

Javaid, M.Y. & Akhtar, N. (1977). Haematology of fishes in Pakistan, II. Studies on fourteen species of teleosts. Biologia, 23, 79-90.

Johansen, K. (1971). Comparative physiology: gas exchange and circulation in fishes. Ann. Rev. Physiol., 33, 569-612.

Johansson, M.-L. (1973). Peroxidase in blood cells of fishes and cyclostomes. Acta Reg. Soc. Sci. Litt. Gotheburg Zool., 8, 53-6.

Johansson-Sjöbeck, M.-L. & Stevens, J.D. (1976). Haematological studies on the blue shark, Prionace glauca L. J. Mar. Biol. Ass. U.K., 56, 237-40.

Johnson, E. & Smith, P.D. (1984). Attachment and phagocytosis by salmon macrophages of agarose beads coated with human C3b and C3bi. Dev. Comp. Immunol., 8, 623-30.

Jordan, H.E. & Speidel, C.C. (1924). Studies on lymphocytes II. The origin, function and fate of the lymphocytes in fishes. J. Morphol., 38, 529-49.

Jordan, H.E. & Speidel, C.C. (1930). Blood formation in cyclostomes. Amer. J. Anat., 46, 355-91.

Jordan, H.E. & Speidel, C.C. (1931). Blood formation in the African lungfish under normal conditions and under conditions of prolonged estivation and recovery. J. Morphol. Physiol., 51, 319-71.

Kamiya, H. & Shimizu, Y. (1980). Marine biopolymers with cell specificity II. Purification and characterization of agglutinins from mucus of windowpane flounder Lophopsetta maculata. Biochim. Biophys. Acta., 622, 171-8.

Kampmeier, O.F. (1969). Evolution and Comparative Morphology of the Lymphatic System. Springfield: C. C. Thomas.

Kanner, J. & Kinsella, J.E. (1983). Lipid deterioration initiated by phagocytic cells in muscle foods: β carotene destruction by a myeloperoxidase-hydrogen peroxide-halide system. J. Agric. Food Chem., 31, 370-6.

Kayama, M., Sado, T., Iijima, N., Asada, T., Igarashi, M., Shiba, T., Yamaguchi, Y. & Hirai, T. (1985). The prostaglandin synthesis in carp thrombocyte. Bull.Jap. Soc. Sci.Fish., 51, 1911.

Kelényi, G. & Larsen, L.O. (1976). The haematopoietic supraneural organ of adult sexually immature river lampreys (Lampetra fluviatilis [L] Gray) with particular reference to azurophil leucocytes. Acta. Biol. Acad. Hung., 27, 45-56.

Kelényi, G. & Németh, A. (1969). Comparative histochemistry and electron microscopy of the eosinophil leucocytes of vertebrates: a study of avian, reptile, amphibian and fish leucocytes. Acta. Biol. Acad. Hung., 20, 405-22.

Kendall, M.D. (1981). The cells of the thymus. In The Thymus Gland, ed. M.D. Kendall. London: Academic Press.

Klontz, G.W. (1972). Haematological techniques and the immune response in rainbow trout. Symp. Zool. Soc. Lond., 30, 89-99.

Kobayashi, K., Tomonaga, S. & Kajii, T.(1984). A second class of immunoglobulin other than IgM present in a cartilaginous fish, the skate, Raja kenojei: isolation and characterization. Mol. Immunol., 5 397-401.

Koler, R.D. (1979). Genetics of hemoglobins. In CRC Handbook Series on Clinical Laboratory Science Section I Hematology Vol. I, ed. R.M. Schmidt. pp. 51-66. Boca Raton, Florida: CRC Press Inc.

Laudano, A.P. & Doolittle, R.F. (1980). Studies on synthetic peptides that bind fibrinogen and prevent fibrin polymerisation. Structural requirements, number of binding sites, and species differences. Biochemistry, 19, 1013-9.

Lester, R.J.G. & Desser, S.S. (1975). Ultrastructural observations on the granulocytic leucocytes of the teleost Catostomus commersoni. Can. J. Zool., 53, 1648-57.

Linna, T.J., Finstad, J. & Good, R.A. (1975). Cell proliferation in epithelial and lympho-hematopoietic tissues of cyclostomes. Amer. Zool., 15, 29-38.

Linthicum, D.S. (1975). Ultrastructure of hagfish blood leucocytes. In Immunologic Phylogeny, eds. W. Hildemann & A.A. Benedict, pp. 241-50. New York: Plenum Press.

Litman, G.W. (1976). Physical properties of immunoglobulins of lower species: a comparison with immunoglobulins of mammals. In Comparative Immunology, ed. J.J. Marchalonis, pp. 239-56. Oxford: Blackwell Scientific.

Lobb, C.J. (1986). Structural diversity of channel catfish immunoglobulins. Vet. Immunol. & Immunopathol., 12, 7-12.

Lobb, C.J. & Clem, L.W. (1981a). Phylogeny of immunoglobulin structure and function. XI. Secretory immunoglobulins in the cutaneous mucus of the sheepshead, Archosargus probatocephalus. Dev. Comp. Immunol., 5, 587-96.

Lobb, C.J. & Clem, L.W. (1981b). The metabolic relationships of the immunoglobulins in fish serum, cutaneous mucus, and bile. J. Immunol., 127, 1525-9.

Lobb, C.J. & Clem, L.W. (1981c). Phylogeny of immunoglobulin structure and function. X. Humoral immunoglobulins of sheepshead, Archosargus probatocephalus. Dev. Comp. Immunol., 5, 271-82.

Lobb, C.J. & Clem, L.W. (1982). Fish lymphocytes differ in the expression of surface immunoglobulin. Dev. Comp. Immunol., 6, 473-80.

Lopez, D.M., Sigel, M.M. & Lee, J.C. (1974). Phylogenetic studies on T cells. I. Lymphocytes of the shark with differential response to PHA and Con A. Cell. Immunol., 10, 287-92.

Lowe-Jinde,L. & Niimi, A.J. (1983). Influence of sampling on the interpretation of haematological measurements of rainbow trout, Salmo gairdneri. Can. J. Zool., 61, 396-402.

MacArthur, J.I. & Fletcher, T.C. (1985). Phagocytosis in fish In Fish Immunology, eds. M.J. Manning & M.F. Tatner, pp. 29-46. Orlando: Academic Press.

MacArthur, J.I., Fletcher, T.C., Pirie, B.J.S., Davidson, R.J.L. & Thomson, A.W. (1984). Peritoneal inflammatory cells in plaice, Pleuronectes platessa L.: effects of stress and endotoxin. J. Fish Biol., 25, 69-81.

MacArthur, J.I., Thomson, A.W. & Fletcher, T.C. (1985). Aspects of leucocyte migration in the plaice, Pleuronectes platessa L. J. Fish Biol., 27, 667-76.

Macey, D.J. & Potter, I.C. (1981). Measurements of various blood cell parameters during the life cycle of the southern hemisphere lamprey, Geotria australis Gray. Comp. Biochem. Physiol., 69A, 815-23.

Maclean, N. (1978). Haemoglobin. Studies in Biology no. 93. London: Edward Arnold.

Mainwaring, G. & Rowley, A.F. (1985a). Separation of leucocytes from the dogfish (Scyliorhinus canicula) using density gradient centrifugation and differential adhesion to glass coverslips. Cell Tissue Res., 241, 283-90.

Mainwaring, G. & Rowley, A.F. (1985b). Studies on granulocyte heterogeneity in elasmobranchs. In Fish Immunology, eds. M.J. Manning & M.F. Tatner, pp. 57-69. Orlando: Academic Press.

Mainwaring, G. & Rowley, A.F. (1985c). The effect of anticoagulants on Blennius pholis L. leucocytes. Comp. Biochem. Physiol., 80A, 85-91.

Manning, M.J. (1981). A comparative view of the thymus in vertebrates. In The Thymus Gland, ed. M.D. Kendall, pp. 7-20. London: Academic Press.

Manning, M.J. & Turner, R.J. (1976). Comparative Immunobiology, Glasgow: Blackie.

Manning, M.J., Grace, M.F. & Secombes, C.J. (1982). Developmental aspects of immunity and tolerance in fish. In Microbial Diseases of Fish, ed. R.J. Roberts, pp.31-46. Orlando: Academic Press.

Marchalonis, J.J. (1977). Immunity in Evolution. London: Edward Arnold.

Marchalonis, J.J., Warr, G.W. & Ruben, L.N. (1978). Evolutionary immunobiology and the problem of the T-cell receptor. Dev. Comp. Immunol., 2, 203-18.

Mattisson, A.G. & Fänge, R. (1977). Light and electron microscopic observations on the blood cells of the Atlantic hagfish, Myxine glutinosa (L). Acta Zool., 58, 205-21.

Mattisson, A. & Fänge, R. (1982). The cellular structure of the
    Leydig organ in the shark, Etmopterus spinax (L.). Biol.
    Bull. Woods Hole, 162, 182-94.
McCumber, L.J., Sigel, M.M., Trauger, R.J. & Cuchens, M.A. (1982).
    RES structure and function of the fishes. In The
    Reticuloendothelial System a Comprehensive Treatise Vol. 3,
    Phylogeny and Ontogeny, eds. N. Cohen & M.M. Sigel, pp.
    393-422. New York: Plenum Press.
McKinney, E.C., Smith, S.B., Haines, H.G. and Sigel, M.M. (1977).
    Phagocytosis by fish cells. J. Reticuloendothelial Soc., 21,
    89-95.
Michel, C. (1982). Progress towards furunculosis vaccination. In
    Microbial Diseases of Fish, ed. R.J. Roberts, pp. 151-169.
    Orlando: Academic Press.
Miller, N.W. & Clem, L.W. (1984). Temperature mediated processes in
    teleost immunity: differential effects of temperatures on
    catfish in vitro antibody responses to thymus-dependent and
    thymus-independent antigens. J. Immunol., 133, 2356-9.
Miller, N.W., Sizemore, R.C. & Clem, L.W. (1985). Phylogeny of
    lymphocyte heterogeneity: the cellular requirements for in
    vitro antibody responses of channel catfish leukocytes. J.
    Immunol., 134, 2884-8.
Moody, C.E., Serreze, D.V. & Reno, P.W. (1985). Non-specific
    cytotoxic activity of teleost leukocytes. Dev. Comp.
    Immunol., 9, 51-64.
Morrow, W.J.W. & Pulsford, A. (1980). Identification of peripheral
    blood leucocytes of the dogfish (Scyliorhinus canicula L.) by
    electron microscopy. J. Fish Biol., 17, 461-75.
Murray, C.K. & Fletcher, T.C. (1976). The immunohistochemical
    localization of lysozyme in plaice (Pleuronectes platessa L.)
    tissues. J. Fish Biol., 9, 329-34.
Nash, K.A., Fletcher, T.C. & Thomson, A.W. (1986). Migration of fish
    leucocytes in vitro: the effect of factors which may be
    involved in mediating inflammation. Vet. Immunol.
    Immunopathol., 12, 83-92.
Nelson, G.L. (1969). Gill arches and the phylogeny of fishes, with
    notes on the classification of vertebrates. Bull. Amer. Mus.
    Nat. Hist., 141, 475-552.
Olivier, G., Eaton, C.A. & Campbell, N. (1986). Interaction between
    Aeromonas salmonicida and peritoneal macrophages of brook
    trout (Salvelinus fontinalis). Vet. Immunol. Immunopathol.,
    12, 223-34.
Ortiz-Muniz, G. & Sigel, M.M. (1971). Antibody synthesis in lymphoid
    organs of two marine teleosts. J. Reticuloendothelial Soc.,
    9, 42-52.
Östberg, Y., Fänge, R., Mattisson, A. & Thomas, N.W. (1976). Light
    and electron microscopical characterization of heterophilic
    granulocytes in the intestinal wall and islet parenchyma of
    the hagfish, Myxine glutinosa (Cyclostomata). Acta Zool.,
    57, 89-102.

Ourth, D.D. (1980). Secretory IgM, lysozyme and lymphocytes in the skin mucus of the channel catfish, Ictalurus punctatus. Dev. Comp. Immunol., 4, 65-74.

Ozaki, H., Ohwaki, M. & Fukada, T. (1983). Studies on lectins of amago (Oncorhyncus rhodurus) I. Amago ova lectin and its receptor on homologous macrophages. Dev. Comp. Immunol., 7, 77-88.

Page, M. (1983). Studies on the Immune System of Lampreys. Ph.D. Thesis, University of Wales.

Page, M. & Rowley, A.F. (1982). A morphological study of pharyngeal lymphoid accumulations in larval lampreys. Dev. Comp. Immunol., Suppl. 2, 35-40.

Page, M. & Rowley, A.F. (1983). A cytochemical, light and electron microscopical study of the leucocytes of the adult river lamprey, Lampetra fluviatilis (L. Gray). J. Fish Biol., 22, 503-17.

Page, M. & Rowley, A.F. (1984). The reticulo-endothelial system of the adult river lamprey, Lampetra fluviatilis (L.): the fate of intravascularly injected colloidal carbon. J. Fish Disease, 7, 339-53.

Papermaster, B.W., Condie, R.M., Finstad, J. & Good, R.A. (1964). Evolution of the immune response. I. The phylogenetic development of adaptive immunologic responsiveness in vertebrates. J. Exp. Med., 119, 105-30.

Parish, N., Wrathmell, A. & Harris, J.E. (1985). Phagocytic cells in the dogfish (Scyliorhinus canicula L.). In Fish Immunology, eds. M.J.Manning & M.F. Tatner, pp 71-83. Orlando: Academic Press.

Parish, N., Wrathmell, A., Hart, S. & Harris, J.E. (1986). The leucocytes of the elasmobranch Scyliorhinus canicula L. - a morphological study. J. Fish Biol., 28, 545-61.

Pederson, T. & Gelfant, S. (1970). Macromolecular synthesis in dogfish peripheral blood cells. J. Cell Biol. 45, 183-7.

Percy, R. & Potter, I.C. (1976). Blood cell formation in the river lamprey, Lampetra fluviatilis. J. Zool. (Lond.), 178, 319-40.

Percy, R. & Potter, I.C. (1977). Changes in haemopoietic sites during the metamorphosis of the lampreys, Lampetra fluviatilis and Lampetra planeri. J. Zool. (Lond.), 183, 111-23.

Percy, R.P. & Potter, I.C. (1981). Further observations on the development and destruction of lamprey blood cells. J. Zool. (Lond.), 193, 239-51.

Perey, D.Y.E., Finstad, J., Pollara, B. & Good, R.A. (1968). Evolution of the immune response. VI. First-and second-set skin homograft rejections in primitive fish. Lab. Invest., 19, 591-60.

Perutz, M.F. (1976). Structure and mechanism of haemoglobin. Br. Med. Bull. 32, 195-208.

Pettey, C.L. & McKinney, E.C. (1981). Mitogen induced cytotoxicity in the nurse shark. Dev. Comp. Immunol. 5, 53-64.

Piavis, G.W. & Hiatt, J.L. (1971). Blood cell lineage in the sea
lamprey, <u>Petromyzon</u> <u>marinus</u> (Pisces: Petromyzontidae).
Copeia, <u>4</u>, 722-8.

Pica, A., Grimaldi, M.C. & Della Corte, F. (1983). The circulating
blood cells of torpedoes (<u>Torpedo</u> <u>marmorata</u> Risso and <u>Torpedo</u>
<u>ocellata</u> Rafinesque). Monitore Zool. Ital (N.S.), <u>17</u>, 353-74.

Piomelli, D. (1985). Leukotrienes in teleost fish gills.
Naturwissenschaften, <u>72</u>, 276-7.

Potter, I.C., Percy, R., Barber, D.L. & Macey, D.J. (1982). The
morphology, development and physiology of blood cells. <u>In</u>
The Biology of Lampreys Vol. 4A, eds. M. W. Hardisty & I. C.
Potter, pp. 233-92. London: Academic Press.

Potter, I.C. Robinson, E.S. & Brown, I.D. (1974). Studies on the
erythrocytes of larval and adult lampreys (<u>Lampetra</u>
<u>fluviatilis</u>). Acta. Zool. (Stockh.), <u>55</u>, 173-7.

Preston, A. (1960). Red blood values in the plaice (<u>Pleuronectes</u>
<u>platessa</u> L.). J. Mar. Biol. Ass. U.K. <u>39</u>, 681-7.

Pulsford, A., Fänge, R. & Morrow, W.J.W. (1982). Cell types and
interactions in the spleen of the dogfish <u>Scyliorhinus</u>
<u>canicula</u> L.; an electron microscopic study. J. Fish Biol.,
<u>21</u>, 649-62.

Pulsford, A., Morrow, W.J.W. and Fänge, R. (1984). Structural studies
on the thymus of the dogfish, <u>Scyliorhinus</u> <u>canicula</u> L. J.
Fish. Biol., <u>25</u>, 353-60.

Rafn, S. & Wingstrand, K.G. (1981). Structure of intestine, pancreas
and spleen of the Australian lungfish, <u>Neoceratodus</u> <u>forsteri</u>
(Krefft). Zool. Scripta, <u>10</u>, 223-39.

Randall, D.J. (1970). The circulatory system. <u>In</u> Fish Physiology,
Vol. IV. eds W.S. Hoar & D. J. Randall. pp. 133-72. London:
Academic Press.

Ratcliffe, N.A., Rowley, A.F., Fitzgerald, S.W. & Rhodes, C.P.
(1985). Invertebrate immunity: basic concepts and recent
advances. Int. Rev. Cytol., <u>97</u>, 183-350.

Reznikoff, P. & Reznikoff, D.G. (1934). Hematological studies in
dogfish (<u>Mustelus</u> <u>canis</u>). Biol. Bull. Woods Hole, <u>66</u>, 115-23.

Rijkers, G.T. (1982). Kinetics of humoral and cellular immune
reactions in fish. Dev. Comp. Immunol., <u>Suppl. 2</u>, 93-100.

Rijkers, G.T. Frederix-Wolters, E.M.H. & van Muiswinkel, W.B.
(1980). The immune system of cyprinid fish. Kinetics and
temperature dependence of antibody-producing cells in carp
(<u>Cyprinus</u> <u>carpio</u>). Immunology, <u>41</u>, 91-7.

Riviere, H.B., Cooper, E.L., Reddy, A.L. & Hildemann, W.H. (1975). In
search of the hagfish thymus. Amer. Zool. <u>15</u>, 38-49.

Roberts, R.J., Young, M. & Milne, J.A. (1972). Studies on the skin of
plaice (<u>Pleuronectes</u> <u>platessa</u>). I. The structure and
ultrastructure of normal plaice skin. J. Fish. Biol., <u>4</u>,
87-98.

Roitt, I., Brostoff, J. & Male, D. (1985). Immunology. Edinburgh:
Churchill Livingstone.

Rola-Pleszczynski, M. & Lemaire, I. (1985). Leukotrienes augment interleukin I production by human monocytes. J. Immunol. 135, 3958-61.

Rosenberg-Wiser, S.. & Avtalion, R.R. (1982). The cells involved in the immune response of fish: III. Culture requirements of PHA-stimulated carp. (Cyprinus carpio) lymphocytes. Dev. Comp. Immunol., 6, 693-702.

Rowley, A.F. & Page, M. (1985a). Ultrastructural, cytochemical and functional studies on the eosinophilic granulocytes of larval lampreys. Cell Tissue Res., 240, 705-9.

Rowley, A.F. & Page, M. (1985b). Lamprey melano-macrophages: structure and function. In Fish Immunology, eds. M. J. Manning & M. F. Tatner. pp. 273-84. Orlando: Academic Press.

Ruben, L.N., Warr, G.W., Decker, J.M. & Marchalonis, J.J. (1977). Phylogenetic origins of immune recognition: lymphoid heterogeneity and the hapten/carrier effect in the goldfish, Carassius auratus. Cell. Immunol., 31, 266-83.

Ruud, J.T. (1954). Vertebrates without erythrocytes and blood pigment. Nature, 173, 848-50.

Sailendri, K. & Muthukkaruppan, V.F. (1975). Morphology of lymphoid organs in the cichlid teleost, Tilapia mossambica (Peters). J. Morphol., 147, 109-22.

Sakai, D.K. (1984). Opsonization by fish antibody and complement in the immune phagocytosis by peritoneal exudate cells isolated from salmonid fishes. J. Fish Disease, 7, 29-38.

Sanchez, G.A., Gajardo, M.K. & De Ioannes, A.E. (1980). IgM-like natural hemagglutinin from ratfish serum: isolation and physio-chemical characterisation. (Callorhynchus callorhynchus). Dev. Comp. Immunol., 4, 667-78.

Satchell, G.H. (1971). Circulation in Fishes. Cambridge: Cambridge University Press.

Saunders, D.C. (1966). Elasmobranch blood cells. Copeia, 1966, 348-51.

Saunders, D.C. (1968). Variations in thrombocytes and small lymphocytes found in the circulating blood of marine fishes. Trans. Amer. Microsc. Soc., 87, 39-43.

Scott, A.L., & Klesius, P.H. (1981). Chemiluminescence: a novel analysis of phagocytosis in fish. In International Symposium on Fish Biologics: Serodiagnostics and Vaccines. Develop. Biol. Standard Vol. 49, pp. 243-54, Basel: Karger.

Scott, A.L., Rogers, W.A. & Klesius, P.H. (1985). Chemiluminescence by peripheral blood phagocytes from channel catfish: function of opsonin and temperature. Dev. Comp. Immunol., 9, 240-50.

Secombes, C.J. (1986). Immunological activation of rainbow trout macrophages induced in vitro by sperm autoantibodies and factors derived from testis sensitized leucocytes. Vet. Immunol. Immunopathol. 12, 193-201.

Secombes, C.J. & Manning, M.J. (1980). Comparative studies on the immune system of fishes and amphibians: antigen localization in the carp, Cyprinus carpio L. J. Fish Disease, 3, 339-412.

Secombes, C.J., van Groningen, J.J.M., van Muiswinkel, W.B. & Egberts, E. (1983a). Ontogeny of the immune system in carp (Cyprinus carpio L.). The appearance of antigenic determinants on lymphoid cells detected by mouse anti-carp thymocyte monoclonal antibodies. Dev. Comp. Immunol., 7, 455-64.

Secombes, C.J., Van Groningen, J.J.M. & Ebgerts, E. (1983b). Separation of lymphocyte subpopulations in carp, Cyprinus carpio L. by monoclonal antibodies: immunohistochemical studies. Immunology, 48, 165-75.

Sekhon, S.S. & Beams, H.W. (1969). Fine structure of the developing trout erythrocytes and thrombocytes with special reference to the marginal band and the cytoplasmic organelles. Amer. J. Anat., 125, 353-74.

Sekhon, S.E. & Maxwell, D.S. (1970). Fine structure of developing hagfish erythrocytes with particular reference to the cytoplasmic organelles. J. Morphol. 131, 211-36.

Shankey, T.V. & Clem, L.W. (1980). Phylogeny of immunoglobulin structure and function - VIII. Intermolecular heterogeneity of shark 19S IgM antibodies to pneumococcal polysaccharide. Mol. Immunol. 17, 365-75.

Sharon, N. (1984). Surface carbohydrates and surface lectins are recognition determinants in phagocytosis. Immunology Today, 5, 143-7.

Shepro, D., Belamarich, F.A. & Branson, R. (1966). The fine structure of the thrombocyte in the dogfish (Mustelus canis) with special reference to microtubule orientation. Anat. Rec., 156, 203-14.

Sherburne, S.W. (1974). Occurrence of both heterophils and neutrophils in the blood of the spiny dogfish, Squalus acanthias. Copeia, 1974, 259-61.

Sigel, M.M., Hamby, B.A. & Huggins, E.M. Jr. (1986). Phylogenetic studies on lymphokines. Fish lymphocytes respond to IL-1 and epithelial cells produce an IL-1 like factor. Vet. Immunol. Immunopathol., 12, 47-58.

Sigel, M.M., Lee, J.C., McKinney, E.C. & Lopez, D.M. (1978). Cellular immunity in fish as measured by lymphocyte stimulation. Mar. Fish. Rev., 40, 6-11.

Sizemore, R.C., Miller, N.W., Cuchens, M.A., Lobb, C.J. & Clem, L.W. (1984). Phylogeny of lymphocyte heterogeneity: the cellular requirements for in vitro mitogenic responses of channel catfish leukocytes. J. Immunol., 133, 2920-4.

Smedsrud, T., Dannevig, B.H., Tolleshaug, H. & Berg. T. (1984). Endocytosis of a mannose-terminated glycoprotein and formaldehyde-treated human serum albumin in liver and kidney cells from fish (Salmo alpinus). Dev. Comp. Immunol., 8, 579-88.

Smith, G.L., Hattingh, J. & Schoonbee, H.J. (1977). Observations on some effects of disodium ethylenediamine tetra-acetate and heparin on fish blood. Comp. Biochem. Physiol., 57C, 35-8.

Smith, M.J.H., Ford-Hutchinson, A.W. & Bray, M.A. (1980).
    Leukotriene B: a potential mediator of inflammation. J.
    Pharm. Pharmacol., 32, 517-8.
Smith. P.D. & Braun-Nesje, R. (1981). Cell mediated immunity in the
    salmon: lymphocyte and macrophage stimulation,
    lymphocyte/macrophage interactions and the production of
    lymphokine-like factors by stimulated lymphocytes. Dev.
    Comp. Immunol., Suppl. 2. 233-38.
Smith, R.T. (1970). Origin of the spindle cell in Myxine. Bull. Mt.
    Desert Isl. Biol. Lab., 9, 60.
Snieszko, S.F. (1960). Microhaematocrit as a tool in fishery
    research. Special Scientific Report - Fisheries No. 341.
    U.S. Department of the Interior, Fish & Wildlife Service.
St. Louise-Cormier, E.A., Osterland, C.K. and Anderson, P.D. (1984).
    Evidence for a cutaneous secretory immune system in rainbow
    trout (Salmo gairdneri). Dev. Comp. Immunol., 8, 71-80.
Stokes, E.E. & Firkin, B.G. (1971). Studies of the peripheral blood
    of the Port Jackson shark, (Heterodontus portusjacksoni) with
    particular reference to the thrombocyte. Br. J. Haematol.,
    20, 427-34.
Stolen, J.S. & Mäkelä, O. (1980). Cell collaboration in a marine
    teleost. A demonstration of specificity in the carrier
    effect. Immunol. Lett., 1, 341-5.
Suzuki, K. (1986). Morphological and phagocytic characteristics of
    peritoneal exudate cells in tilapia, Oreochromis niloticus
    (Trewavas), and carp, Cyprinus carpio. J. Fish Biol., 29,
    349-64.
Tanaka, Y., Saito, Y. & Gotoh, H. (1981). Vascular architecture and
    intestinal hematopoietic nests of two cyclostomes.
    Eptatretus burgeri and ammocoetes of Entosphenus reissneri: a
    comparative morphological study. J. Morphol., 170, 71-93.
Tatner, M.F. (1985). The migration of labelled thymocytes to the
    peripheral lymphoid organs in the rainbow trout, Salmo
    gairdneri Richardson. Dev. Comp. Immunol., 9, 85-91.
Taussig, M.J. (1984). Processes in Pathology and Microbiology, 2nd
    edn. Oxford: Blackwell Scientific Pubs.
Thoenes, G.H. & Hildemann, W.H. (1970). Immunological responses of
    Pacific hayfish. II. Serum antibody production to soluble
    antigen. In Developmental Aspects of Antibody Formation and
    Structure, pp. 711-22. Prague: Czech.Acad.Sci.
Tomonaga, S. & Kobayashi, K. (1985). A second class of immunoglobulin
    in the cartilaginous fishes. Dev. Comp. Immunol., 9, 797-802.
Tomonaga, S., Hirokane, T., Shinohara, H. & Awaya, K. (1973a). The
    primitive spleen of the hagfish. Zool. Mag., 82, 215-7.
Tomonaga, S., Shinohara, H. & Awaya, K. (1973b). Fine structure of the
    peripheral blood cells of the hagfish. Zool. Mag., 82, 211-4.
Tomonaga, S., Kabayashi, K., Hagiwara, K., Sasaki, K. & Sezaki, K.
    (1985). Studies on immunoglobulin and immunoglobulin-forming
    cells in Heterodontus japonicus, a cartilaginous fish. Dev.
    Comp. Immunol., 9, 617-26.

Tomonaga, S., Kobayashi, K., Kajii, T. & Awaya, K. (1984). Two populations of immunoglobulin-forming cells in the skate, Raja kenojei: their distribution and characterization. Dev. Comp. Immunol., 8, 803-12.

Vogel, W.O.P. & Claviez, M. (1981). Vascular specialization in fish, but no evidence for lymphatics. Z. Naturforsch., 36, 490-2.

Ward, J.W. (1969). Hematological studies on the Australian lungfish, Neoceratodus forsteri. Copeia, 1969, 633-5.

Ward, P.D. (1982). The development of bacterial vaccines for fish. In Microbial Diseases of Fish, ed. R.J. Roberts, pp. 47-58. Orlando: Academic Press.

Wardle, C.S. (1971). New observations on the lymph system of the plaice Pleuronectes platessa and other teleosts. J. Mar. Biol. Ass. U.K., 51, 977-90.

Warr, G.W. (1981). Evolution of the lymphocyte. Immunology Today, 2, 63-8.

Warr, G.W. & Marchalonis, J.J. (1977). Lymphocyte surface immunoglobulin of the goldfish differs from its serum counterpart. Dev. Comp. Immunol., 1, 15-22.

Warr, G.W. & Marchalonis, J.J. (1978). Specific immune recognition by lymphocytes: an evolutionary perspective. Quart. Rev. Biol., 53, 225-41.

Warr, G.W. & Marchalonis, J.J. (1982). Molecular basis of self/non-self discrimination in the ectothermic vertebrates. In The Reticuloendothelial System A Comprehensive Treatise Vol. III eds. N. Cohen & M.M. Sigel, pp.541-67. New York: Plenum Press.

Warr, G.W., De Luca, D. & Anderson, D.P. (1983). Thymocyte plasma membrane of the rainbow trout, Salmo gairdneri: associated immunoglobulin and heteroantigens. Comp. Biochem. Physiol., 76B, 515-21.

Warr, G.W., De Luca, D., Decker, J.M., Marchalonis, J.J. & Ruben, L.N. (1977). Lymphoid heterogeneity in teleost fish: studies on the genus Carassius. In Developmental Immunology eds. J.B. Solomon & J.D. Horton, pp. 241-8. Amsterdam: Elsevier.

Warr, G.W., De Luca, D. & Griffin, B.R. (1979). Membrane immunoglobulin is present on thymic and splenic lymphocytes of the trout Salmo gairdneri. J. Immunol., 123, 910-7.

Warr, G.W., De Luca, D. & Marchalonis, J.J. (1976). Phylogenetic origins of immune recognition: lymphocyte surface immunoglobulins in the goldfish, Carassius auratus. Proc. Natl. Acad. Sci. USA, 73, 2476-80.

Warr, G.W., De Luca, D. & Marchalonis, J.J. (1980). Phylogeny and ontogeny of antigen-specific T cell receptors. In Development and Differentiation of Vertebrate Lymphocytes, ed. J. Horton, pp. 99-111. Amsterdam: Elsevier, North/Holland.

Warr, G.W., Vasta, G.R., Marchalonis, J.J., Allen, R.C. & Anderson, D.P. (1984). Molecular analysis of the lymphocyte membrane. Dev. Comp. Immunol., 8, 757-72.

Watson, L.J., Shechmeister, I.L. & Jackson, L.L. (1963). The hematology of goldfish, Carassius auratus. Cytologia, 28, 118-30.

Watson, M.E., Guenther, R.W. & Royce, R.D. (1956). Hematology of healthy and virus-diseased sockeye salmon, Onchorhynchus nerka. Zoologica (N.Y.), 41, 27-37.

Wedemeyer, G.A., Gould, R.W. & Yasutake, W.T. (1983). Some potentials and limits of the leucocrit test as a fish health assessment method. J. Fish Biol., 23, 711-6.

Weinreb, E.L. (1963). Studies on the fine structure of teleost blood cells. I. Peripheral blood. Anat. Rec., 147, 219-38.

Weinreb, E.L. & Weinreb, S. (1969). A study of experimentally induced endocytosis in a teleost. I. Light microscopy of peripheral blood cell responses. Zoologica (New York), 54, 25-34.

Weiss, L., Powell, R. & Schiffman, F.J. (1985). Terminating arterial vessels in red pulp of human spleen: a transmission electron microscopic study. Experientia, 41, 278-83.

Weissmann, G., Bloomgarden, D., Kaplan, R., Cohen, C., Hoffstein, S., Collins, T., Gotlieb, A. & Nagle, D. (1975). A general method for the introduction of enzymes, by means of immunoglobulin-coated liposomes, into lysosomes of deficient cells. Proc. Natl. Acad. Sci. USA, 72, 88-92.

Weissmann, G., Finkelstein, M.C., Csernansky, J., Quigley, J.P., Quinn, R.S., Techner, L., Troll, W. & Dunham, P.B. (1978). Attack of sea urchin eggs by dogfish phagocytes: a model of phagocyte-mediated cellular cytotoxicity. Proc. Natl. Acad. Sci. USA, 75, 1825-9.

White, J.G. (1972). Uptake of latex particles by blood platelets. Phagocytosis of sequestration? Am. J. Pathol., 69, 439-52.

Wilkinson, P.C. (1982). Chemotaxis and Inflammation, 2nd edn. Edinburgh: Churchill Livingstone.

Williams, R.W. & Warner, M.C. (1976). Some observations on the stained blood cellular elements of channel catfish, Ictalurus punctatus. J.Fish Biol., 9, 491-7.

Wintrobe, M.M. (1933). Variations in the size and hemoglobin content in erythrocytes in the blood of various vertebrates. Folia Haematol., 51, 32-49.

Wrathmell, A.B. & Parish, N.M. (1980). Cell surface receptors in the immune response in fish. In Phylogeny of Immunological Memory, ed. M.J. Manning, pp. 143-52. Amsterdam: Elsevier/North-Holland Biomedical Press.

Yamaga, K.M., Kubo, R.T. & Etlinger, H.M. (1978). Studies on the question of conventional immunoglobulin on thymocytes of primitive vertebrates. II. Delineation between Ig-specific and cross-reactive membrane components. J. Immunol., 120, 2074-9.

Zapata, A. (1979a). Ultrastructural study of the teleost fish kidney Dev. Comp. Immunol., 3, 55-65.

Zapata, A. (1979b). Estudio ultraestructural de la mielopoiesis en peces teleosteos. Morf. Norm. y patol., 3, 737-47.

Zapata, A. (1980). Ultrastructure of elasmobranch lymphoid tissue. 1.
    Thymus and spleen. Dev. Comp. Immunol., 4, 459-72.
Zapata, A. (1981a). Ultrastructure of elasmobranch lymphoid tissue.
    2. Leydig's and epigonal organs. Dev. Comp. Immunol., 5,
    43-52.
Zapata, A. (1981b). Lymphoid organs of teleost fish. I.
    Ultrastructure of the thymus of Rutilus rutilus. Dev. Comp.
    Immunol., 5, 427-36.
Zapata, A. (1982). Lymphoid organs of teleost fish. III. Splenic
    lymphoid tissue of Rutilus rutilus and Gobio gobio. Dev.
    Comp. Immunol., 6, 87-94.
Zapata, A. (1984). Phylogeny of the fish immune system. Bull. Inst.
    Pasteur, 81,
Zapata, A. & Carrato, A. (1980). Ultrastructure of elasmobranch and
    teleost thrombocytes. Acta Zool. (Stockho), 61, 179-82.
Zapata, A., Ardavin, C.F., Gomariz, R.P. & Leceta, J. (1981). Plasma
    cells in the ammocoete of Petromyzon marinus. Cell Tissue
    Res., 221, 203-8.
Zucker-Franklin, D. & Grusky, G. (1979). Platelet ultrastructure. In
    CRC Series in Clinical Laboratory Science Section I:
    Hematology Vol. I, ed. R.M. Schmidt, pp.313-28. Boca Raton:
    CRC Press Inc.

Fig. 1. Protospleen in larval <u>Lampetra</u> with central
mesenteric artery (MA), sinusoidal spaces (SS) and gut wall (W).
Bar=20μm.

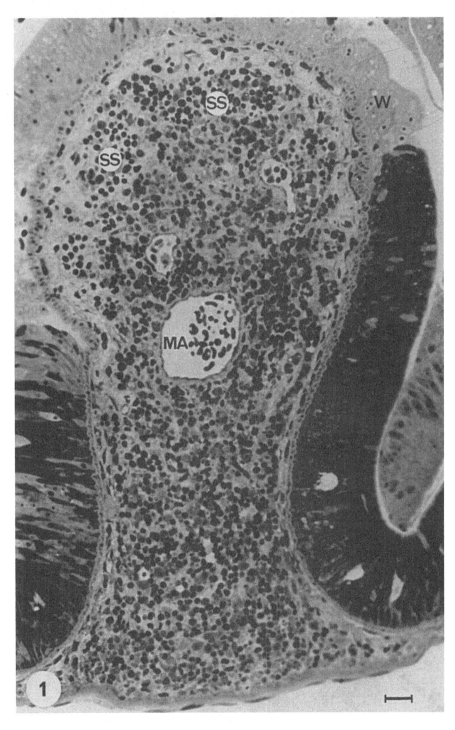

Fig. 2. Macrophage in sinusoidal space of the protospleen of Lampetra attached to endothelial lining (unlabelled arrows). Bar=1μm.

Fig. 3. Migration of granulocyte (G) from the intersinusoidal parenchyma (IP) to a sinusoidal space (SS) in the protospleen of Lampetra. Note immature granulocyte (IG) and mitotic erythroblast (EB). Bar=5μm

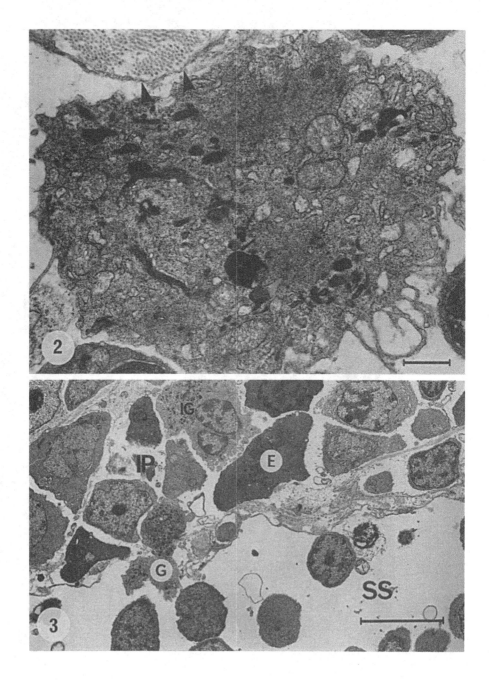

Fig. 4. Intersinusoidal parenchyma of protospleen from Lampetra with immature granulocyte (IG), mitotic cells (MC), putative lymphocytes (L) and supporting reticular cells (RC). Bar=5μm.

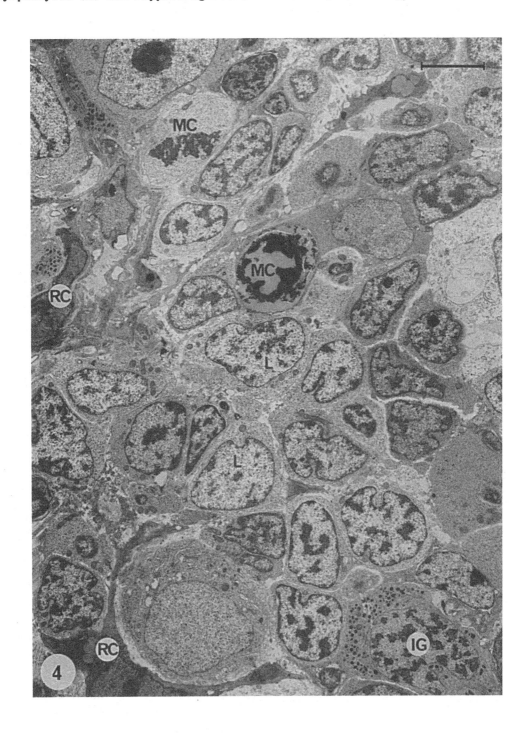

Fig. 5. Kidney of larval <u>Lampetra</u> with haempoietic tissue between the excretory tubules (T). Bar=20μm.

Fig. 6. Supraneural haemopoietic tissue of adult <u>Lampetra</u> <u>fluviatilis</u> with prominent fat vacuoles (V). Bar=50μm.

Fig. 7. Part of supraneural haemopoietic tissue of adult <u>L.</u> <u>fluviatilis</u> containing developing granulocytes (G) and vascular area (VA) with melano-macrophages (MM). Dotted line marks edge of vascular area. Bar=10μm.

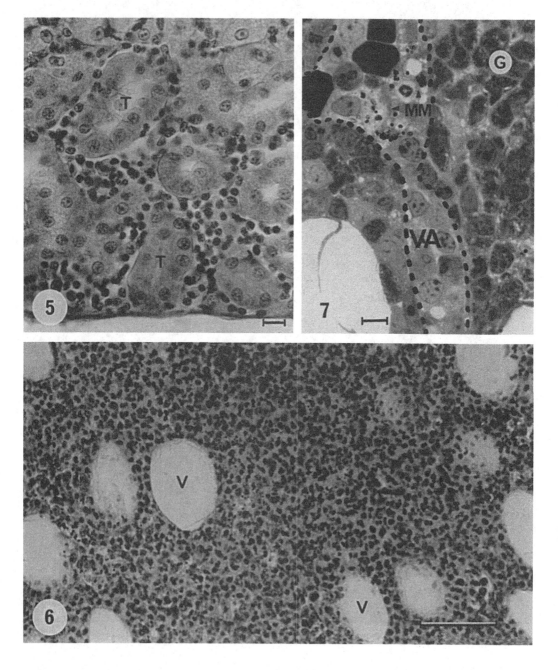

Fig. 8. Lymphoid accumulations (unlabelled arrows) in the pharyngeal region of larval Lampetra. Bar=0.5mm. From Page & Rowley (1982), courtesy of Pergamon Journals Ltd.

Fig. 9. Structure of pharyngeal lymphoid accumulations in larval Lampetra with lymphocytes (L), lymphoblasts (LB) and a supporting reticular cell (RC). Bar=10μm.

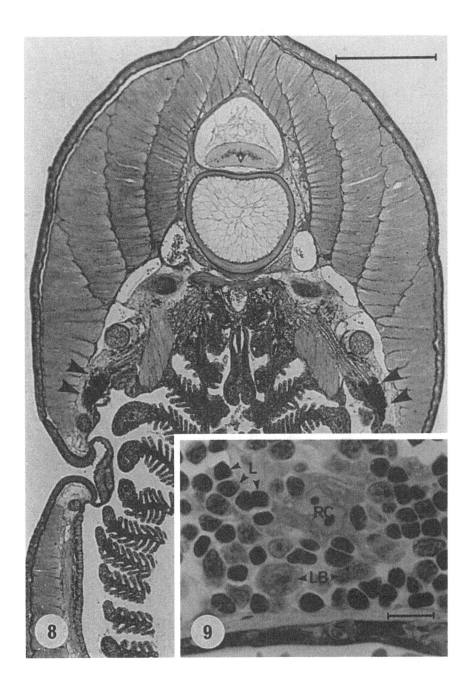

Fig. 10. Lobulate thymus (T) of young dogfish, Scyliorhinus canicula. Immature gill filaments (F). Bar=100μm.

Fig. 11. Lobulate thymus of Raja clavata. Note the cortical (CO) and medullary (ME) zones. Bar=100μm. Micrograph courtesy of Professor A. Zapata.

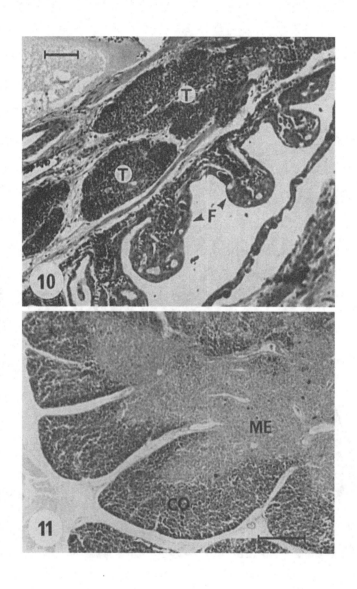

Fig. 12. Spleen of dogfish, Scyliorhinus canicula, with white (W) and red (R) pulp. Ellipsoids (E). Bar=100μm.

Fig. 13. Ellipsoid of dogfish with central capillary (CA), sheath of macrophages (MA) and reticular cells (RC). Red pulp (RP). Bar=5μm.

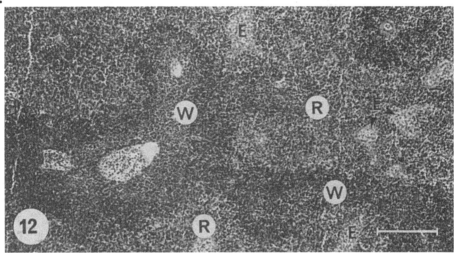

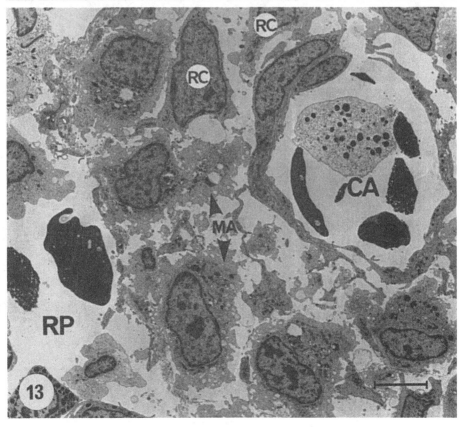

Fig. 14. Leydig's organ of a young dogfish, <u>S. canicula</u> situated as lobulate tissue (unlabelled arrows) around the oesophagus (OE). Bar=100μm.

Fig. 15. Showing the lobular appearance of the epigonal organ in <u>S. canicula</u>. Bar=100μm.

Fig. 16. Developing G1 (G1) and G3 (G3) granulocytes in the epigonal organ of <u>S. canicula</u>. Bar=10μm.

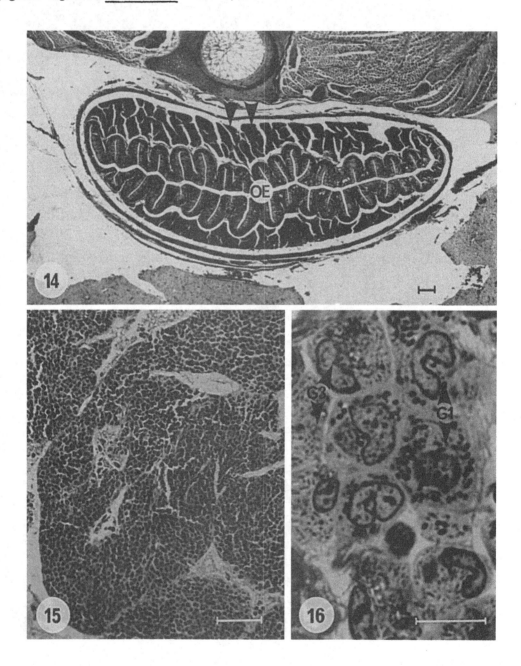

Fig. 17. Thymus of young blenny, <u>Blennius pholis</u>. Note proximity with the pharyngeal epithelium (EP). Bar=10µm.

Fig. 18. Thymus of young rainbow trout, <u>Salmo gairdneri</u>. Pharyngeal epithelium (EP), outer (OZ) and inner (IZ) zones of thymus. Bar=200µm. From Chilmonczyk (1983) with permission of Pergamon Journals Ltd.

Fig. 19. SEM of surface of pharyngeal epithelium above the thymus of young rainbow trout. Note pores in the epithelium. Bar=10µm. From Chilmonczyk (1985) with permission of Academic Press Inc.

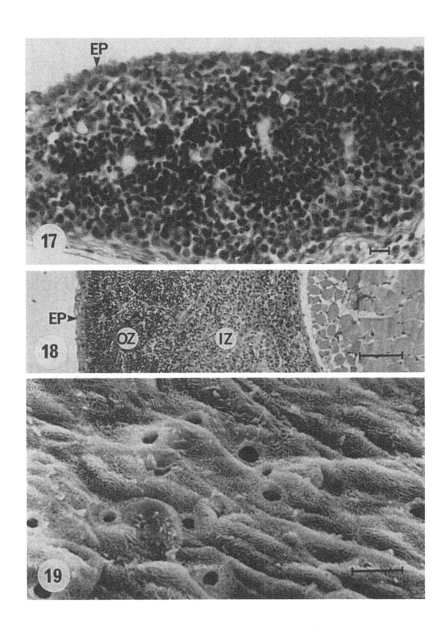

Fig. 20. Low power micrograph of the anterior kidney in the blenny, Blennius pholis. Note haemopoietic tissue (HT), melano-macrophage centre (MC) and spinal column (S). Bar=100μm.

Fig. 21. Detail of haemopoietic tissue in the kidney of the blenny, B. pholis. Nephric tubules (T). Bar=20μm.

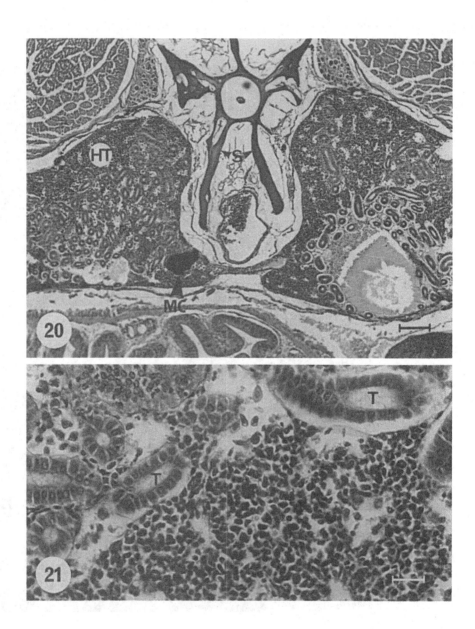

Fig. 22. Heterophilic (H) and basophilic (B) granulocytes at different stages of maturation in the anterior kidney of <u>Tinca tinca</u>. Bar=5μm. From Bielek (1981) with permission of Springer-Verlag.

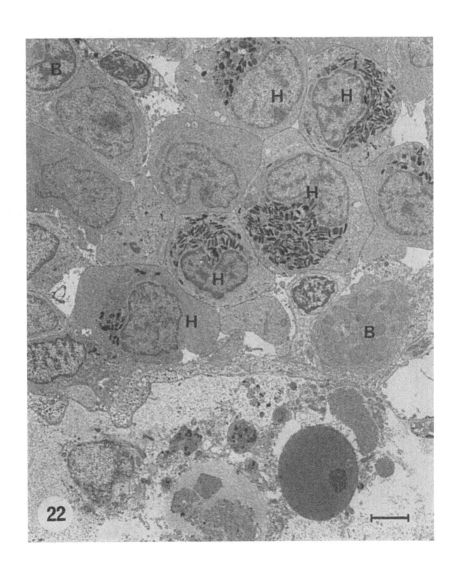

Fig. 23. Erythrocytes at various stages of maturation in the
blood of the dogfish, S. canicula. Note late erythroblast (EB),
erythrocytes with basophilic inclusions (arrows) and eosinophilic zone
(EZ) around nucleus. Bar=10μm.

Fig. 24. Putative haemoblast in S. canicula. Bar=1μm.

Fig. 25. Mature erythrocytes in ammocoete larva of Lampetra
with mitochondrial remnants (R) in the cytoplasm. Bar=1μm.

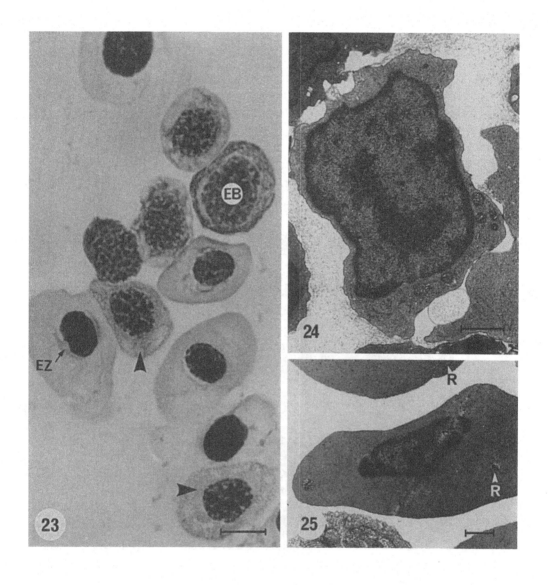

Fig. 26. Lymphocyte in the rudd, Scardinius
erythrophthalmus. Bar=5μm.

Fig. 27. Lymphocyte in blenny, Blennius pholis with poorly
differentiated cytoplasm containing mainly monoribosomes (R). Bar=1μm.

Fig. 28. Lymphocyte of lamprey, Lampetra fluviatilis.
Bar=1μm. From Page & Rowley (1983) with permission of the Fisheries
Society of the British Isles.

Fig. 29. Lymphoblast of L. fluviatilis with cytoplasmic
polyribosomes (PR) and prominent nucleolus (NU). Bar=1μm. From Page &
Rowley (1983) with permission.

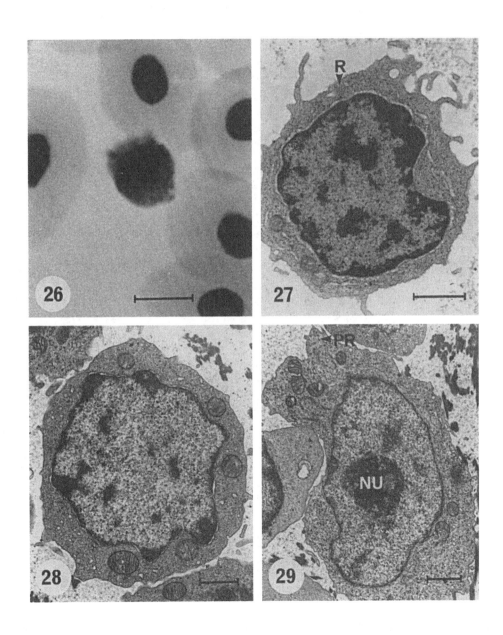

Fig. 30. Plasma cell in the peripheral blood of the dogfish, Scyliorhinus canicula. Note characteristic prominent Golgi complex (GO) and swollen rough endoplasmic reticulum (RER). Bar=1µm.

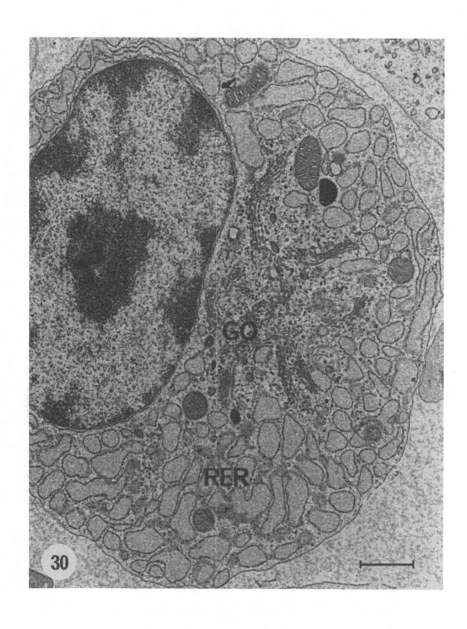

Fig. 31. Spindle-shaped thrombocyte from the blenny,
Blennius pholis. Bar=5μm.
Fig. 32. Clump of thrombocytes from the lamprey, Lampetra
fluviatilis. Note ragged appearance of the cells and characteristic
nuclear clefts (C). Bar=10μm. From Page & Rowley (1983) with permission.
Fig. 33. Spreading thrombocytes of the dogfish, Scyliorhinus
canicula. Note small cytoplasmic granules (G). Bar=10μm.

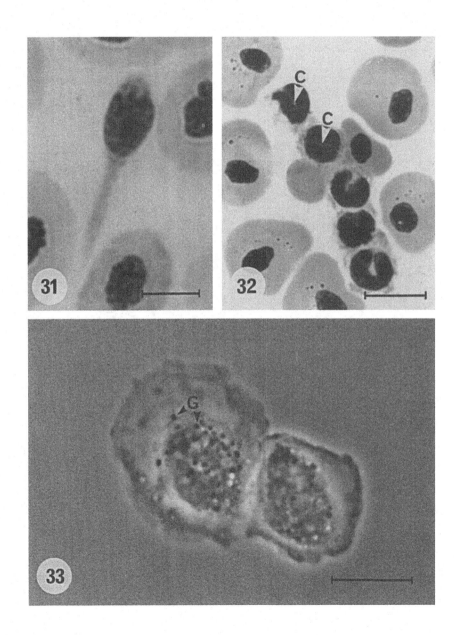

Fig. 34. Thrombocytes from the blenny, <u>Blennius pholis</u> with
canalicular system (CA) opening at one point to the extracellular
environment (arrow), occassional granules (G) and bundles of
microtubules (MT). Bar=1µm.

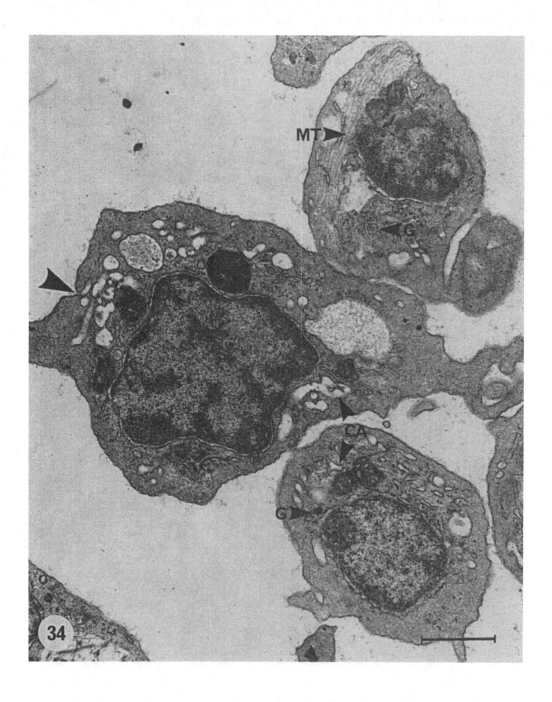

Fig. 35. Thrombocytes from the dogfish, <u>Scyliorhinus</u>
<u>canicula</u> with numerous small granules (G), canalicular system (CA),
<u>myelin</u> whorls (MW) and Golgi complexes (GO). Bar=1μm.

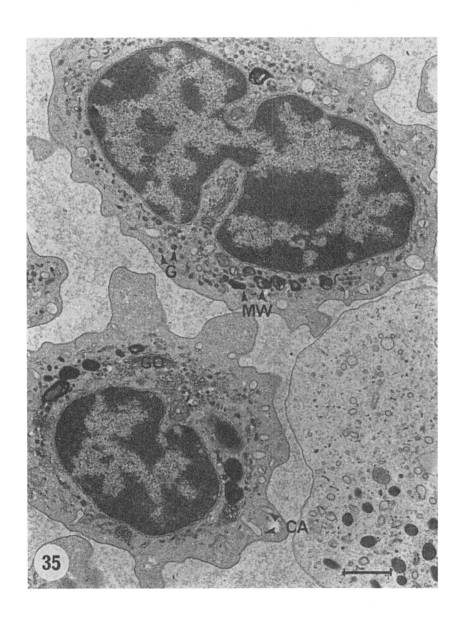

Fig. 36. Neutrophilic/heterophilic granulocyte of adult lamprey, Lampetra fluviatilis with bilobed nucleus. Bar=5µm. From Page & Rowley (1983) with permission.

Fig. 37. Eosinophilic granulocyte from larval Lampetra with coarse granules. Bar=5µm. From Rowley & Page (1985a) with permission of Springer-Verlag.

Fig. 38. Immature neutrophilic/heterophilic granulocyte of L. fluviatilis with characteristic Golgi complex (GO), rough endoplasmic reticulum (RER) and predominant Type I (T1) and Type II (T2) granules. Bar=1µm.

Fig. 39. Mature neutrophilic/heterophilic granulocyte of L. fluviatilis with mainly Type II (T2) and Type III (T3) granules in the cytoplasm. Bar=1µm.

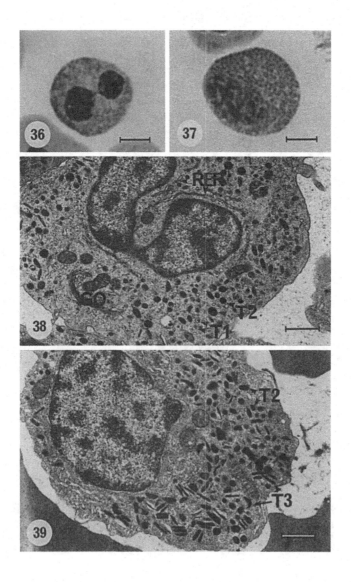

Fig. 40. Granule types (T1-T3) in a
neutrophilic/heterophilic granulocyte of Lampetra fluviatilis. Note
crystalline substructure (arrows) in the Type III granules. Bar=0.2µm.
From Page & Rowley (1983) with permission.

Fig. 41. Eosinophilic granulocyte in larval Lampetra. Note
numerous electron-dense granules. Bar=1µm. From Rowley & Page (1985a)
with permission.

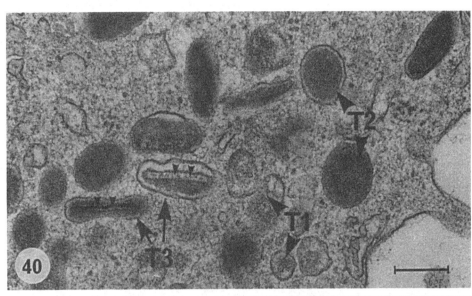

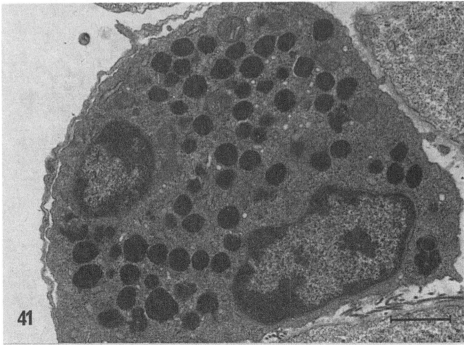

Figs. 42-44. G1 eosinophilic granulocytes of the dogfish,
<u>Scyliorhinus canicula.</u> <u>Fig. 42.</u> G1 granulocyte in smear with prominent
cytoplasmic granules. Bar=10μm. <u>Fig. 43.</u> Phase contrast micrograph
showing flattened amoeboid forms. Bar=10μm. <u>Fig. 44.</u> Electron micrograph
showing characteristic ovoid electron-dense granules in a G1
granulocyte. Bar=1μm. From Mainwaring & Rowley (1985a) with permission.

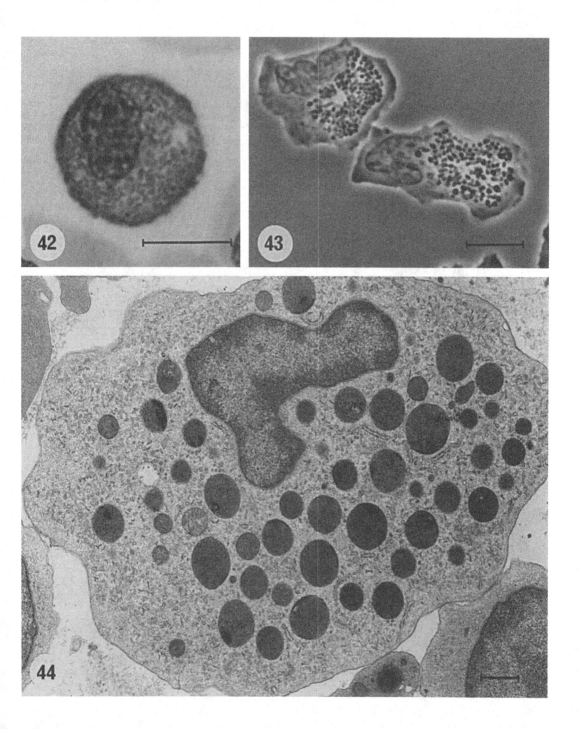

Figs. 45-47. G4 eosinophilic granulocytes of <u>Scyliorhinus</u>
<u>canicula</u>. <u>Fig. 45</u>. G4 granulocyte in smear showing elongate shape.
Bar=10μm. <u>Fig. 46</u>. Nomarski interference micrograph showing granules and
fine protoplasmic extensions (PE). Bar=10μm. <u>Fig. 47</u>. Electron
micrograph of G4 granulocytes with extensive heterochromatin (H),
angular granules and protoplasmic extensions (PE). Bar=1μm. Figs. 46 &
47 from Mainwaring & Rowley (1985a) with permission.

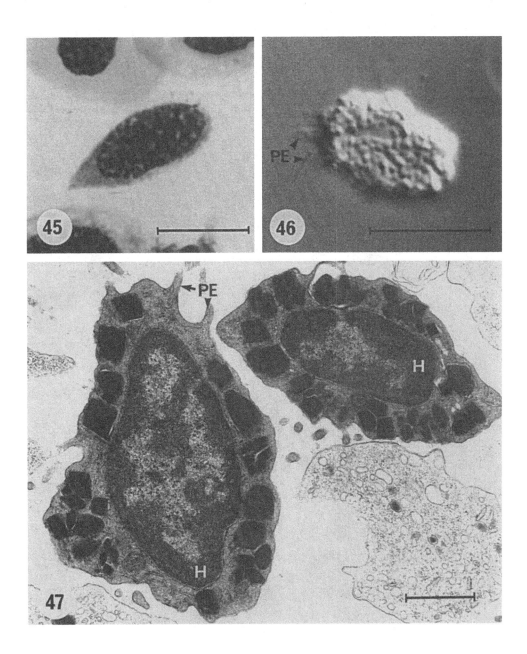

Figs. 48-50. G3 eosinophilic granulocytes of <u>Scyliorhinus</u> <u>canicula</u>. <u>Fig. 48.</u> G3 granulocyte in smear with eccentric lobate nucleus. Bar=10μm. <u>Fig. 49.</u> Phase contrast micrograph showing flattened amoeboid form with elongate granules and lobate nucleus (N). Bar=10μm. <u>Fig. 50.</u> Electron micrograph showing the characteristic eccentric nucleus and elongate electron-dense granules. Bar=1μm. From Mainwaring & Rowley (1985a) with permission.

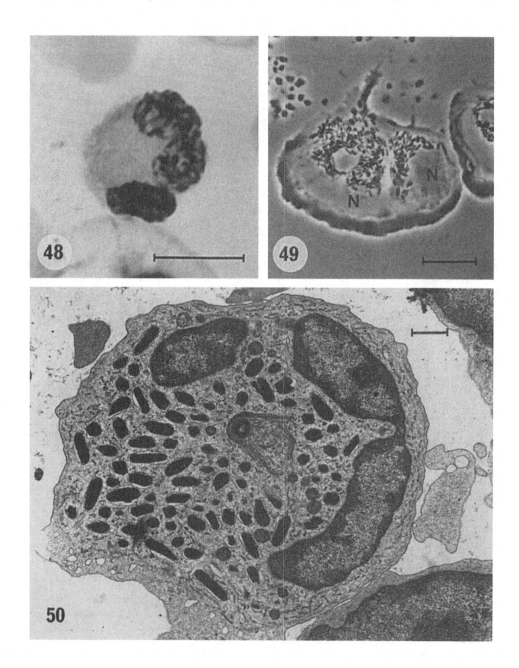

Figs. 51-52. G2 granulocytes of Scyliorhinus canicula. Fig. 51. Vacuolated neutrophilic appearance of a G2 in smears. Bar=10μm. Fig. 52. Electron micrograph showing presumptive G2 granulocyte with numerous granules and eccentric nucleus. Bar=1μm. From Mainwaring & Rowley (1985a) with permission.

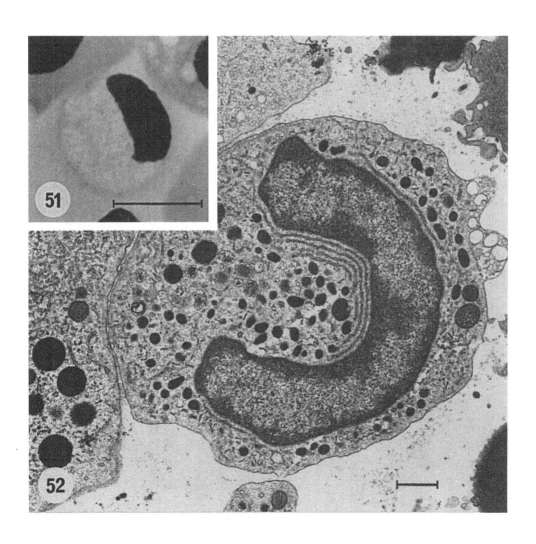

Figs. 53a,b,c. Structure of the granules in eosinophilic granulocytes of the nurse shark, Ginglymostoma cirratum. Note central crystalline component (C) in granules and matrix consisting of paracrystalline arrays (arrows). Bars=1μm (a) and 0.1μm (b,c). From Hyder et al. (1983) with the permission of Churchill-Livingstone.

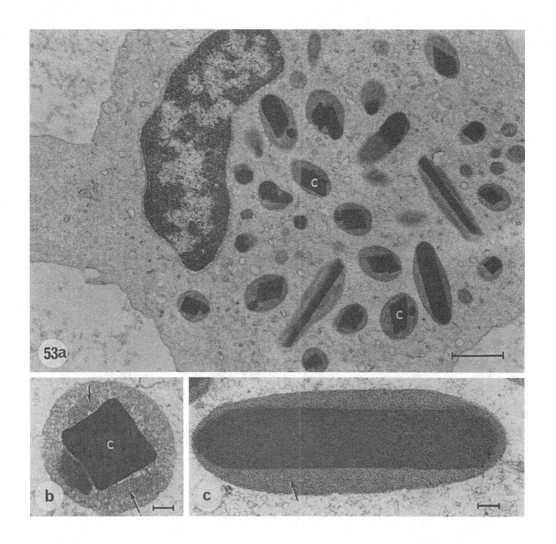

Fig. 54. Immature heterophilic granulocyte of <u>Cyprinus</u> <u>carpio</u> with rod-shaped granules containing crystalline inclusions. Bar=1μm. Micrograph courtesy of Dr. E. Bielek.

Fig. 55. Immature eosinophilic granulocyte of <u>C. carpio</u> with granules at different stages of maturation. Granules in stages 1-3 (1-3) show the appearance of the maturation sequence. Bar=1μm. Micrograph courtesy of Dr. E. Bielek.

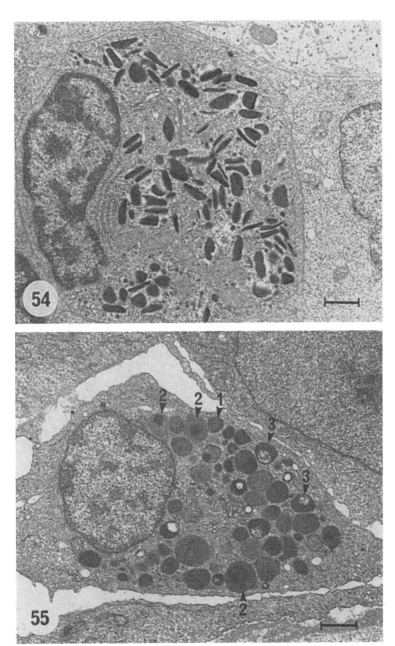

Fig. 56. Immature basophilic granulocyte of <u>Cyprinus carpio</u>. The largest granules with a heterogeneous substructure are the mature forms (arrows). Note granule packaging by Golgi complex (GO). Bar=1μm. Micrograph courtesy of Dr. E. Bielek.

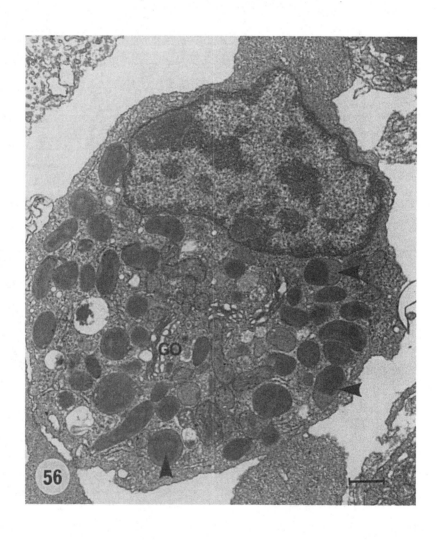

Fig. 57. Neutrophilic/heterophilic granulocyte of the blenny, Blennius pholis with a cytoplasm containing a single population of granules. Bar=1μm.

Fig. 58. Detail of granules in the heterophilic/neutrophilic granulocyte of B. pholis with a fibrillar substructure. Bounding membrane of granules (BM). Bar=0.1μm.

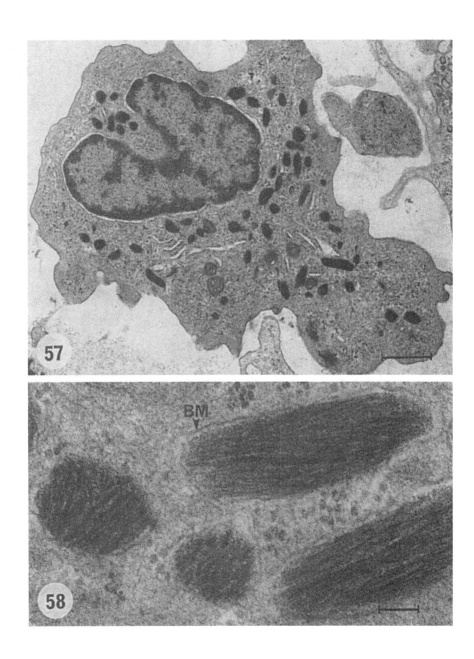

Fig. 59. Granulocyte of the rudd, <u>Scardinius</u>
<u>erythropthalmus</u>. Note granules with three different morphologies; Type I
(T1), Type II (T2) and Type III (T3). Bar=1μm.
        Fig. 60. Part of a rudd granulocyte showing Type I (T1) and
Type III (T3) granules with a crystalline substructure. Insert shows
detail of crystalline rods (arrows). Bars=0.1μm.

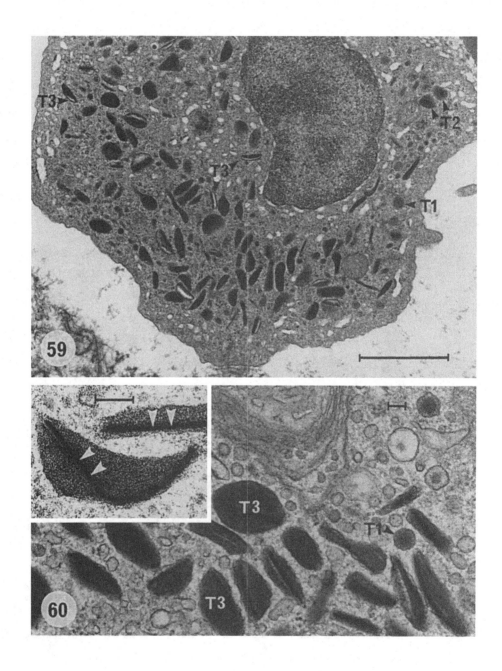

Fig. 61. Monocyte from the blenny, <u>Blennius pholis</u> with
eccentric non-lobate nucleus, irregular outline and vacuolated
cytoplasm. Bar=10μm.

Fig. 62. Monocyte of <u>B. pholis</u> with characteristic prominent
Golgi (GO), primary lysosomes (PL), vacuoles (V), protoplasmic
extensions (PE) and pinocytotic vesicles (PV). Bar=1μm.

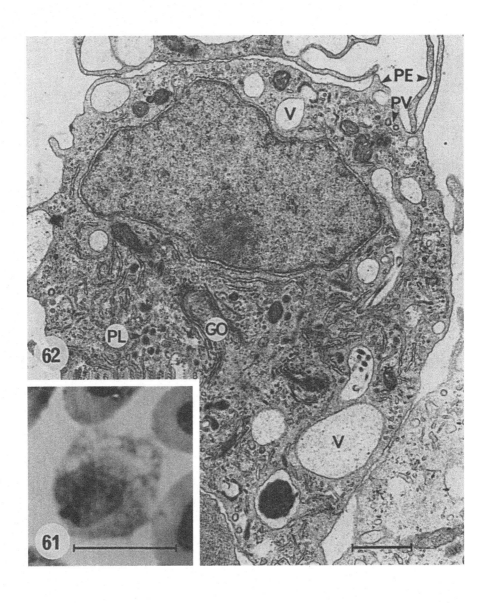

Fig. 63. Melano-macrophage in the kidney interstitium of an
adult lamprey, <u>Lampetra fluviatilis</u> following injection of bacteria.
Note melanin granules (ME) and intracellular bacteria (arrows). Bar=1μm.
Fig. 64. Scanning electron micrograph of a blood clot from
the dogfish, <u>Scyliorhinus</u> canicula. Note thrombocytes (T) and associated
fibrin strands (F) with enmeshed erythrocytes (E). Bar=10μm.

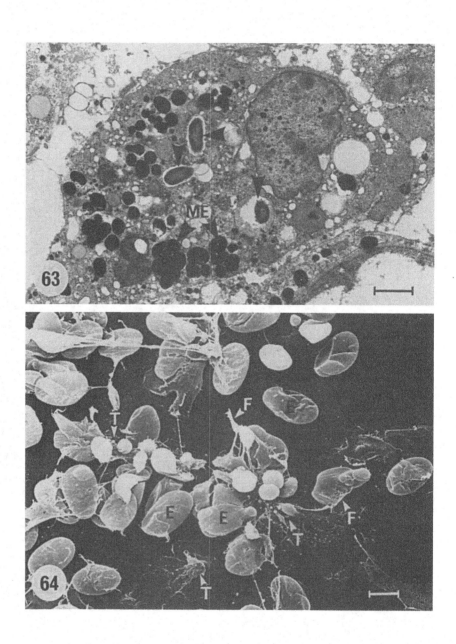

Figs. 65, 66. Neutrophilic (Fig. 65) and eosinophilic (Fig. 66) granulocytes from the peritoneal cavity of the bass, Morone saxatilis elicited by the injection of Bacillus cereus. Bar=1μm. Micrographs courtesy of Dr. J. Bodammer.

Fig. 67. Aggregation of G1 granulocytes (G1) in the cavernous body of the gills of the dogfish, Scyliorhinus canicula following the intravascular injection of bacteria. Note monocytes (M) and specialised endothelial cells called cavernous body cells (CB). Bar=10μm.

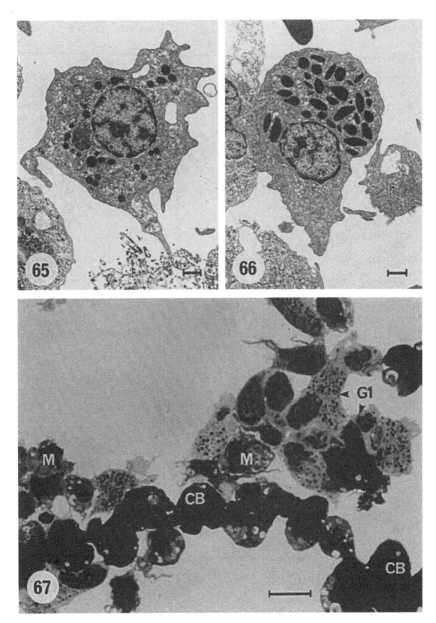

3 Amphibians

R.J. Turner
Department of Zoology, University College of Wales, Aberystwyth
SY23 3DA, UK.

INTRODUCTION

The developmental history of the frogs and toads (order Anura) is unique among the vertebrates, involving as it does the emergence of a free-living aquatic animal at a very early stage, followed by a delayed but rapid metamorphosis into an adult with quite different form and diet and a more terrestrial habit. Amphibian embryos and larvae have proved ideal for experimental studies on the sites of haemopoiesis, since removal or exchange of tissues is possible before the blood system is established: the young animals are accessible, robust, easy to manipulate and available in large numbers. Such operations are obviously more difficult to perform in viviparous animals, and their outcome may be influenced by maternal factors. Older tadpoles with their leisurely development and attainment of a reasonable size (at least in some species) have facilitated study of the respiratory and immunological characteristics of independent but still immature animals, whilst the move from water to land by newly metamorphosed adults invites comparisons about the layout of their blood system and its associated organs, and about the molecular and physiological properties of their haemoglobin. Moreover, it seems reasonable to suppose that metamorphosis has major immunological repercussions, since not only will there be new species of pathogen and modes of attack to be countered by the adult, but also new "self" structures to be tolerated.

Frogs of the family Ranidae are the most successful and widely distributed of the amphibians; they are also the most studied. Gibbs et al. (1971) estimated that U.S. suppliers alone were collecting 9 million ranid frogs (360 tons) a year for educational and research purposes. Such predations, coupled with the effects of pollution, land drainage, interference with migratory routes and so on, have caused a decline in wild populations of "common frogs", and the use of animals bred and reared in controlled colonies instead (which is already a trend) is obviously desirable. Apart from any ecological considerations, the latter will be less variable, less stressed and healthier (see also Elkan 1976). The most successful of existing laboratory-adapted amphibians, the clawed toad Xenopus (family Pipidae) is easy to maintain, can be bred any time of the year, its generation time of about 12 months is shorter than that of many other anurans, and clones of genetically identical heterozygous animals can be produced (Kobel and Du Pasquier 1977). These isogeneic animals are ideal for studies involving cell transfer. Furthermore, pressure-treatment (Tompkins 1978) can be used to create polyploid Xenopus, whose cells are easily distinguishable from normal diploid cells on the basis of size, DNA content and number of nucleoli as well as chromosome number. Methods for producing isogeneic and polyploid Rana in the laboratory have also been described (Dasgupta 1962; Volpe and McKinnell 1966). These techniques have paved the way for detailed studies on blood cell migrations.

Amphibians are a specialised and diverse group of vertebrates: not all are amphibious, and some are only distantly related to frogs. This should be borne in mind when particular species are presented as models. The newts and salamanders (order Urodela) are generally more fish-like than the Anura in their appearance and physiology since the adult terrestrial stage is often suppressed; where metamorphosis does occur it is restricted in scope, with the most obvious changes taking place in the respiratory apparatus and skin (Etkin 1964). The immunobiology of urodeles and the distribution and histology of their haemopoietic tissues are also reminiscent of fishes (Manning and Turner 1976; Manning 1978). The remaining amphibian order, the Apoda, is an obscure assemblage of tropical worm-like, burrowing creatures possessing small or vestigial eyes and no limbs. Our knowledge of their blood systems is limited. Some apodans have aquatic larvae which metamorphose, others do not (see Dodd and Dodd 1976).

Besides their remarkable development and ability to exploit both aquatic and terrestrial environments, amphibian species have other attributes which should be mentioned, even though these are not exclusive to the class. Firstly, since the animals are ectothermic, the speed of biological events can be manipulated. Secondly, the nucleated red blood cells of amphibians and other non-mammalian vertebrates (notably chickens) have provided useful material for a variety of cell and molecular studies: the cells can be obtained easily as relatively uniform populations, they have a well-defined end product, and show progressive and selective inactivation of the genome during differentiation (Sinclair and Brasch 1975). Amphibian erythrocytes have been employed for ultrastructural and biochemical analysis of microtubules (see, for example, Monaco et al. 1982), for studies on histones and chromatin (Nelson and Yunis 1969; Brown et al. 1981) and for work on drug and hormone receptors. In this latter area, frog β adrenoceptors have received the most attention, although it is claimed that they do not correspond strictly to mammalian $\beta 1$ or $\beta 2$ receptors (Dickinson and Nahorski 1981). Finally, the large size of urodele erythrocytes (up to 70μm across) presents special opportunities. Lassen et al. (1978), for example, have inserted microelectrodes into Amphiuma cells to obtain direct measurements of membrane potential.

ORGANIZATION OF THE CIRCULATORY SYSTEM

Blood circulation
Amphibians use gills, lungs, skin and the buccal cavity for gaseous exchange. The importance of a given respiratory surface

Figure 1.

Amphibian blood circulations.

A. <u>Anuran</u>. Note the blood supply
   to the skin via special cutaneous
   arteries (ca).

pca pulmocutaneous arch
sa   systemic   "
RA   right atrium
LA   left   "
V   ventricle

—— deoxygenated blood
········ oxygenated   "
—·—· mixed   "

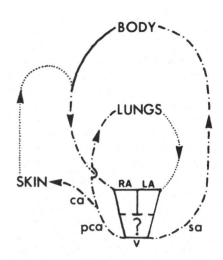

B. <u>Urodele</u>. The special cutaneous
   circuit seen in anurans is missing.
   However, the dorsal section of aortic
   arch no.6 is often retained as a ductus
   Botalli (dB) (= ductus arteriosus) so
   that the pulmonary arch (pa) and
   systemic arch remain connected.
   (sa+5 = systemic arch plus arch no.5).
   The interatrial septum is sometimes
   perforated.

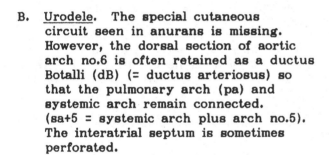

C. <u>Urodele (lungless)</u>. Note the
   absence of an interatrial septum.

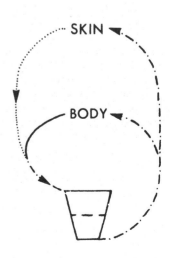

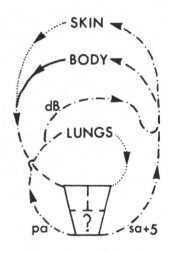

differs between species, between developmental stages of the same species, and in a given individual under different environmental conditions (Noble 1931; Foxon 1964): the group has kept its options open as it were, and generalizations are hard to make. These diverse respiratory mechanisms in turn affect the layout of blood vessels, the organization of the heart, and the composition of the blood.

In larval stages, aortic arches provide vascular loops which form a capillary network in the gills. The capillaries rejoin the aortic arches for dissemination of the oxygenated blood to the head and body. Blood reaching the gills will already be partially oxygenated following its passage through the skin and (in air-gulping tadpoles) through the lungs. The gills and their blood connections are, of course, lost at metamorphosis; aortic arch no.3 remains to supply the head region, no.4 (systemic arch) to supply the rest of the body, whilst no.5 persists only in adult urodeles (arches 1 and 2 disappear early in development). With the appearance of lungs, aortic arch no.6 has become a pulmonary artery; the pulmonary veins bypass the sinus venosus and empty directly back into the heart. The logical outcome of this development is the familiar double circulation of "higher" land vertebrates, where systemic and pulmonary circulations are kept apart by atrial and ventricular septa. In amphibians, however, where the skin may be a major site of oxygen uptake, such a division is counterproductive because although the same arterial system may be used to supply the lungs and large areas of the skin (i.e. the pulmocutaneous arch of anurans), returning blood drains into atria of opposite sides (Fig. 1A and 1B). Thus, although the atria are divided in anurans and some urodeles, the ventricle remains undivided and the destination of blood from a given atrium can vary. Any separation of blood which is maintained through the ventricle is thought to be accomplished by projections of the ventricular wall, by pressure differential (Toews 1971) and/or by a spiral valve located in the conus arteriosus (Shelton 1976).

The plethodont urodeles provide a striking illustration of the potential of the skin as a respiratory surface. In the adults there are no gills or lungs, and the buccal cavity provides less than 11% of the respiratory capillaries (see Foxon 1964). These animals possess a single blood circulation (Fig. 1C). This trend is also seen in urodeles where the lungs are poorly developed or assume a hydrostatic function: the interatrial septum becomes perforated and the spiral valve lost. A similar situation is found in the apodan, Siphonops annulatus, which obtains most of its oxygen through the skin, whereas Typhlonectes compressicauda, which relies mostly on its lungs has completely divided atria and retains the ability to separate blood flowing through the ventricle (Toews and Macintyre 1978).

The effects of these different modes of respiration on the red cells themselves will be examined later.

Lymphatic circulation

For a detailed consideration of amphibian lymphatics the reader is referred to Kampmeier (1969). A major feature in anurans is the presence of extensive subcutaneous lymph sacs (for the experimentalist, these provide an easy, leak-proof injection route into the circulation). The sacs form a "water jacket" around the animal which could assist it in gaseous exchange. According to Kampmeier, the copious volumes of lymph produced may be associated with the relatively easy movement of plasma proteins from blood capillaries into the tissues. Unlike birds and mammals with their valved thoracic duct, the movement of lymph back to the blood is achieved by a series of two-chamber pumps (lymph hearts) which communicate directly with lymph cavities at one end and veins at the other. Anurans typically possess one pair of anterior lymph hearts; posterior hearts vary in number among species and individuals. Modifications of the lymphatic channels are less extreme in urodeles and apodans, and the ancestral series of segmental lymph hearts has survived to a greater degree.

Haemopoietic and lymphoid tissues

Leaving aside for the moment the question of stem cell production and migration, two main groups of tissues which are associated with the blood can be delineated: (i) those which produce the different kinds of blood cells and where developing stages can be identified, i.e. haemopoietic tissues and primary lymphoid tissues; (ii) tissues which house mature leucocytes responsible for reacting against foreign material, i.e. secondary lymphoid tissues. In practice, such distinctions are frequently difficult to make in amphibians: the association between a particular function and a particular site tends to be less consistent or clear-cut in this group (and in fishes) than in "higher" vertebrates.

Descriptive accounts of erythropoietic, granulopoietic and thrombopoietic sites are summarized in Table 1 (reports on monocyte production are fragmentary and hence not included). Cell accumulations likely to be involved in lymphopoiesis are listed separately in Table 2. Note that references to granulocytes usually describe neutrophils and eosinophils, not basophils. In adult anurans, the basophils may arise in the spleen alongside neutrophils and eosinophils (Kapa et al. 1970), but in urodeles (e.g. Notophthalmus and Amphiuma) granulocyte production is split: the liver produces eosinophils and neutrophils but not basophils, whilst the spleen is an important source of basophils but not neutrophils or eosinophils (Cowden et al. 1964; Cowden 1965; Hightower and Haar 1975). In all three amphibian orders, the spleen is also considered to be a major site of red cell production and destruction and, as one might expect, splenectomy disturbs the balance and numbers of circulating blood cells. However, the operation is not usually life-threatening: splenectomized Xenopus, ranid frogs and newts (Notophthalmus) have survived for prolonged or indefinite periods (Jordan 1938; Fey 1965a). Following removal of the spleen in R. esculenta, the thymus appears to take over basophil production (Kapa &

Table 1 : SITES OF HAEMOPOIESIS IN AMPHIBIANS

Animal	Stage	Erythrocytes	Granulocytes[a]	Thrombocytes	Reference
(variety of genera)	early	ventral blood islands			1,2,3,4
ANURA					
Rana pipiens	larva	kidney, liver (spleen)[b]	kidney, liver (lymphomyeloid nodes)[b]		1,2,3,4,5,6
	adult	liver, spleen bone marrow	spleen, bone marrow, gut wall, kidney (lymphomyeloid nodes)	spleen	
R. catesbeiana	larva	kidney, liver (spleen)	(lymphomyeloid nodes) +?		2,3,7,8
	adult	liver, spleen bone marrow	spleen, kidney (lymphomyeloid nodes)	spleen	
R. temporaria (=fusca)	larva	kidney, liver	kidney, lungs, (lymphomyeloid nodes)	kidney liver	2,6,7
	adult	bone marrow	bone marrow (lymphomyeloid nodes)	bone marrow	
Xenopus	larva	kidney	kidney, liver	spleen	6,9,10
	adult	spleen, liver kidney	liver bone marrow		
URODELA					
Notophthalmus[c]		spleen	liver	spleen	1,2,11
Proteus		spleen (liver)	kidney	spleen (liver)	1,2
Necturus		spleen	liver, kidney, heart, (gut wall)	spleen	1,2,3
Amphiuma		spleen	liver		2
Plethodon		spleen	liver, bone marrow	spleen	1,2,12
APODA					
Ichthyophis		spleen	liver (spleen)	spleen	13,14

Note: (a) Granulocytes = neutrophils + eosinophils. Basophils may arise elsewhere (see text).

(b) Brackets indicate possible/minor role.

(c) The newt Notophthalmus viridescens is called Diemyctilus or Triturus in older literature.

References: 1, Jordan (1938); 2, Foxon (1964); 3, Le Douarin (1966); 4, Hollyfield (1966); 5, Turpen et al. (1979); 6, Manning and Horton (1982); 7, Cooper (1976); 8, Broyles et al. (1981);9, Maclean and Jurd (1972); 10, Fey (1965b); 11, Hightower and Haar (1975); 12, Curtis et al. (1979a); 13, Welsch and Storch (1982); 14, Zapata et al. (1982).

Csaba 1972). Since other amphibian species show significant basophil populations in or around the thymus and often lack basophils in major eosinophil- and neutrophil-producing areas, it has been suggested that the former are specialized lymphoid cells rather than true granulocytes (see Curtis et al. 1979a,b).

The kidneys are granulopoietic centres in a wide variety of amphibians during larval and adult stages (Table 1). Some authorities assert that they are also the main erythropoietic sites in anuran larvae, but according to others, the larval liver is more important in this respect (see Manning and Horton 1982). In Rana pipiens, the larval (pronephric) kidney is responsible for monopoiesis (Turpen & Knudson 1982).

A feature of special interest in amphibians is the appearance of bone marrow for the first time as a haemopoietic tissue. This is seen chiefly in adult anurans. In some species e.g. R. temporaria the marrow apparently becomes the sole source of erythrocytes and thrombocytes, whereas in others e.g. R. pipiens it shows seasonal fluctuations and the spleen remains active (see Le Douarin 1966). In the aquatic Xenopus, bone marrow is poorly developed. Attempts to assess the contribution of bone marrow to the amphibian blood system by experimental means are very limited, although Cooper and Schaefer (1970) have shown that by shielding one hind limb of frogs (R. pipiens) during whole-body irradiation, survival rates are improved and changes in blood leucocyte counts are diminished. Bone marrow is lacking in the apodans examined so far, and in urodeles it is restricted to a single family, the Plethodontidae. Studies on Plethodon glutinosus suggest that the bone marrow provides significant numbers of neutrophils and eosinophils, but unlike ranid anurans, there is no erythropoietic activity here (Curtis et al. 1979a). As in other urodeles, red cells and thrombocytes are produced in the spleen, and eosinophils and neutrophils continue to be produced in the liver.

It is thus evident that no single tissue has the monopoly of blood cell production in amphibians: requirements are met by different sites depending on species and developmental stage. Bone marrow at this level could be seen simply as an additional site where space became available following the appearance of lighter, hollow bones needed to support the vertebrate body on land. On the other hand, the use of bony cavities to house delicate blood-forming cells in terrestrial amphibians and especially in "higher" vertebrates is seen by Cooper et al. (1980) as a response to increased radiation hazards on land.

A comparison of Tables 1 and 2 shows that a number of organs with erythropoietic or granulopoietic sites also house lymphoid tissues, notably the spleen, liver, kidney, and bone marrow. The spleen (Fig.2) is the most highly organized and sophisticated of these and resembles its mammalian counterpart, at least in anurans: the lymphocytes are concentrated in a regular pattern around arterioles (white pulp), and separated clearly from the erythrocytic sinusoidal

zone (red pulp); in apodans and particularly some urodeles, the
demarcation is less obvious (see Manning and Turner 1976; Welsch and
Storch 1982). The amphibian spleen has the hallmarks of a secondary
lymphoid organ: circulating lymphocytes "home" into it, it traps
antigen, its lymphoid cells proliferate following antigenic
triggering, and it eventually produces antibody-forming cells (Manning
and Horton 1982). Other lymphocytic sites such as in the liver,
kidney and gut, are simpler and more scattered than in the spleen, but
they may be equally important to the animal: splenectomy does not
appear to have   major immunological repercussions either in anurans

Table 2 : LYMPHOCYTE ACCUMULATIONS IN AMPHIBIANS

Site	Anura (Rana, Bufo)	Anura (Xenopus)	Urodela	Apoda (Ichthyophis, Nectrocaecilia)
Thymus	+	+	+	+
Spleen	+	+	+	+
Kidney	+	+	(+)	?
Lymphomyeloid nodes	+	-	-	-
Lymphoepithelial GALT[a]	+	-	+	?
Nodular GALT[a]	+	+	-	-
Liver	(+)	+	+	+
Bone marrow	+	(+)	+ or -[b]	-

Sources: Manning & Horton, (1982); Zapata et al. (1982)

Notes: + = present; - = absent; (+) = minor contribution.

[a]GALT = gut-associated lymphoid tissue

[b]Lymphocytes found in bone marrow of Plethodontidae
   but not in other urodele families.

Figures 2-5. Amphibian lymphoid tissues (courtesy of J.D. Horton).
Fig.2. Circular white pulp "island" in the spleen of a Xenopus
toadlet, silver-stained to show network of reticular fibres. Note
also the central arteriole (arrows).
Fig.3. Mesonephric kidney of larval R. pipiens (H&E). Haemopoietic
tissue (arrows) is extensive in the intertubular regions, where
lymphocytes, granulocytes and other developing blood cell types can be
found.
Fig.4. Larval lymph gland of R. pipiens (H&E). The gland (gl) shows
dense accumulations of lymphoid cells with intervening blood sinuses,
and projects into the large anterior lymphatic channel (lc).
(gc = gill chamber).
Fig.5. Thymus of Xenopus toadlet (H&E). The cortex is densely
populated with lymphocytes. The paler medulla contains fewer
lymphocytes but a variety of other cell types including basophils and
macrophages and large numbers of epithelial cells. Trabeculae with
associated blood vessels can be seen penetrating the organ from the
capsular region.

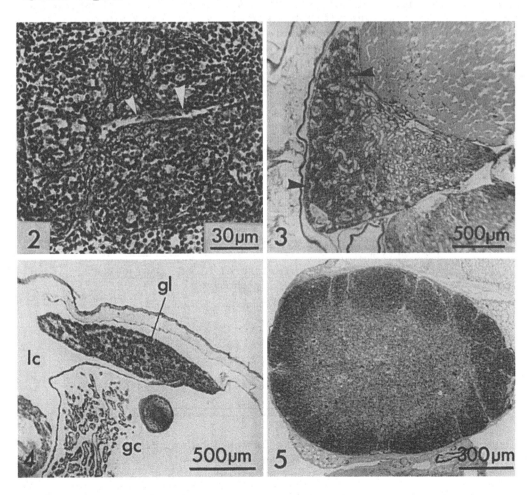

or urodeles.   In the liver, lymphoid cells may occur in the
subcapsular region or in small, deeper foci, whilst in the kidney they
are found in intertubular regions (Fig.3).   Presumably, the sluggish
blood flow in these organs allows lymphocytes to cluster and respond
to antigen reaching them in the circulation or (in the case of the
kidney) from the body cavity (Turner 1973).   Elsewhere, especially
along the gut (including the pharynx), the foci may act as
"sentinels", monitoring antigens coming into the body.   The exact
status of many such accumulations has not been adequately determined:
their small size and diffuse nature make it difficult to conduct
depletion experiments or in vitro tests, whilst there is a need to
devise experiments which provide more appropriate and varied antigenic
challenges to the animal in vivo.

The role of discrete nodular lymphocytic accumulations in the gut of
anurans (Table 2) is not clear.   These are somewhat similar
histologically to Peyer's patches, and they house phagocytes and
antibody-forming cells (Goldstine et al. 1975; Chin and Wong 1977).
However, a secretory antibody akin to IgA has not been identified, and
the proportions of lymphocyte subsets found within the nodules are
very similar to those in the peripheral blood (Jurd 1977 and Table 3).
Thus there is no evidence as yet to suggest that the mucosal surfaces
in amphibians are protected by distinct populations of lymphocytes
with their own special secretions and circulatory pathways.

Table 3 : SURFACE IMMUNOGLOBULINS IN ADULT XENOPUS

LYMPHOCYTES (from Jurd, 1977)

Antiserum (conventionally raised)	% of cells labelled	
	Peripheral blood lymphocytes	GALT cells
anti-M (heavy chain)	62-75	51-68
anti-IgRAA (" ")[a]	1-3	1-4
anti-L chain	64-74	54-66

[a] A distinct, 7S IgG-like immunoglobulin, found in
a number of non-mammalian tetrapods
(RAA = Reptilia, Aves, Amphibia).

The significance of different organs as blood filters has been
assessed by Kent (1966) in adult frogs (R. temporaria) by injecting
carbon particles into the femoral vein. The K value, which describes
the overall rate of clearance of particles from the circulation, was
comparable with several mammalian species. Seventy to eighty five
per cent of each dose was subsequently recovered in the liver,
although on a weight-for-weight basis the spleen showed similar
uptake, and some carbon was detected also in the lungs, kidneys and
heart. These organs together accounted for 95% of the injected
material. In larval and adult Xenopus, blood- and lymph-borne
particles are again removed mainly by cells of the liver and spleen
(Turner 1969).

On Kent's criteria, other phagocytic sites in frogs such as the bone
marrow and lymphomyeloid organs must play only a minor role in
removing non-biotic particles, but this does not reflect their
importance as trappers of antigenic material, nor as sites of immune
responses. In adult ranids, the bone marrow can house large
populations of lymphocytes, cells from this source are capable of
graft rejection, and the number of antibody-forming cells found there
following immunization can exceed their numbers in the spleen (Eipert
et al. 1979; Cooper et al. 1980).

The lymphomyeloid organs are small nodular structures which take up
carbon from the circulation and can thus be pinpointed. They occur in
the neck and thoracic regions of ranid and bufonid anurans (see Cooper
1976 for listing and classification) but are absent in Xenopus,
urodeles and apodans. They have sometimes been described as lymph
nodes, but the term "lymphomyeloid organ" is usually preferred since
they have a minor granulopoietic function, and they are not strictly
comparable with mammalian lymph nodes: unlike the latter they filter
particulate material mainly from the blood, not the lymph.
Nevertheless, lymphatic vessels have recently been described within
adult nodes (Villena and Zapata 1981), and in R. pipiens certain
larval nodes (lymph glands) actually project into a lymphatic sinus
(Fig. 4). Horton (1971) suggests that particulate uptake occurs
across the surface of the node from this lymphatic channel as well as
from blood sinuses within.

The major role of anuran lymphomyeloid nodes is clearly an
immunological one: they trap antigen, house proliferating
immunocompetent cells and secrete antibodies. Moreover, they are able
to retain antigen in the form of immune complexes on the surface of
specialized dendritic cells for a prolonged period, a phenomenon which
may be important in the development of immunological memory and/or in
feedback inhibition of the immune response (van Rooijen 1980). In
anurans such as Xenopus which lack lymphomyeloid nodes, dendritic
cells are located in the spleen. The appearance of these lymph node-
like structures in the "higher", more terrestrial amphibians, and
their further development in the homoiothermic vertebrates (together
with the eventual immunological redundancy of the kidney and liver)
may perhaps be correlated with increased efficiency of the circulatory

system. Circulatory improvements inevitably eliminated a number of sinusoidal sites which had previously allowed lymphocytes to settle out and make contact with antigen and new, more specialised secondary lymphoid organs were needed to replace them (see also Manning and Turner 1976; Manning 1978).

The primary lymphoid organs of amphibians have been the focus of considerable effort and speculation in recent years, and it is now established that the thymus plays a central immunological role. In anurans, the gland consists of a pair of single-lobed structures with a dense cortex and paler medulla (Fig.5); in urodeles and apodans there are minor differences in its embryological derivation, and cortico-medullary differentiation is less obvious (Manning and Horton 1982; Zapata et al. 1982). Invariably, however, thymic development precedes that of other lymphoid organs.

Thymectomy experiments have been performed on a number of species, including R. pipiens, R. catesbeiana, the midwife toad Alytes obstetricans, and the urodeles Triturus alpestris, Ambystoma tigrinum and Pleurodeles waltlii (see reviews of Cooper 1973 and Du Pasquier 1976), but the most detailed and productive work has come from Xenopus. Its thymus lies in a superficial position on each side in the transparent pharyngeal region during larval life, and the rudimentary organs can be removed by microcautery when the animal is only a few days old (Horton and Manning 1972). The operated areas can easily be checked for regeneration, and there are no later problems of runting. Moreover, these athymic Xenopus can later be reconstituted with chromosomally marked histocompatible cells, e.g. using 3n donors and 2n recipients (Katagiri et al. 1980). Experiments show that thymus-derived (T) lymphocytes participate directly in allograft rejection (Katagiri et al. 1980; Kaye and Tompkins 1983); they are involved indirectly in antibody production against some (T-dependent) antigens (Manning and Collie 1975; Katagiri et al. 1980); and they seed other areas including parts of the spleen, liver, kidney, gill- and gut-associated lymphoid tissues (Tochinai, 1976a). Nevertheless, a population of lymphocytes (putative B cells) can be identified in athymic Xenopus, even after the earliest operation (in 4-day old larvae), and some antibody responses (to thymus-independent antigens) can still be mounted. There is thus a functional and developmental dichotomy of the immune system similar to that in "higher" vertebrates. Thymectomy in other anurans and in urodeles confirms the need for thymus-derived cells in cell-mediated immune reactions; their role in antibody production is less certain, however, and there may be differences between urodeles and anurans in this respect. In the axolotl, for example, thymectomy actually increases antibody responses to horse erythrocytes (Charlemagne and Tournefier 1977a) (see section on functions of lymphocytes).

The identity of organs responsible for the maturation of B lymphocytes in amphibians continues to intrigue investigators. There is no bursa of Fabricius, and the bone marrow seems an unlikely candidate at this phylogenetic level since it appears only in adults or not at all, whereas antibodies can be demonstrated in larvae (see Du Pasquier, 1976; Manning & Horton, 1982). Using fluoresceinated anti-Rana IgM, Zettergren (1982) has shown that the first immunoglobulin-carrying cells of larval R. pipiens appear in the pronephric kidney at 8-9 days after fertilization. These cells resemble the pre-B cells of birds and mammals in that their IgM is exclusively cytoplasmic. Two days later, cells bearing surface IgM (i.e. B cells) could be detected in the pronephros and liver. During the larval period, the frequency of pre-B cells and the pre-B:B cell ratio remained higher in the pro- and meso-nephros than elsewhere. In adult frogs, pre-B cells were seen only in the bone marrow. The main sites of B cell maturation in this species thus appear to be the kidneys, switching to the bone marrow at metamorphosis. In Xenopus, Zettergren et al. (1977) found pre-B cells in the subcapsular region of the larval liver, whilst Bruning et al. (1981) claim that IgM-synthesising cells are present in Xenopus as early as 2 days post-fertilization (stage 35) when the embryos are first emerging from their jelly coats and well before the thymus differentiates. Other experiments on 2-day old Xenopus embryos (Brown 1981) have detected terminal de-oxynucleotide transferase, an enzyme which is said to be exclusive to immature lymphoid cells.

The question of primary lymphoid organs has been complicated by claims in the literature that not all functioning T cells arise in the thymus, and that the thymus might produce B cells as well as T cells. The possibility of a non-thymic source of functioning (alloreactive) T cells can now be discounted, at least in Xenopus, on the grounds that some thymectomy experiments were not being carried out early enough, thus allowing an escape of cells to other sites (Kaye and Tompkins 1983). The production of B cells in the amphibian thymus is more contentious. Firstly, antigen-stimulated anurans do sometimes house antibody-forming cells in their thymus (see Manning and Horton 1982; Hsu et al. 1983), but this is not to say that the organ has necessarily produced those B cells. Secondly, not all responses classified as T-independent in mammals are unimpaired in thymectomized Xenopus (Horton et al. 1980; see also section on functions of lymphocytes). Thirdly, thymic cells from Xenopus respond in culture to a mammalian B cell mitogen, LPS (Horton et al. 1980), but Bleicher and Cohen (1981) have cast doubt on the identity of the reacting cells in such experiments. The problems of distinguishing amphibian T and B lymphocytes, and the validity of "borrowing" probes perfected in mammalian systems for use in amphibian work (especially anti-Ig and mitogens) will be examined further in the section on lymphocyte structure, cytochemistry and development.

ORIGINS OF BLOOD CELLS
Two key questions are considered here: (i) Do the tissues which house maturing blood cells produce their own haemopoietic stem cells, or do these migrate in from elsewhere? (ii) If the latter is the case, where do the stem cells come from? Undifferentiated precursors are obviously more difficult to locate than definitive blood cells, and amphibian studies in this field have

relied largely on embryological manipulations involving the removal of suspected stem cell sources, or grafting procedures which use cytogenetically labelled tissues.

During amphibian development there is a succession of erythropoietic sites (Table 1). The first of these is a Y-shaped structure, the blood island, extending along the mid-ventral surface of the embryo (see Le Douarin 1966). Haemoglobin can be detected in this region at early tailbud stages, and radioactive iron concentrates in cells of the lateral and ventral regions in late gastrulae and neurulae (Pantelouris et al. 1963). Surgical removal of presumptive blood island areas of urodele and anuran embryos deprives the animals of red blood cells. Leucocytes are spared, however, and the early experiments on blood island function were usually of short duration, giving no opportunity for other erythropoietic sites to come into play (see Hollyfield 1966; Le Douarin 1966). Hollyfield exchanged blood islands between 3n and 2n embryos of R. pipiens at a stage (15-16) before blood cells have differentiated or the circulation has become established, and found that blood island derivatives were limited to circulating red cells of early larvae : the red cells of mid-larval stages did not originate from the graft. Thus red cells from the ventral blood islands are short-lived, and the region cannot provide stem cells for erythrocytes produced elsewhere later in development. Turpen (1980) has confirmed Hollyfield's studies on R. pipiens, and shown that circulating white cells and leucopoietic sites in the larval liver and kidney were all of host origin. Deparis (1968) has likewise shown that the spleen of newts (Pleurodeles waltlii) is not dependent on cells from ventral blood islands. On the other hand, Kau and Turpen (1983) have reported that cells from the ventral blood islands of Xenopus contribute more than a temporary population of erythrocytes, since they detected a colony from this source in the thymus gland of young (Stage 49) larvae. The longevity and eventual fate of these blood island-derived thymocytes are unknown, but Kau and Turpen speculate that they could form the subset of T cells which lingers only for a brief period in the Xenopus thymus early in development, before moving out to the periphery as alloreactive cells (see section on lymphocyte functions and Table 10). Lymphocytes in the thymus of older Xenopus (regulatory T cells?) are not derived from the ventral blood islands (Tompkins et al. 1980; see also Fig. 6).

These amphibian findings are of interest in a wider context. The view expounded by Moore and Owen (1967) was that pluripotent stem cells which seed the haemopoietic tissues of "higher" vertebrates originate in the blood islands of the extra-embryonic yolk sac. However, products of the equivalent structures in amphibians – the ventral blood islands – are prominent only in early life. Moreover, avian studies now argue in favour of an intra-embryonic source of stem cells, and erythrocytes derived from the yolk sac disappear completely from the circulation around the time of hatching (Dieterlen-Lièvre et al. 1980; see Chapter 5). A new, intra-embryonic source of lymphocytes has also been described in mice (Kubai and Auerbach 1983). Turpen's group (Kau and Turpen 1983; Turpen and Smith 1985) suggests that vertebrates may have two stocks of haemopoietic stem cells: the first, an extraembryonic one, would respond to early pressures, particularly by producing erythrocytes; the second, an intraembryonic one, would be protected from these early pressures and instead go on to produce in a more leisurely fashion the animal's definitive blood cells.

Notwithstanding arguments about intra- versus extra- embryonic stem cell sources, workers on mammals and birds are generally agreed that haemopoietic tissues (including primary lymphoid tissues) are seeded with stem cells arriving from elsewhere. Claims by Turpen et al. (1973) that the thymus gland of frogs (R. pipiens) was responsible for producing its own stem cells were thus greeted with scepticism, and their results could not be repeated in other amphibians (Pleurodeles, Xenopus) nor, indeed, in R. pipiens (Deparis and Jaylet 1976; Volpe et al. 1979). In these experiments, thymic primordia were removed from 2n embryos at the tail-bud stage and placed orthotopically or heterotopically in 3n or 4n recipients, followed later by analysis of lymphocyte ploidy in the differentiated gland. The more recent results all argue conclusively in favour of a non-thymic source of lymphopoietic cells in amphibians, as in birds and mammals.

In pursuing the origins of lymphopoietic cells which populate the Xenopus thymus in late larval stages, Tompkins et al. (1980) and Volpe et al. (1981) performed a series of orthotopic grafts using 2n:3n segments of different sizes and from different parts of the embryo (see Fig. 6). In this way the source of T cells was narrowed down to a region containing the presumptive mesonephros and closely adjacent structures (proximal lateral mesoderm, dorsal mesentery and genital ridge): 2n grafts of this tissue placed on each side of a 3n embryo resulted in a larva carrying T cells which were almost all 2n (99%). Some, but not all erythrocyte precursors must have come from the same site, since 75% of definitive red cells were 2n. Volpe suggests that the remaining red cells came from a more anterior position in the embryo, possibly the pronephros: in 1/2:1/2 chimaeras (Fig. 6), significant numbers of red cells were derived from the anterior half.

A similar picture has emerged in R. pipiens. The embryonic pro- and meso-nephric regions (also known as the nephrogenic band) both supply red cell precursors, since grafts carried out at the tail-bud stage (15-16) affect erythrocyte ploidy in mid-larval animals (Hollyfield 1966). These erythrocytes replace the transient population supplied by the ventral blood islands. In addition, the mesonephric region donates cells which appear later in the thymus, spleen, liver and bone marrow as well as the kidneys (Turpen 1980; Turpen et al. 1981). The mesonephros, incidentally, is itself subject to colonization - a unilaterally transplanted piece of presumptive mesonephros will seed its opposite number - and Turpen suggests that the precursor cells derive from lateral plate mesoderm in the region of the presumptive mesonephros rather than the mesonephros itself. The main point of interest which emerges from these experiments, however, is the ability of this one particular

Figure 6. <u>Localization of lymphopoietic stem cells using 3n:2n grafts</u>
           <u>in Xenopus embryos</u> (adapted from Volpe <u>et al</u>. 1981).

Thick line = position of graft. Stippled area = 2n. Clear area = 3n.
Small circle = position of presumptive thymus. Area within circle
indicates subsequent ploidy of thymic lymphocytes (in late-stage
larvae).

<u>¹/₂ : ¹/₂ chimaera</u>   <u>³/₄ : ¹/₄ chimaera</u>

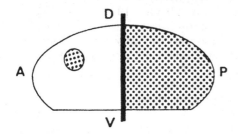   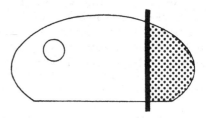

<u>grafted ventral blood island</u>   <u>grafted neural area</u>

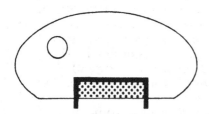   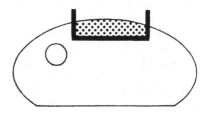

<u>presumptive mesonephros</u>

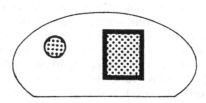

embryonic region to supply organs which are known to produce all kinds
of blood cells: T cells in the thymus, B cells and granulocytes in the
kidneys, erythrocytes in the liver, for example. There is, then, at
least a suggestion that the precursors which emigrate from the lateral
mesoderm/nephrogenic region are pluripotent, although the apparent
stem cell zonation within the nephrogenic region of Xenopus as
described by Volpe et al. (1981) needs to be explored further. The
relationship between the nephrogenic site and the ventral blood islands
is also problematic: are there two distinct populations of precursors,
or has a common precursor arisen and migrated to the two sites before
transplants were carried out? Recent studies by Deparis & Jaylet
(1984) suggest that distant mesodermal regions of Pleurodeles embryos
contain similar populations of blood cell precursors: the ability to
produce a transient versus permanent (and more diverse) population of
differentiated blood cells seems to depend on the differing inducer
properties of associated endoderm.

Precursor cells appear to use interstitial routes and the circulation
during their migration. In experiments on R. pipiens, Turpen and
Knudson (1982) produced posterior-anterior chimaeras at successive
embryonic stages (14-17), but always bisected just behind the
pronephros (or presumptive pronephros). Larval organs were tested
later for blood-forming cells of "posterior" versus "anterior"
derivation. Their results suggested an anterior migration of
precursor cells between stages 14 and 17; other experiments using
posterior nephrogenic segments transplanted at stage 14-15 showed that
this migration continues beyond stage 17 and leads to colonization of
the aortae and pronephros at stages 19-20. All of this has taken
place before the circulation becomes operational. Subsequent
migrations over longer distances presumably rely on the circulation, a
point best illustrated by parabiosis ("Siamese twin") experiments,
where haemopoietic precursors from one animal are able to colonize the
partner's organs (e.g. Deparis and Jaylet 1976). Colonizing cells
which have reached their destination retain their proliferative
capacity (Turpen et al. 1981).

Based on examinations made of the embryonic aortae and pronephros,
Turpen's group has tentatively identified the blood cell precursors in
R. pipiens. They are large, macrophage-like cells with excentric oval
nuclei, prominent nucleoli, diffuse chromatin and a foamy cytoplasm
containing yolk and pigment granules. According to Goldschneider et
al. (1980), mammalian stem cells are much smaller, blast-like cells,
but their nuclear features resemble those described by Turpen.
Embryonic haemopoietic cells from Xenopus are shown in Fig. 7.

Although evidence is limited, the differentiation of amphibian blood
cells appears to be controlled by a variety of micro-environmental and
hormonal signals. Embryonic induction of Pleurodeles mesoderm by
associated endoderm (Deparis and Jaylet 1984) has already been noted,
and dorsal lateral plate mesoderm of Xenopus influences the
differentiation of ventral-derived haemopoietic cells in vitro (Turpen
and Smith 1985). Proteinaceous secretory material has been observed

in epithelial cells of amphibian thymuses (Curtis et al. 1979b), but
it has not been convincingly shown by experiment that such products
are comparable to mammalian thymic hormones. Thyroxine has an
erythropoietic effect on some species of adult anurans but not others
(Kaloustian and Dulac 1982), whilst mammalian erythropoietin
apparently has no effect on amphibian cells in vivo or in vitro. On
the other hand, cultured spleen cells from frogs (R. pipiens) do show
increased proliferative and erythropoietic activity in the presence of
serum from R. pipiens or R. catesbeiana with experimentally induced
anaemia (Carver and Meints 1977), thus indicating that an erythropoietin-like factor exists. Work with R. catesbeiana has also
provided indirect evidence for short-range stimuli during erythropoiesis (Broyles et al. 1981): the larval liver and mesonephros
produce morphologically and biochemically distinct types of erythrocytes, and contacts between stromal, parenchymal and erythroid
cells in the erythropoietic foci of the two organs are quite
different. In organ culture, the kidney can markedly alter the
haemoglobin profile of an adjacent liver. Production of red cells is
limited by negative feedback: Aleporou and Maclean (1978) have shown
that factors released from mature erythrocytes inhibit proliferation
of the immature stages, probably in a tissue-specific fashion.

Figure 7. Photomicrograph of haemopoietic precursor cells
of Xenopus, harvested from 5-day lateral plate cultures by
cytocentrifugation (courtesy of J.B. Turpen & P.B. Smith).
Lateral plate mesoderm and overlying ectoderm were
explanted from embryos 20 h old and cultured as vesicles.
The cells collected from these vesicles form small
colonies of granulocytes or macrophages when plated out for
10 days in methyl cellulose medium. Supernatant from Con A-
treated adult spleen cells was used as a source of colony-
stimulating activity.

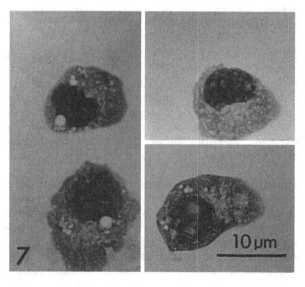

In summary, we have to date what might best be described as a sketch of amphibian haemopoietic stem cells: a population with migratory and proliferative properties, possibly pluripotent, possibly recognisable, susceptible to humoral signals during development. The regulation of these stem cells and their capacity for self-renewal have yet to be explored in depth.

BLOOD CELL STRUCTURE, CYTOCHEMISTRY AND DEVELOPMENT

Erythrocytes (Figs. 8,9,10,13,14)
    Amphibian erythrocytes are usually elliptical discs with a nuclear bulge in the centre, looking rather like fried eggs. Sizes vary considerably. The largest, indeed the largest of all vertebrate erythrocytes, are found in aquatic neotenous urodeles: up to 70μm diameter in Amphiuma. Anurans have much smaller, less elongated and more numerous red cells, and this is most noticeable in high-altitude forms: the smallest red cells in the class are found in South American frogs of the genus Telmatobius, several species of which live at altitudes over 4000m. Ruiz et al. (1983) report a mean size of 17 x 12μm for four species of Telmatobius, compared with a mean of 22 x 14μm for a variety of other anuran genera (Bombina, Hyla, Microhyla, Rana, Rhacophorus: see Kuramoto 1981). In apodans, the aquatic form Typhlonectes compressicauda has large red cells as in urodeles (up to 55μm diameter), whereas in Boulengerula taitanus, a terrestrial apodan, red cell dimensions are within the anuran range (see Toews and Macintyre 1978).

Anuran and urodele red cells in the circulation are also distinguishable on the basis of ultrastructural characteristics. In anurans, the cytoplasm is homogeneous and packed with haemoglobin; organelles such as lysosomes and mitochondria are rarely seen. Many species of urodele, on the other hand, show clusters of granular and vacuolar bodies in the cytoplasm . The presence of lysosomes in circulating urodele erythrocytes (confirmed by acid phosphatase cytochemistry) is of particular interest, since they appear to be responsible for degrading the other organelles (Tooze and Davies 1965; Grasso 1973b). Sean et al. (1980) have also described cytoplasmic vacuoles with a variety of contents which can be extruded from the erythrocyte (Figs. 8,9). These observed differences between anuran and urodele cells seem to be due largely to timing: anuran erythrocytes are released from haemopoietic organs only after their cytoplasmic organelles are eliminated, whereas urodele cells complete their maturation in the circulation.

Investigations on the cellular changes which occur during erythropoiesis are usually impeded by the presence of erythroid cells of several stages at a given site, and by the fact that the differentiating stages are often inaccessible. Grasso (1973a,b) has overcome this in newts (Triturus spp.) by combining acetylphenylhydrazine injection with splenectomy: the former destroys mature red cells, the latter ensures that the erythropoiesis which follows occurs in the peripheral circulation. Thus successive, relatively homogeneous "waves" of increasingly mature erythroid cells can easily be examined. The first "wave", which Grasso tentatively described as erythroid precursors, accounts for 50-60% of the total blood cell population at 11-14 days after treatment with acetylphenylhydrazine, and consists of undifferentiated, rounded lymphocyte-like cells of various sizes with weakly basophilic or unstained cytoplasm and large nuclei. Ultrastructurally, the latter show light-staining chromatin clumps and large compact nucleoli, and there is a paucity of cytoplasmic organelles: Golgi apparatus and endoplasmic reticulum are poorly developed and ribosomes few in number, although more ribosomes can be seen in cells at 13-14 days after treatment. Labelling experiments indicated that these early cells are engaged in DNA, RNA, haem and protein synthesis. Similar cells can be identified during naturally induced erythropoietic bursts in the springtime.

Figures 8-9. Erythrocytes from the circulation of the urodele, _Pleurodeles waltlii_ (courtesy of K.E. Sean and colleagues). In Fig. 8, note the cytoplasmic vacuoles with granular and membranous contents. In Fig. 9, these seem to be extruded from the cell surface.

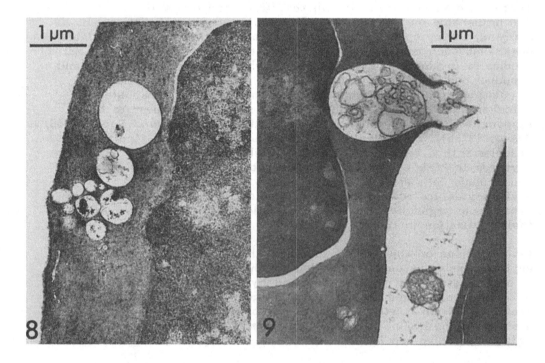

populations of larval red cells, one originating from the liver, one from the mesonephros. The cells from the liver have a central nucleus, whereas those from the kidney have a peripheral nucleus and are somewhat longer (27 versus 24μm, with little overlap), and they contain different larval haemoglobins (Broyles et al. 1981). A third population of red cells, of different density and containing adult-type haemoglobin, appears during metamorphosis (see section on erythrocyte functions). According to Tyler et al. (1982), larval and adult-type red cells of R. catesbeiana (at least from some geographic locations) can be distinguished using dark-field illumination: larval cells usually showed white/grey granular luminescence in their cytoplasm, whilst adult cells were non-luminescent.

The development and maintenance of the flat, ellipsoidal shape of the amphibian erythrocyte is achieved by a marginal band of microtubules, together with a fibrillar meshwork known as the transmarginal band which extends across the cell (Bertolini and Monaco 1976; Cohen 1982). A third cytoskeletal component which links the peripheral fibrillar network to the nucleus has been identified in Xenopus erythrocytes (Gambino et al. 1981). The nuclear skeleton itself consists almost entirely of a nuclear pore-lamina; an intra-nuclear matrix is absent.

Not all amphibian red cells are nucleated. Anucleate forms are most evident in lungless urodeles, where they usually account for less than 5% of the circulating red cell population. The slender salamander (Batrachoceps), however, is remarkable (and unique among the non-mammalian vertebrates) in that over 90% of its erythrocytes lack nuclei (Foxon 1964; Cohen 1982). This is thought to be a special device for increasing its oxygen-carrying capacity in the absence of lungs by improving the surface-volume ratio of the erythrocyte. The loss of the nucleus contributes to respiratory efficiency only to a limited extent in terms of haemoglobin content since the condensed chromatin in mature conventional amphibian erythrocytes is surrounded by nuclear haemoglobin at the same concentration as in the cytoplasm – the two regions appear to be continuous via nuclear pores – whilst lower concentrations (about 10% of cytoplasmic values) are present

Figure 10.   Erythrocyte shapes in Batrachoceps.
For explanation see text.

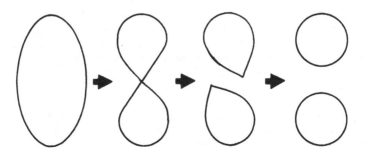

The second stage (days 15-18) consists of basophilic erythroblasts, which show many chains and clusters of ribosomes but otherwise little ultrastructural change. Beyond this stage, through polychromatophilic blasts (days 19-25), reticulocytes and mature erythrocytes (day 32 onwards) there is a continual decrease in basophilia, increased eosinophilia and a change from a rounded to oblong cell shape. The nuclei and nucleoli become smaller, and chromatin increases in density. RNA synthesis ceases in the polychromatophilic erythroblasts. Haemoglobin accumulates progressively, and isolated or clustered ribosomes (now reduced in number) become embedded in it. Cells with significant amounts of haemoglobin still undergo cell division (through intermediate and reticulocyte stages) but at a less rapid rate. Micropinocytotic vesicles and diverse vacuoles persist throughout intermediate and late erythroid stages; abnormal mitochondria and other organelles (probably undergoing autolytic attack) are encountered in polychromatophilic blasts onwards.

Similar experiments to those described above have been carried out on Xenopus (Chegini et al. 1979). At about 6-10 days following treatment of adults with acetylphenylhydrazine, basophilic erythroblasts with a large nucleus, light-staining chromatin, dense ribosomal clusters and large mitochondria (but little endoplasmic reticulum) are released into the circulation. They are replaced by polychromatophilic blasts and reticulocytes. The polychromatophilic cells stain slightly with o-dianisidine, indicating a little haemoglobin; the nucleus shows some condensation, but the nucleolus is still prominent (Thomas and Maclean 1975). The non-dividing, mature Xenopus erythrocytes have very condensed chromatin, a high haemoglobin content, and have lost their cytoplasmic inclusions and nucleoli.

Information on apodan erythropoiesis is sketchy, and no studies on animals with experimentally induced anaemia have been carried out. However, the observations of Boschini (1979) and Zapata et al. (1982) suggest that normal apodan blood contains red cells at different stages of maturation. Boschini has distinguished three types of erythrocyte in the circulation of Typhlonectes compressicauda: (i) a rounded cell with a high nucleo-cytoplasmic ratio, flocculent chromatin and visible nucleolus; (ii) a large rounded cell possessing considerably more cytoplasm (homogeneous or with small, rare azurophilic granules), round or oval nucleus with more condensed chromatin blocks and no visible nucleolus; (iii) a large discoid cell with abundant cytoplasm containing small but prominent clusters of azurophilic granules, plus a dense oval nucleus (Fig. 13). Electron-micrographs of Typhlonectes erythrocytes revealed a variety of cytoplasmic organelles. Zapata et al. (1982) similarly report a mixture of large oval erythrocytes and smaller cells with large rounded nuclei in the peripheral blood of Ichthyophis. Red cells undergoing division were also noted.

Within a species, the structure of a circulating cell may be governed not only by its state of maturity but also its organ of origin and the age of the animal. In Rana catesbeiana, for example, there are two

within the chromatin (Small and Davies 1970). The anucleate cells of
Batrachoceps are usually flattened elliptical structures of different
sizes, but some are pointed, round or dumb-bell shaped (Cohen 1982).
Unlike mammalian erythrocytes the marginal band system is retained,
and Cohen attributes the variety of cell shapes to a twisting and
nipping-off by the cytoskeleton during amitotic cell division (Fig.
10).

Thrombocytes (Figs, 11,19,20,21)
        These are cells with a dense round or oval nucleus and two
spindle-shaped extensions of the cytoplasm. Where spindles are not
visible under the light microscope, thrombocytes may be mistaken for
small lymphocytes. For this reason, and because of their adhesive and
aggregating properties which cause non-random distribution in smears,
thrombocyte counts tend to be unreliable. Anucleate forms which
resemble mammalian platelets have also been described (see Foxon 1964).

In the older literature, the production of thrombocytes has been
linked with that of erythrocytes (Table 1) and with "lymphoid blast"
cells (see Foxon 1964). Immature stages do not normally occur in the
blood, but will do so following splenectomy, and can thus be examined
more readily (Fey 1965a, 1966). According to Fey (1966),
thromboblasts of Xenopus are distinguishable from lymphoblasts on the
basis of a large irregular excentrically placed nucleolus. Their
chromatin is fine and the cytoplasm weakly basophilic. Strikingly
large, round cells identified as prothrombocytes were also described.
By this stage, the nucleus has assumed a more elongated shape, whilst
the cytoplasm is vacuolated and contains granules staining pale blue
with Giemsa.

More recent work has focussed on the ultrastructure of mature
thrombocytes, and enabled clearer distinctions to be drawn between
them and lymphocytes (Figs. 19-23). In particular, there is an
extensive system of canaliculi and vesicles which join up within the
cytoplasm and open to the exterior. The canaliculi have a smaller
diameter (250nm) in Rana spp. than in Xenopus (Daimon et al. 1979).
Campbell (1970), working on R. pipiens, also described a deep but
narrow indentation in the nucleus, free ribosomes in the cytoplasm, a
small Golgi complex, dense cytoplasmic granules with homogeneous
content, and small particles which appeared to be glycogen. Apodan
(Ichthyophis) thrombocytes too show a cleft nucleus, prominent
cytoplasmic vesicles and dense granules (Welsch and Storch 1982).

On the basis of tests reported by Fey (1962), the most distinctive
cytochemical feature of Xenopus and Rana thrombocytes appears to be
their moderately strong alkaline phosphatase reaction (lymphocytes were
negative), although results for other amphibians are less clear-cut
(see Table 4). Boschini (1980) has employed the PAS test to distinguish
thrombocytes in Typhlonectes blood (Fig. 11). Immunofluorescence tests
are also useful in this respect: Charlemagne and Tournefier (1975)
showed, for example, that some (i.e. "B") lymphocytes in Pleurodeles
emitted a bright specific membrane fluorescence with an anti-Ig reagent

Figures 11-14. Peripheral blood smears of the apodan, <u>Typhlonectes</u>
<u>compressicauda</u> (courtesy of J. Boschini Filho). Figs. 11 and 12:
PAS/haematoxylin stain. Figs. 13 and 14: Leishman's.
<u>Fig. 11.</u> Thrombocyte. The cell is slightly elongate. Its cytoplasm
is full of fine PAS +ve granules; denser granulation is seen at each
pole.
<u>Fig. 12.</u> Lymphocyte. The cell is rounder than a thrombocyte, and
granulation less regular.
<u>Fig. 13.</u> Neutrophil, with multilobed nucleus and indistinct
cytoplasm. Note also the small but prominent clusters of azurophilic
granules in the cytoplasm of the neighbouring erythrocyte.
<u>Fig. 14.</u> Eosinophil. The nucleus is less highly lobed than the
neutrophil, its granules more prominent.

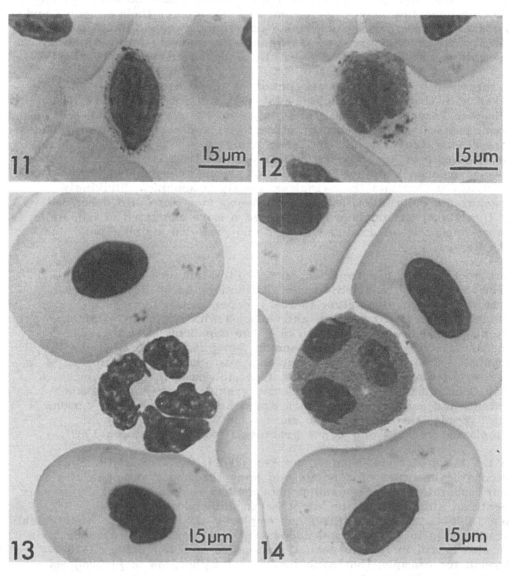

Figures 15-18.   Peripheral blood smears of the apodan, Typhlonectes
compressicauda   (courtesy of J. Boschini Filho).
Fig. 15.   Basophil (PAS).   Note "lymphoid" nucleus.
Fig. 16.   Basophil, showing coarse granulation (toluidine blue).
Fig. 17.   Monocyte, showing nuclear morphology (Leishman's).
Fig. 18.   Monocyte (peroxidase).

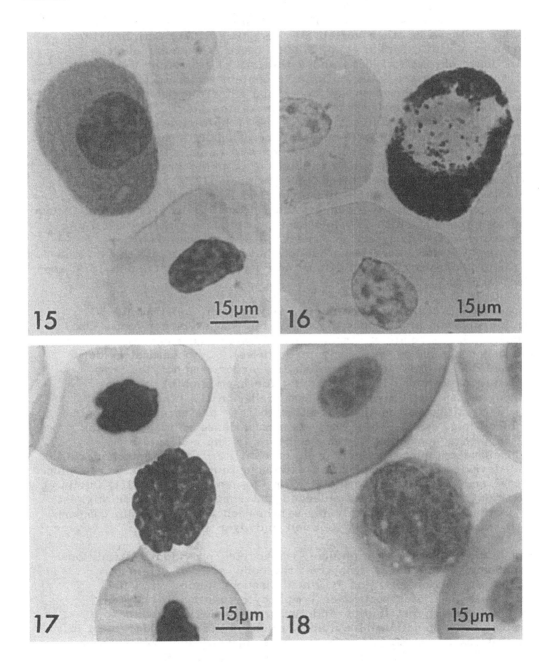

(see also next section), whereas thrombocytes produced only a non-specific, diffuse autofluorescence in their cytoplasm and nuclei. Finally, it is worth noting that doses of irradiation which cause a dramatic decline in the numbers of circulating lymphocytes (in R.pipiens) have little or no effect on thrombocytes (Stearner 1950).

Lymphocytes (Figs. 12,22,23)
Small lymphocytes of amphibians are morphologically similar to those in other vertebrates (e.g. Fey 1967b; Campbell 1970; Curtis et al. 1979; Zapata et al. 1982). There is a round nucleus with dense chromatin blocks and a thin rim of basophilic cytoplasm. Ultrastructurally the latter is seen to contain moderate to large numbers of free ribosomes and occasional mitochondria; there is little development of endoplasmic reticulum or of the Golgi complex (Figs. 22,23). Supravitally stained lymphocytes of Bufo alvarius showed a few mitochondria (Janus green stain), occasionally a single nucleolus, and 10-24 neutral red granules (Cannon and Cannon 1979). In lymphocytes of ranid frogs, distinctive azurophilic granulation of the cytoplasm has often been noted, and Campbell (1970) reports more cytoplasmic granules at the ultrastructural level than in mammalian lymphocytes. The neutral red granules described by Cannon and Cannon appear to coincide with PAS and acid phosphatase-positive sites. As in mammalian lymphocytes, non-specific esterases can also be demonstrated and peroxidases cannot, but unlike mammals there is no β-glucuronidase or aryl sulphatase activity (Cannon & Cannon 1979; see also Table 4).

These small lymphocytes are particularly common constituents of amphibian blood and lymphoid tissues, but similar larger forms also occur. (Size is relative: "small" leucocytes can be large by mammalian standards, especially in urodeles). Cytochemical evidence is desirable in order to distinguish the larger lymphocytes from immature erythroid cells (by staining for haemoglobin) or from monocytes (see next section). Plasma cells, on the other hand, are easily identified in the tissues, although they are rare constituents of blood (e.g. Fey 1967b). They possess a dense rounded nucleus (often with a nucleolus) at the periphery of the cell, and a voluminous cytoplasm containing significant numbers of mitochondria, a well-developed Golgi complex and extensive rough endoplasmic reticulum (Tooze and Davies 1968; Campbell 1970; Cowden and Dyer 1971; Curtis et al. 1979a). The presence of enlarged cisternae in some cases suggests an increased storage capacity for synthesised immunoglobulin compared with mammalian plasma cells (Cowden and Dyer 1971).

The uniform picture of amphibian lymphocytes provided by traditional morphological studies is deceptive: the cells' sites of origin (see section on haemopoietic and lymphoid tissues), their functional properties (see later section on lymphocyte functions), enzyme content, density distribution and membrane biology all point to a complex cell population. Inevitably, much of the technology used is mammal-based and not all of it has been adapted successfully or exploited fully; nevertheless, considerable progress has been made.

Figures 19-23. Thrombocytes and lymphocytes of R. catesbeiana (courtesy of T. Daimon). Figs.19 and 20. Thrombocytes cut in longitudinal and transverse section, respectively. The nucleus is indented, and numerous electron-lucent vesicles are distributed throughout the cytoplasm. These vesicles are transverse sections of a surface-connected canicular system which is clearly shown in a ruthenium red-treated semi-thin section (Fig. 21).
Figs. 22 and 23. Lymphocytes for comparison. Note the spherical shape, smooth rounded nucleus, microvilli at the cell surface, and absence of canaliculi.

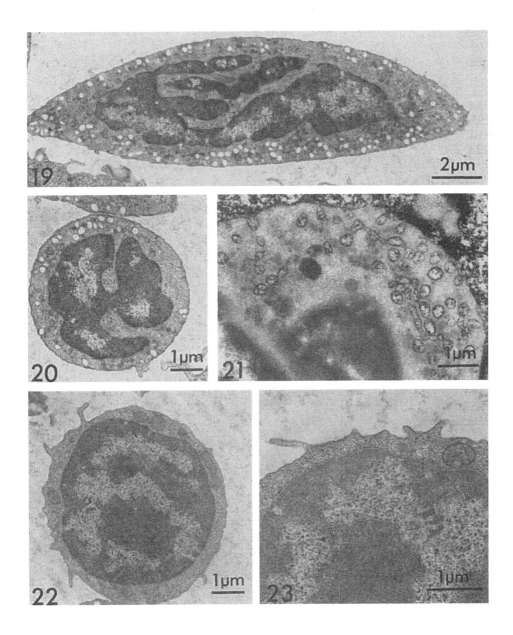

Table 4 : <u>CYTOCHEMICAL FEATURES OF AMPHIBIAN LEUCOCYTES</u>

	thrombo-cyte	lympho-cyte	mono-cyte	neutro-phil	eosino-phil	baso-phil	reference
A. PEROXIDASE							
Rana	−	−	+	+	+	−	1,2,3,4
Bufo	−	−	+	+	+	−	1,2,5
Hyla				+	+		1
Pelobates	−	−	+	+	−	−	2
Bombina	−	−	+	+	+	−	2
Xenopus	−	−	+	+	−	−	2
Notophthalmus				+	+		1
Triturus	−	−	+	+	−	−	1,2
Pseudotriton				+	+		1
Hynobius				+	−		1
Onychodactylus				+	−		1
Megalobatrachus				+	−	−	1,6
Necturus				+			1
Amphiuma				+	−		1
Plethodon				+	−		1,7
Ambystoma	−	−	+	+	−	−	1,2
Rhyacotriton				+	−		1
B. ALKALINE PHOSPHATASE							
Rana	+	−	(+)	+	+	+/(+)	2
Bufo	(+)	+/−	(+)/−	+	(+)/−	+	2,5
Pelobates	(+)	−	(+)	+	+	+	2
Bombina	(+)	−	−	+	+	−	2
Xenopus	+	−	−	(+)	−	+	2
Triturus	+	+/(+)	+	(+)/+	(+)/+	+	2
Ambystoma	+	(+)	+	+	+	+	2
Megalobatrachus				−			6
C. ACID PHOSPHATASE							
Rana	+	+/(+)	+	(+)/+	+	+/(+)	2
Bufo	+	+/(−)	+	(+)/+	−/+	(+)	2,5
Pelobates	−	−	(+)	(+)	−	−	2
Bombina	+	+	+	+	+	+	2
Xenopus	−	−	−	−	−	−	2
Triturus	+	+	+	(+)/+	+	+	2
Ambystoma	−	−	+	+	(+)	−	2

	thrombo-cyte	lympho-cyte	mono-cyte	neutro-phil	eosino-phil	baso-phil	reference
D. ESTERASE							
Rana	+	+	+	+	+	+	2
Bufo	+	+	+	+/(+)	+/(+)	+	2,5
Pelobates	+	+	+	+	+	+	2
Bombina	+	+	+	+	+	+	2
Xenopus	+	+	+	+	+	+	2
Triturus	+	+	+	+	+	+	2
Ambystoma	+	+	+	+	+	+	2
E. PAS							
Rana	+	+	+	+	+	+	2,8
Bufo	+	(+)/+	+	+	+	+	2,5
Pelobates	+	(+)	+	+	+	+	2
Bombina	+	(+)	+	+	+	+	2
Xenopus	+	(+)	+	+	+	+	2
Notophthalmus				+	+	+	9
Triturus	+	+	+	+	+	+	2
Ambystoma	+	+	+	+	+	+	2
Amphiuma				+		+	10
F. β-GLUCURONIDASE							
Bufo		-	-	-	-		5
G. ARYL SULPHATASE							
Bufo		-	-	-	-		5

+ = positive; - = negative; ( ) = inconstant or weak reaction (within a species); / = species difference.

References: 1, Mitsui (1965); 2, Fey (1966, 1967a,b); 3, Gangwar and Untawala (1971); 4, Rogovin et al. (1978a); 5, Cannon and Cannon (1979); 6, Mitsui and Fukui (1964); 7, Curtis et al. (1979a); 8, Kapa et al. (1970); 9, Cowden et al. (1964); 10, Cowden (1965).

The use of fluorescent-labelled anti-immunoglobulin (Ig) for anuran studies has been reported by several laboratories. Labelled and unlabelled subpopulations of circulating lymphocytes can be detected which, by mammalian criteria, correspond to B and T cells respectively. Figures for the blood lymphocytes of adult Xenopus obtained by Jurd (1977) are shown in Table 3. The labelled molecules are products of the lymphocyte itself, since capping, loss and regeneration can be induced as in mammalian B cells. Zettergren (1982), working on Rana pipiens, has shown that the sequence of Ig isotype expression during maturation is also comparable: pre-B cells with cytoplasmic IgM only were replaced by B cells bearing surface IgM; cells which co-expressed surface IgM and IgRAA ("IgG") appeared later and increased in number; finally, cells carrying only IgRAA could be detected. Jurd and Stevenson (1976), on the other hand, were unable to show co-expression of IgM and IgRAA on the surface of Xenopus lymphocytes: the great majority of labelled cells bore only IgM, the remainder only IgRAA (Table 3).

Reports of high percentages of Ig-positive cells in the thymus gland of anurans (Du Pasquier et al. 1972; Jurd and Stevenson 1976) raised doubts about the derivation of labelled and unlabelled lymphocytes in the circulation. However, Mattes and Steiner (1978), working on R. catesbeiana and Hadji-Azimi and Schwager (1982) using Xenopus, concluded that anti-carbohydrate antibodies which cross-reacted with non-Ig molecules were responsible for the fluorescence of thymic lymphocytes: antisera raised in animals of a different class (e.g. frog to rabbit) are more likely to show these additional specificities than mammal-to-mammal immunization. Hadji-Azimi and Schwager (1982) also failed to isolate Igs from solubilized membranes of thymic lymphocytes. These more recent findings are supported by Bleicher and Cohen (1981), who used a fluoresceinated monoclonal anti-Xenopus IgM with anti-protein specificity: 43-58% of blood lymphocytes from adult Xenopus were labelled - a somewhat lower figure that that reported by Jurd (1977) - and thymic cells were all unlabelled. Moreover, the percentage of unlabelled lymphocytes in the spleen was reduced after early thymectomy. Thus peripheral Ig-ve lymphocytes in anurans do appear to be thymus-derived, and Ig+ve cells do not.

The picture in urodeles is less detailed. Two groups of lymphocytes have been distinguished in newts (Pleurodeles) by means of conventional anti-IgM: a weakly fluorescent population (T?) which included almost all thymic lymphocytes and about a third of splenic lymphocytes; and brightly labelled lymphocytes (B?) from the rest of the splenic population (Charlemagne and Tournefier 1975).

"T cell-specific" phytolectins and "B cell-specific" bacterial mitogens have been widely employed to investigate lymphocyte heterogeneity in amphibians. Phytohaemagglutinin (PHA), concanavalin A (Con A) and bacterial lipopolysaccharide (LPS) all induce blastogenesis in urodele and anuran blood cells (see Edwards and Ruben 1982), but the conclusions that can be drawn from these reactions are not always clear-cut or particularly informative. The use of PHA and

Con A as T cell probes is justified, at least in frogs such as R. pipiens and Xenopus, since early thymectomy severely impairs or eliminates lymphocyte responses to these mitogens (Manning et al. 1976; Rollins-Smith and Cohen 1982), and the responding population is Ig-ve (Bleicher and Cohen 1981; Schwager and Hadji-Azimi 1982). The value of LPS as a B cell probe is more doubtful, however. Although thymectomized animals show unimpaired responses to LPS (Manning et al. 1976; Rollins-Smith and Cohen 1982), Ig-ve cells still respond to a limited extent in serum-free medium and to a normal extent in the presence of foetal calf serum supplement (Bleicher and Cohen 1981). Thus, LPS appears to stimulate amphibian T as well as B cells. Moreover, high concentrations of commercial LPS (up to 2 mg ml^{-1}) are invariably needed in order to obtain measurable stimulation of amphibian cultures, and rigorously purified LPS is completely non-mitogenic (Bleicher et al. 1983). A second "B cell" mitogen, purified protein derivative of tuberculin (PPD), elicits good responses by cells from thymectomized Xenopus and at lower concentrations than LPS (Manning et al. 1976), but has not (to the author's knowledge) been tested in conjunction with anti-Ig preparations.

Attempts to obtain mitogen responses in urodeles have proved interesting, and sometimes vexing. For example, although blood and splenic lymphocytes of Ambystoma respond to Con A and PHA, thymic cells do not (good responses to these mitogens by cells from the anuran thymus are possible, but are not always obtained - see Horton et al. 1980). In general, urodele cells respond slowly and undramatically to mitogens (Edwards and Ruben 1982), and whilst poor or negative responses may have a technical basis - inappropriate media or supplements, imbalance or loss of accessory cells, for example - there appear to be genuine phylogenetic differences in this respect between the two orders. In any case, it seems prudent to correlate mitogen reactivity where possible with appropriate functional assessments such as antibody assays, cell-mediated immune reactions and tests for regulatory cells (see section on lymphocyte functions).

Attempts have recently been made to discriminate subpopulations of amphibian lymphocytes on the basis of erythrocyte (E) rosetting, nylon wool adherence and α-naphthyl acid esterase (ANAE) staining (Klempau and Cooper 1983, 1984). The percentage of E-rosetting cells in adult R. pipiens was higher in the blood than in various lymphoid organs, including the bone marrow and thymus; the rosetting population was enriched after passage through nylon wool; there was some overlap, but not identity, between E-rosetting and ANAE-positive cells; and two monoclonal antibodies known to interfere with E-rosetting by human T-cells, had similar effects on R. pipiens lymphocytes. Late thymectomy in Xenopus produced a decline in circulating rosetting and ANAE-positive cells. Other studies on Xenopus (Koppenheffer and Stadig 1981) have shown that when spleen cells are incubated with sheep erythrocytes (SRBC) which have been coated with Xenopus anti-SRBC, 10% of them (cells bearing Fc receptors?) form rosettes, whereas only 1% of thymus cells react. These approaches need further, critical attention.

Guillet and Tournefier (1981) have examined isozymes in axolotl
(Ambystoma mexicanum) lymphocytes from different organs. Supernatants
from homogenized lymphocytes of thymus, spleen and blood were tested
for isozymes of lactate dehydrogenase (LDH) by gel electrophoresis.
Six isozymes were represented, which revealed different activities
according to source:
        LDH 1,2,5 & 6 activity: spleen > blood > thymus
        LDH 3 & 4 activity: thymus > spleen and blood
Distinct cell subpopulations could not be discerned within a given
tissue on this basis, however: spleen or thymic lymphocytes separated
on a Ficoll gradient showed the same zymograms as the whole
unseparated population. The density distribution of these axolotl
cells is complex (Tournefier 1982).

Xenopus lymphocytes are denser than their rodent or avian equivalents,
and lymphocytes from the blood have a narrower distribution on BSA
gradients than splenic cells (Donnelly et al. 1976). Some low density
lymphocytes in the spleen do not therefore appear to recirculate.
Unlike "higher" vertebrates, thymic lymphocytes in Xenopus are no
denser than peripheral T cells, nor are the latter significantly
different from B cells in this respect: thymectomy in Xenopus did not
affect the density profile of the blood.

Monocytes and Macrophages (Figs. 17,18,24,25)
        Large mononuclear cells are the first leucocytes to appear
in the blood of bullfrogs (Rana catesbeiana) during larval
development: from 15 days post-hatching at 25°C (Hildemann and Haas
1962). Cells of this type with a linear chromatin pattern were taken
to be immature members of the monocyte series, and cells with clumped
chromatin were taken to be lymphocytes. Large cells described as
"mature monocytes" with rounded nuclei first appeared at day 22 in R.
catesbeiana, and similar forms with kidney-shaped or lobulated nuclei
were seen 10 days later (Hildemann and Haas 1962). A variety of nuclear
shapes has been described in monocytes of other amphibians: kidney-
shaped in Bufo alvarius (Cannon and Cannon 1979) and Ichthyophis
(Zapata et al. 1982); rounded in Notophthalmus (Jordan 1938); horse-
shoe shaped in Xenopus (Fey 1962). In conventionally stained blood
smears (e.g. Wrights) the cytoplasm is grey-blue, may be foamy or
vacuolated (Jordan 1938; Fey 1962, 1967b) and may contain numerous
small azurophilic granules (Jordan 1938; Cannon and Cannon 1979). In
supravitally stained preparations of B. alvarius blood, numerous
mitochondria and small neutral red granules can be seen (Cannon and
Cannon 1979). The presence of pseudopodia in monocytes has often been
noted.

Conventionally stained blood smears are not ideal for distinguishing
monocytes from other large mononuclear cells, especially lymphocytes:
positive identification requires evidence of phagocytic activity (e.g.
uptake of carbon or latex) and/or a characteristic enzyme content.
Amphibian monocytes contain at least some of the hydrolytic enzymes
present in their mammalian counterparts (see Table 4), and the enzyme-
positive granules, azurophilic granules and neutral red granules

Figures 24-25. Cells tentatively identified as macrophages from lymphoid tissues of <u>Plethodon</u> (courtesy of S.K. Curtis). Note the large size, irregular (but not multilobed) nucleus, and prominent cytoplasm with heterogeneous inclusions.

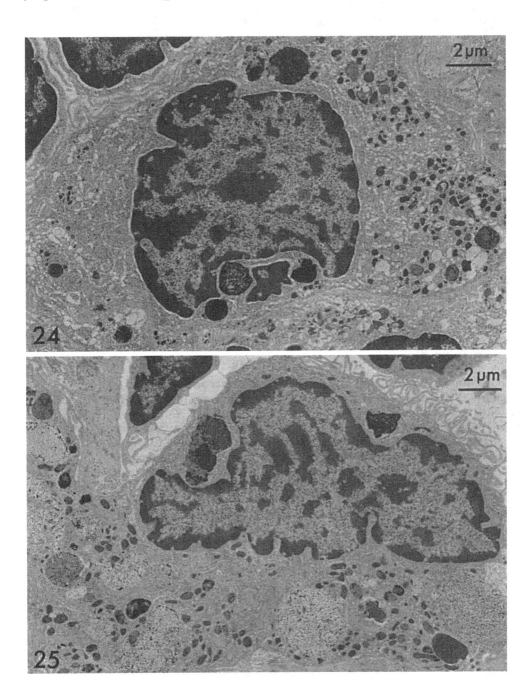

appear to correspond with each other in number and distribution
(Cannon and Cannon 1979). The monocytes invariably show peroxidase
activity, whereas lymphocytes are always negative (Table 4). Splenic
macrophages of Bufo and Xenopus also show a more diffuse pattern of
esterase activity than lymphoid cells (Garavini 1981; Baldwin and
Cohen 1981). A positive reaction with anti-Ig preparations does not
necessarily exclude cells of the phagocytic series, since
subpopulations carrying cytophilic antibody have been described (e.g.
Ruben et al. 1981).

Granulocytes (Figs. 13-16,26-32)
Classification of granulocytes in amphibians has proved to
be a less hazardous exercise than in fish, at least from a morphological standpoint: cells resembling the neutrophils, eosinophils and
basophils of "higher" vertebrates can all be recognised.

Amphibian neutrophils (heterophils) are abundant and conspicuous cells
of "moderate" size (10-25µm diameter in most species; 40µm in
Amphiuma) (Jordan 1938; David & McMullen 1972). There is a distinctive
multilobed nucleus with a dense band of chromatin around the nuclear
periphery (Fig.26). The cytoplasmic granules are small compared with
other granulocytes, but they vary in size, shape and substructure
within and between species (Table 5 and Figs. 26, 29, 30). This cell
type (unlike basophils and some eosinophils) is always peroxidase-
positive, but phosphatase reactions are less consistent (Table 4).
The giant granules sometimes found in neutrophils of the giant
salamander, Megalobatrachus, have been compared with those seen in
human patients with Chediak-Higashi disease (Mitsui and Fukui 1964;
see also Table 5).

Eosinophils are usually of similar size to neutrophils, and less
frequent. They have prominent round or oval granules (Table 6), and
the cell nucleus tends to be less highly lobulated than in neutrophils
(Fey 1967a and Figs. 14, 27). The eosinophilic granules may or may
not show substructures, depending on species (Table 6 and Figs. 27,
31). As noted above, these cells often fail to show peroxidase
activity, particularly in urodeles (Table 4). Phosphatase reactions
are again inconsistent. Aryl sulphatase and β-glucuronidase are
absent from the eosinophils and neutrophils of Bufo alvarius, but
other amphibians have not been tested (Cannon and Cannon 1979).

Circulating basophils are rare in some amphibian species, common in
others. Cannon and Cannon (1979), for example, note that only 1% or
fewer of blood leucocytes are basophils in Bufo alvarius, whereas
Cowden (1965) reports up to 20% in Amphiuma and Takaya (1968) about
50% in Triturus pyrrhogaster (see also Jordan 1938 for differential
counts). Their size relative to other granulocytes also depends on
the species. The cells usually show an unsegmented nucleus and large,
metachromatic granules, although granule ultrastructure is variable
(Table 7 and Figs. 28, 32). In Rana esculenta, basophils can be
distinguished from mast cells on both morphological and cytochemical

grounds: the latter show elongated or kidney-shaped nuclei, pronounced cytoplasmic processes and are metachromatic with azure A, whereas basophils from blood (and from the spleen) are "lymphoid" with a round nucleus and little cytoplasm and fail to react with azure A. The granules of both forms, however, react with alcian blue and stain metachromatically with toluidine blue (Kapa et al. 1970; Csaba et al. 1970). In Notophthalmus the differences are less dramatic: basophils are again "lymphoid" in shape with coarse granules, whilst mast cells show more cytoplasm and finer granulation (but rounded nuclei); cytochemically they do not differ significantly (Cowden et al. 1964). The two types of cell may represent different developmental stages of the same population, but this seems unlikely since they do not co-exist in all amphibians (Cowden 1965).

The characteristic metachromatic reactions of mammalian mast cell granules are normally attributed to heparin, a highly sulphated acid mucopolysaccharide with a peptide backbone (proteoglycan), whilst related substances (chondroitin sulphate and heparan sulphate) have been detected in blood basophils (Orenstein et al. 1977).
Studies by Cowden et al. (1964), Cowden (1965), Kapa et al. (1970) and Chiu and Lagunoff (1972) have indicated that amphibian basophil and mast cell granules too contain acid mucopolysaccharides (glycosaminoglycans), but that these are less highly sulphated than in mammals: metachromasia is more limited and safranin staining is weak or negative. Moreover, PAS reactivity is stronger in the amphibian cells, and the proteins which form complexes with the polysaccharides appear to be different. These results suggest an intermediate status between the "typical" metachromatic, PAS-negative mast cell/basophil of "higher" vertebrates, and the non-metachromatic, PAS-positive equivalent (PAS-GL cell) of fishes (Barber & Westermann 1978). However, the overall histamine content of amphibians is low (Reite 1972), and very little has been detected in tissues such as the tongue which contain large numbers of mast cells. The mast cells themselves fail to fluoresce following o-phthalaldehyde treatment (Takaya et al. 1967; Takaya 1968; Chiu and Lagunoff 1972). These experiments were carried out on several ranid species and on the newt, Triturus pyrrhogaster. Blood from the latter was also tested since it is particularly rich in basophils, and found to be negative (Takaya 1968). Clearly, the association between heparin (or related mucosubstances) and histamine in basophilic granules is not a universal one: the amphibians have either "lost" their histamine as suggested by its presence in lungfish (Reite 1972), or the "higher" vertebrates acquired these agents at different stages of their phylogeny.

Biogenic amines other than histamine have received comparatively little attention. Kapa et al. (1970) have detected serotonin (5-hydroxytryptamine) in the mast cells and basophils of Rana esculenta by means of paraldehyde-induced fluorescence, but negative results were reported for R. pipiens and R. catesbeiana (Takaya et al. 1967; Chiu and Lagunoff 1972). Similar searches in a wider variety of amphibians would be of interest here.

Table 5 : GRANULATION AND NUCLEAR FEATURES OF AMPHIBIAN NEUTROPHILS

Genus/Species	Site	Nucleus	Staining characteristics of cytoplasmic granules	Granule Sizes (diameter in μm)	Granule Shapes	Granule Substructure	Reference
Rana pipiens	blood	multi-lobed	Wright's: fine, neutrophilic				Jordan (1938)
"	bone marrow	dense chromatin at nuclear periphery		small	round or oval	crystalloid	Campbell (1970)
Rana esculenta	bone marrow spleen, blood		(i) EM: dense (ii) less dense	(0.1-0.3) x (0.5-0.9) 0.1-0.3	elongated	some with crystalloid no	Kelényi and Nemeth (1969)
"		mostly 3-lobed (some 4-lobed)					Fey (1967a)
Rana temporaria	bone marrow	lobed	(i) peroxidase +ve (ii) peroxidase -ve	0.1 0.05	round,oval or rod-like	some with crystalloid. dense centre, light matrix	Rogovin et al. (1978a)
Bufo vulgaris	blood	lobed, excentric. Dense chromatin at nuclear periphery (central area less dense) Nucleolus not seen.	Giemsa: red-purple azurophilic. EM: (i) dense (ii) dense (iii) dense	0.25(max.) x 0.7 0.7(max.) x 0.07 0.03 x 0.2	oval elongated club- or dumbbell	membranes may form myelin figures or vesicles long, denser regions some: long, faint bands	Setoguti et al. (1970)
Bufo bufo		usually bilobed (a few with 3-4 lobes)		0.1(av.)			Fey (1963, 1967a)
Bufo marinus	blood sinusoids of adrenal	multilobed. Most of the dense chromatin is at nuclear periphery. Nucleolus not seen.	(i) Azurophilic, EM dense (ii) Less dense	large smaller: range 0.13-0.42	round,oval or elongated		Surbis (1978)
Bufo alvarius	blood	2-4 lobes	(i) Wright-Giemsa: lavender Acid phosphatase +ve PAS +ve Peroxidase +ve (ii) Neutral red +ve				Cannon and Cannon (1979)

Triturus cristatus	spleen	multi-lobed. Dense chromatin along edge. Small nucleoli	May Grunwald-Giemsa: unstained	0.08-0.2	elongated or dumbbell	almost homogeneous	Tooze and Davies (1968)
"	liver, spleen, blood		(i) EM: dense	(0.1-0.3) x (0.5-0.9)	elongated	some with crystalloid	Kelényi and Nemeth (1969)
(2)			(ii) EM: less dense			no	
Triturus vulgaris	liver, spleen, blood						Fey (1967a)
"		usually 3-4 lobes					
Triturus pyrrhogaster	liver	lobed	(i) EM: pale (ii) EM: dense (iii) EM: dense	0.17 0.4 small	round spindle or rod-like	crystalloid	Ishizeki (1980)
Notophthalmus viridescens	blood, liver	4-6 lobes, dense chromatin at nuclear periphery	Wright's: fine, light lilac (i) EM: dense (ii) EM: less dense	small larger}	variable no dumbbells		Jordan (1938) Hightower and Haar (1975)
Plethodon glutinosus	bone marrow	lobed, chromatin condensed	Peroxidase +ve (largest granules only)	small	usually oval	fine, faint, long filaments	Curtis et al. (1979a)
Batrachoceps attenuatus	liver	lobed, dense chromatin concentrated at nuclear periphery	EM: dense		variable	long fibres (5nm diameter)	Campbell (1969)
Amphiuma tridactylum	blood	lobed	May Grunwald-Giemsa: light and metachromatic	0.1-0.5 0.6-1(a few)	round to rod-like round		David and McMullen (1972)
Megalobatrachus japonicus			peroxidase +ve, Alkaline phosphatase -ve, Giemsa -ve	6-10 (i.e. sometimes single, central giant granule)			Mitsui and Fukui (1964)
Ichthyophis kohtaoensis	blood	lobed	Giemsa: pale				Zapata et al. (1982)
" + Afrocaecilia sp.	liver				elongated or spindle		Welsch and Storch (1982)

Table 6 : <u>GRANULATION AND NUCLEAR FEATURES OF AMPHIBIAN EOSINOPHILS</u>

Genus/Species	Site	Nucleus	Staining characteristics of cytoplasmic granules	Granule Sizes (diameter in um)	Granule Shapes	Granule Substructure	Reference
Rana pipiens	blood	bilobed, often peripheral	eosinophilic	large	round		Jordan (1938)
"	bone marrow		EM: dense			no	Campbell (1970)
Rana esculenta	bone marrow	mostly 2-3 lobes		1	oval	no	Fey (1967a)
" "	bone marrow spleen,blood		Basic protein +ve	0.6-1.4 (some smaller and denser)	round or oval	no	Kelényi and Nemeth (1969)
Rana catesbeiana	bone marrow blood			1-2	round	no	Gouchi (1982)
Rana temporaria	bone marrow		(i) peroxidase +ve (ii) peroxidase -ve (iii) peroxidase -ve phosphatase +ve	0.3-0.5 0.7-0.8 0.05-0.06	round or oval round,oval or elongated	no a few have dense inclusions	Rogovin et al. (1978b)
Bufo marinus	blood sinusoids of adrenal			0.4-0.625	round	dense round core	Surbis (1978)
Bufo alvarius	blood	round- to bean-shaped or bilobed. Dense chromatin.	(i) 50-100 refractile peroxidase +ve (ii) 30-50 neutral red +ve PAS +ve, acid phosphatase +ve	large small			Cannon and Cannon (1979)
Triturus cristatus	spleen	"highly" lobed; dense chromatin at edges	May Grunwald-Giemsa: orange EM: dense	2-3	oval		Tooze and Davies (1968)
" " Triturus vulgaris	liver, spleen, blood		basic protein +ve	0.6-1.4 (some smaller and denser)	round or oval	no	Kelényi and Nemeth (1969)
" "		mostly bilobed					Fey (1967a)
Triturus pyrrhogaster	liver	lobed	dense	large (a few smaller)	round (small may be elongated)	occasional white spot-like structure or crystalloid core	Ishizeki (1980)

Species	Tissue	Nucleus	Staining	Size	Shape	Structure	Reference
Triturus pyrrhogaster	liver		EM: Variable density	0.8-1.6	round or oval	No structure like human crystalloid	Mitsui (1965)
Notophthalmus viridescens	blood	bilobed	strongly acidophilic	large	round		Jordan (1938)
"	liver		EM: dense			some denser rounded areas occasional circular membranous substructure	Hightower and Haar (1975)
"	liver		peroxidase +ve	0.7-0.9		no structure like human crystalloid	Mitsui (1965)
Plethodon glutinosus	bone marrow	lobed (less than neutrophil)	EM: dense peroxidase -ve	large	round	some: light irregular core; crystalloids with 10mm spacing; vesicular material	Curtis et al. (1979a)
Battrachoceps attenuatus	liver			large	round or oval	some: internal membranes or interfaces; no crystalloid core	Campbell (1969)
Ambystoma maculatum	liver		EM: medium or variable density	0.8-1.5	oval	no structure like human crystalloid	Mitsui (1965)
Ichthyophis kohtaoensis	blood		acidophilic	large	large		Zapata et al. (1982)
"	spleen	lobed	(i) EM: dense	large	oval	crystalloid core	Welsch and Storch (1982)
+ _Afrocaecilia_ sp.			(ii) EM: dense	small	round	no	

Table 7 : GRANULATION AND NUCLEAR FEATURES OF AMPHIBIAN BASOPHILS

Genus/Species	Site	Nucleus	Staining characteristics of cytoplasmic granules	Granule Sizes (diameter in μm)	Granule Shapes	Granule Substructure	Reference
Rana pipiens	blood	type (i) faint, irregular; type (ii) round	(i) metachromatic (ii) orthobasophilic	variable; uniform	variable; round		Jordan (1938)
"	blood sinusoids in bone marrow			large	irregular	variable density: particulate	Campbell (1970)
Rana esculenta	blood	round	Alcian blue +ve, PAS +ve, Serotonin fluorescence +ve, Azure A -ve, Safranin -ve, (some) metachromatic with toluidine blue	0.5-2.4	oval	uniform density	Kapa et al. (1970)
"	spleen						Csaba et al. (1970)
Bufo marinus	blood sinusoids of adrenal	usually oval, not indented		0.55 (av.) 0.8 (max.)	round	usually uniform; some irregular, lighter areas	Surbis (1978)
Bufo alvarius	blood	round; clumped chromatin	Wright-Giemsa approx. 150 blue-purple, water soluble granules; 50-60 neutral red granules				Cannon and Cannon (1979)
Triturus cristatus	spleen	bilobed	May Grunwald-Giemsa: deep bluish red. EM: dense	large		crystalloid: 6-9nm spacings	Tooze and Davies (1968)
Notophthalmus viridescens	blood	often bilobed, pale	Azure or metachromatic	as eosinophils			Jordan (1938)
"			PAS weak +ve, Astrablau +ve, Toluidine blue +ve (mainly orthochromatic)				Cowden et al. (1964)
Plethodon glutinosus	bone marrow	irregular but unsegmented	Metachromatic (red) with basic toluidine blue; EM: very dense	large	often disc-shaped	prominent crystalloid: 10nm spacings	Curtis et al. (1979a)
Amphiuma means	blood	round	metachromatic Toluidine blue +ve, Azure A +ve, PAS +ve, Astrablau +ve, Safranin weak +ve	variable			Cowden (1965)
Ichthyophis kohtaoensis	blood			smaller than eosinophil granules	round or oval		Zapata et al. (1982)

Examination of larval bullfrogs (Rana catesbeiana) at successive stages of development has shown that definitive granulocytes are late arrivals in the blood - only small lymphocytes appear later - and all three granulocyte types appear simultaneously (Hildemann and Haas 1962). This suggests a common developmental pathway, but balanced against this is the "lymphoid" appearance of some amphibian basophils, their association with the thymus gland, and the separation of basophil-producing areas from sites of neutrophil and eosinophil production, especially in urodeles (see section on haemopoietic and lymphoid tissues). Beyond this, we know little about the development of basophils in the group, and the remaining discussion will concentrate on neutrophil and eosinophil maturation, based on descriptive accounts of cells in the liver or bone marrow. Since bone marrow only occurs in some species, terms such as "myeloblast" and "myelocyte" are not ideal, but have nevertheless been used by a number of authors.

Granulocyte precursors (progranulocytes, myeloblasts) have not been positively identified in amphibians, but the most likely candidate is a lymphocyte-like cell found nestling amongst developing granulocytes (Campbell 1969; Hightower and Haar 1975; Curtis et al. 1979a). The Golgi complex and endoplasmic reticulum are poorly developed or absent at this cell stage, whilst the cytoplasmic granules, where seen, are small and few in number, and are not diagnostic.

The granulation of developing neutrophils (myelocytes) has been described for a number of urodele and anuran species. It is important to establish whether the diverse granules which have been listed in Table 5 represent distinct subpopulations, or are mature and immature forms of the same population: do the cells acquire one set of granules (i.e. primary) early on, and a second set (secondary) later; and/or are the granules divisible into azurophilic (peroxidase-positive) and specific (peroxidase-negative) types? Unfortunately, few authors have combined ultrastructural and cytochemical observations, and the ontogenetic and phylogenetic implications of the data need to be clarified.

Investigations on bone marrow of the urodele, Plethodon, by Curtis et al. (1979a) have suggested that the neutrophils produce only a single population of granules. Immature cells showed small granules of various shapes, and these appeared to increase in size and density during maturation; only the larger granules showed peroxidase activity. In neutrophils from the liver of Batrachoceps, a urodele closely related to Plethodon, Campbell (1969) found that a limited population of small granules with a fine particulate content appeared first; this was then accompanied and later replaced by small, dense fibrous granules of a single morphological type. The substructure of mature granules was more obvious than in Plethodon (Table 5). In neither genus was there an early population corresponding to the large, dense primary/azurophilic granules of mammalian neutrophils. A population of dense granules which appears early has been described in the liver of another urodele, Notophthalmus (Hightower and Haar 1975), although these primary granules were smaller than the less dense

secondary population which appeared later. Light-microscopic
observations of immature neutrophils in Notophthalmus have shown a
mixture of basophilic and acidophilic granules (Jordan 1938).
Ishizeki (1980) has distinguished three kinds of granule in another
newt, Triturus pyrrhogaster: relatively large rounded granules of
moderate electron density which decline in number in more mature
neutrophils; small, dense elongated granules which also decline in
numbers; small, less dense granules which predominate in the mature
cell, and which possibly differentiate from type 2 granules.

Campbell (1970) claims that differentiating neutrophils of frogs (Rana
pipiens) possess only a single population of granules, whereas reports
on other anuran species describe two or more types (Table 5).   The
investigation by Rogovin et al. (1978a) on Rana temporaria is of
particular interest here: peroxidase-positive and negative granules
are both present in the mature neutrophils as in mammals, but their
sequence of production appears to be reversed; cells which were
manufacturing peroxidase-containing granules already housed a
population of morphologically distinct peroxidase-negative granules in
their cytoplasm. The sequence in other anurans, especially bufonids,
would be worth investigating. Surbis (1978) has described a mammal-
like distribution of granules in mature neutrophils of B. marinus,
whilst Cannon and Cannon (1979) have noted peroxidase activity in
mature neutrophils of B. alvarius but not in myelocytes (using light
microscopy).

The synthesis and packaging of the granules described above are
probably similar to events in higher vertebrates. Rogovin et al.
(1978a) have detected peroxidase in the granules, perinuclear space
and endoplasmic reticulum of immature neutrophils, and observed direct
connections between granules and the endoplasmic reticulum. There was
no clear evidence of a reaction product in the Golgi complex at this
time, but other work suggests that the organelle is important: it
becomes elaborate during neutrophil maturation (Campbell 1969, 1970;
Hightower and Haar 1975), it associates closely with profiles of
endoplasmic reticulum (Campbell 1969; Curtis et al. 1979a), and it
sometimes contains material similar to that seen in immature granules
(Campbell 1969, 1970). In some regions of the cell, the rough
endoplasmic reticulum becomes organized into characteristic stacks of
parallel cisternae (Tooze and Davies 1968; Campbell 1969, 1970;
Setoguti et al. 1970; Curtis et al. 1979a). According to Curtis et
al. (1979a), this arrangement persists throughout neutrophil
development, but in general the endoplasmic reticulum and Golgi
complex decline in mature cells (Campbell 1969, 1970; Setoguti et al.
1970; Surbis 1978). Various authors have also described a distinct
region consisting of centrioles and radiating microtubules (Campbell
1969, 1970; Setoguti et al. 1970; David and McMullen 1972; Ishizeki
1980). The microtubules apparently attach to granules or are closely
associated with them. Mitochondria become abundant in regions of
granule production (Curtis et al. 1979a) but decline in mature cells

(Campbell 1970; Setoguti et al. 1970). Glycogen particles, on the
other hand, become abundant in the late stages (Tooze and Davies 1968;
Setoguti et al. 1970; Curtis et al. 1979a). These cytoplasmic changes
are accompanied by progressive and conspicuous alterations in the
nucleus: the nucleus itself becomes lobed, the chromatin denser, and
the nucleoli smaller in size and number (Campbell 1969, 1970;
Hightower and Haar 1975; Curtis et al. 1979a). The main difference
from mammals is that nuclear indentation begins earlier (at least in
Notophthalmus) in relation to granule formation (Hightower and Haar
1975).

Developing eosinophils are much less common than neutrophils in the
granulopoietic centres, and knowledge of their maturation is
correspondingly more sketchy. In the urodeles Notophthalmus
viridescens (Hightower and Haar 1975) and Triturus pyrrhogaster
(Ishizeki 1980), dense round and conspicuous granules have been seen
in early stages of eosinopoiesis, when the cell has only a small
cytoplasmic area and a rounded nucleus. Later, there is a mixture of
large homogeneous granules and granules with dense or crystalloid
inclusions (which presumably mature from the primary population). In
a number of species, however, crystalloid substructures fail to appear
at all (Table 6), and those which do appear sometimes have no close
counterpart in other amphibians nor in "higher" vertebrates (see e.g.
Curtis et al. 1979a). Inclusions of a membranous nature (Mitsui 1965;
Campbell 1969; Hightower and Haar 1975; Curtis et al. 1979a; Ishizeki
1980) may result from incomplete fusion of smaller granules during the
maturation process.

Campbell (1969) interpreted all differences in the ultrastructure of
granules in Batrachoceps eosinophils as differences in developmental
stages of a single population; enlarged vesicles with a pale-staining
mass were taken to represent very early stages of granule production.
Similar conclusions were drawn from his (1970) study on Rana pipiens.
However, Rogovin et al. (1978b) described chemically distinct subpopulations of granules in eosinophils from R. temporaria: (i) large
peroxidase-positive granules formed by merger of smaller granules;
(ii) large peroxidase-negative granules; and (iii) an abundant
population of small peroxidase-negative granules. In preliminary
experiments the latter did not show phosphatase activity at low pH,
but did so at pH 7.4 and 9. Eosinophils of the blood of Bufo alvarius
(Cannon and Cannon 1979) showed a population of large peroxidase-
positive granules under the light microscope, whilst acid phosphatase
activity was confined to a population of smaller granules and
developed at later stages of maturation. Thus there is some evidence
for the existence of "secondary" or microgranules in amphibian
eosinophils. Further work is needed to establish the developmental
and chemical credentials of these granule subpopulations, to examine
critically the species differences in peroxidase and phosphatase
content (Table 4), and to search more widely for other chemical
constituents such as β-glucuronidase and aryl sulphatase.

Figure 26. Neutrophil in the bone marrow of Plethodon (courtesy of
S.K. Curtis). This cell appears to be at a fairly advanced stage of
differentiation since its nucleus is multilobed and contains
considerable amounts of condensed chromatin at the periphery. The
cytoplasm is voluminous (see also Figs. 29 and 30).

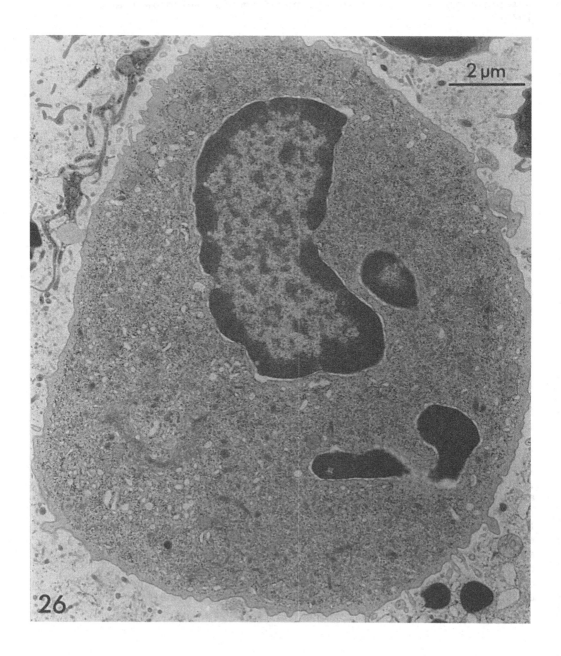

Figure 27. Developing eosinophil in the bone marrow of <u>Plethodon</u> showing large dense granules, of which some display lightly stained regions of assorted shapes and sizes (courtesy of S.K. Curtis) (see also Fig.31).

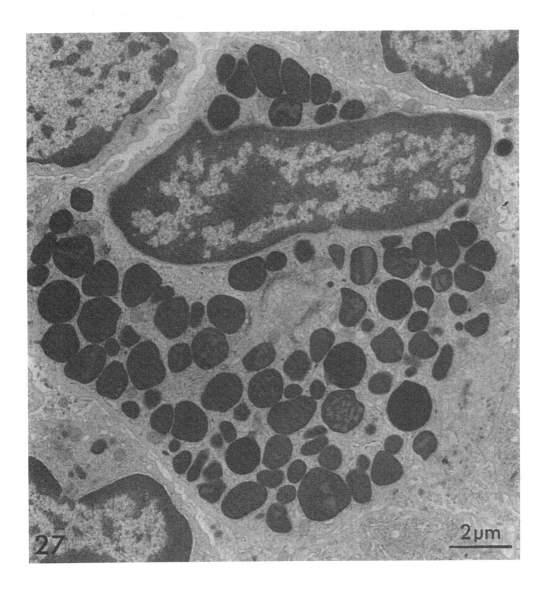

Figure 28. Basophil in the bone marrow of <u>Plethodon</u>, with densely stained cytoplasmic granules of various shapes and sizes (courtesy of S.K. Curtis). Some of the granules are coin-like while others are more rounded or irregularly shaped. The nucleus of the cell is irregular in outline but not multilobed, and it contains considerable amounts of condensed chromatin. Basophils are found only rarely in <u>Plethodon</u> marrow and apparently do not develop there (see also Fig.32).

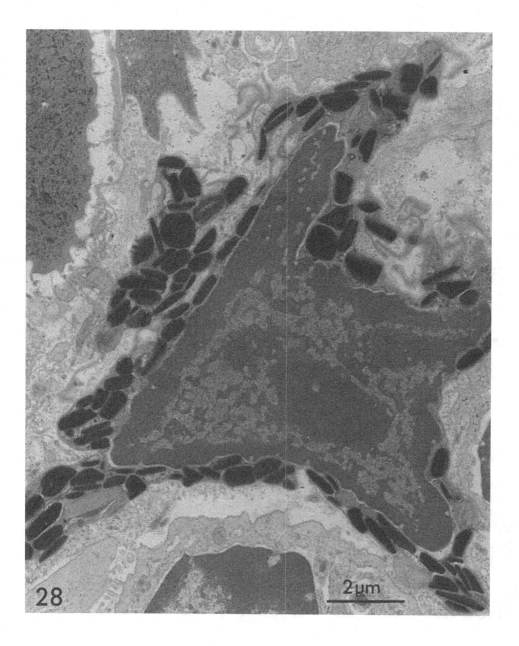

28   2µm

Figures 29-32. Granulocytes of <u>Plethodon</u>: granule substructures
(micrographs supplied by S.K. Curtis). All figures to same scale.
<u>Figs 29 and 30</u>. Neutrophil granules are small, round or elongated. A
fine filamentous substructure may be seen, but is indistinct even in
optimally orientated granules. The dot-like particles are glycogen.
<u>Fig. 31</u>. Eosinophil granule. Its crystalline substructure is always
less electron-dense than the remainder of the granule in this animal.
<u>Fig. 32</u>. Basophil granules. The crystalline substructure is more
prominent than in the eosinophil granules of <u>Plethodon</u>. The
inclusions are orientated at various angles.

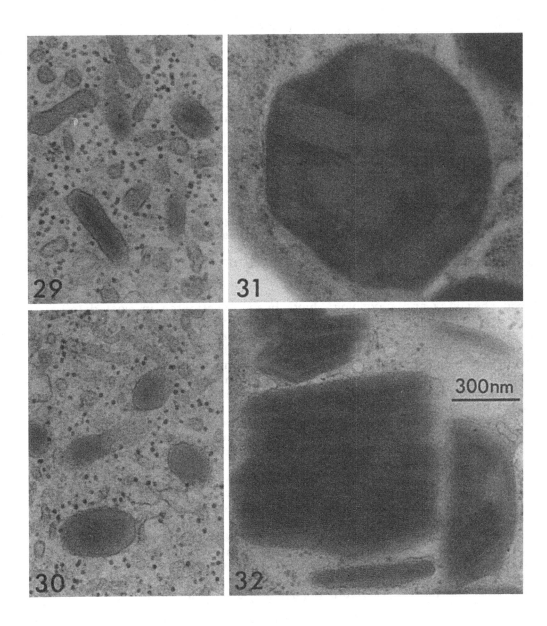

300nm

The Golgi complex becomes prominent during eosinophil maturation, and it remains so in some species (Tooze and Davies 1968; Hightower and Haar 1975) but not in others (Campbell 1970; Surbis 1978). The organelle appears to be involved in granule production, although direct evidence is lacking (Campbell 1969, 1970; Hightower and Haar 1975; Curtis et al 1979a). Abundant mitochondria and an extensive RER may also appear (and later decline), but unlike neutrophils the RER does not form stacked parallel arrays (Mitsui 1965; Curtis et al 1979a). The two cell types show similar developmental changes within their nuclei, even though the number of lobules differs.

FUNCTIONS OF BLOOD CELLS

Erythrocytes
The phenomenon of metamorphosis presents special opportunities to the student of respiratory physiology, and there is an impressive range of respiratory adaptations in adult amphibians: aquatic, terrestrial, fossorial and alpine forms are all represented. The effects of temperature on amphibian respiration are also of interest, bearing in mind the animals' ectothermic status and the loss of oxygen from water as the temperature rises. Changes in the availability of oxygen will affect not only respiratory surfaces and the distribution of blood in these animals (see "blood circulation"), but will also make new demands on the blood itself. Amphibians maintain a flexible and efficient oxygen-transporting system to cope with environmental changes partly by synthesising haemoglobins with different oxygen affinities; in addition, various intracellular modulators of Hb-O_2 affinity have been identified, including organic phosphates and inorganic ions.

Blood chemistry and physiology have been investigated in most detail in the bullfrog Rana catesbeiana, a well-known species with suitably large larvae. Indeed, this was the first reported example of haemoglobin switching in vertebrates (McCutcheon 1936). The oxygen affinity of blood in the gilled larvae was found to be much higher than in the air-breathing adult bullfrogs. This is not surprising, since the oxygen content of water is lower than that of air, and the animal has to work harder to obtain it. Electrophoretic and chromatographic studies have revealed four larval and four adult haemoglobins, all of which are tetramers containing two a-like and two β-like globin chains, but no chains are shared between the adult and larva (see Broyles et al. 1981). According to Just and Atkinson (1972), the larval haemoglobins begin to disappear and adult haemoglobins appear during tail regression (stage XXII) and the transition is nearly complete by the late metamorphic stage (XXIV), two weeks after emergence of the forelimbs. Benbassat (1970) has reported a much slower transition. These haemoglobin changes are under the partial control of thyroid hormones (see Just and Atkinson 1972) and involve a switch in red cell production by the liver, the main erythropoietic organ in late larval and early adult stages: new cells with the new haemoglobin appear while the old larval cells are cleared from the circulation (Forman and Just 1981; Dorn and Broyles 1982; Okazaki et al. 1982). During the switch, large stores of iron hitherto held in circulating larval erythrocytes as ferritin are utilized for the production of adult haemoglobin (Theil 1981). The adult red cells may represent a new cell line, but the continuing erythropoietic role of the liver over the metamorphic period, together with immunofluorescence studies, suggest that this is not the case: antisera raised against larval versus adult haemoglobins showed double labelling in about 16% of erythrocytes at metamorphic climax (Benbassat 1974).

Changes of haemoglobin content in the blood of bullfrogs also occur either side of metamorphosis, but in these instances haemoglobin replacement is more limited and the adaptive advantage of the replacement is not clear. Watt and Riggs (1975) reported that larval haemoglobins I and II predominate in young tadpoles and that components III and IV (which have substantially lower oxygen affinities) become more important in older tadpoles. Futamura et al. (1982), however, reported that component I precedes II as a dominant haemoglobin. The former has unique globin chains and, unlike components II, III and IV, is produced by the mesonephros (Broyles et al. 1981). In adults, the haemoglobin (mainly B) of sexually mature adults has a lower oxygen affinity than that of young froglets (mainly C), and though components B and C occur in the same cells, haemoglobin production shifts during this period from the liver to the spleen and/or bone marrow (Broyles et al. 1981). Such changes in the balance of haemoglobin types during larval or adult life could be just a consequence of relocating erythropoietic activity as development proceeds, having no adequate functional explanation. On the other hand, haemoglobins with different functional properties in the blood at any given stage could be of value to the animal in a fluctuating environment: with rising temperature, for example, haemoglobins with higher oxygen affinity could become more useful and vice versa.

The low oxygen affinity of haemoglobin in air-breathing adults with good oxygen supplies promotes unloading at tissues but raises questions about the level of oxygen uptake which is possible in the lungs. Lykkeboe and Johansen (1978) reported a striking increase in cooperativity (steepening of the O_2 equilibrium curve) at oxygen saturations above the 50% mark in R. catesbeiana and especially in R. temporaria, an effect which promotes oxygen loading. Moreover, it has recently been established (Tam and Riggs 1984) that the two major haemoglobin tetramers in adult bullfrogs (B and C) can combine to form a mixed aggregate with much lower oxygen affinity than either tetramer alone, and that aggregates formed by deoxygenated haemoglobin dissociate as oxygenation proceeds. These B-C links are probably made via α-globin chains; organic phosphates (which bind β-chains) are not involved (Araki et al. 1974). Aggregation increased at low pH and should thus enhance the Bohr effect in bullfrogs (Tam and Riggs 1984). Deoxyhaemoglobin also aggregates in other amphibians, but this is not always pH-dependent (Elli et al. 1970).

Apart from the switch to lower affinity haemoglobins, metamorphosis in R. catesbeiana is accompanied by an increase in haemoglobin concentration, oxygen capacity and haematocrit (Pinder and Burggren 1983). No developmental changes in cooperativity or the Bohr effect (in whole blood) were observed by these authors, and organic phosphates did not contribute to the lowered oxygen affinity at metamorphosis: concentrations of nucleoside triphosphates (ATP and GTP) showed a marked decrease, whilst 2,3-diphosphoglycerate (DPG) - a minor component in this amphibian - was virtually unchanged. The high levels of organic phosphate in larvae may be important in releasing oxygen at the tissues, given the high oxygen affinity of the haemoglobin itself and the low concentrations of CO_2 in larval blood.

Pinder and Burggren (1983) also showed that bullfrog larvae and adults react in different ways to conditions of chronic hypoxia: larvae showed little or no change in the blood itself but a hypertrophy of the gills and improvements in the capillary network at respiratory surfaces; adults on the other hand showed no morphological adjustments for increasing gaseous exchange, whereas the blood showed much increased oxygen capacity and affinity. The latter, it was speculated, may be due to a renewed output of larval-type haemoglobins or to co-factors other than organic phosphates. The costs in energy of the above countermeasures are much lower than devices such as increased ventilation or cardiac output which are seen in acute cases of hypoxia (West and Burggren 1982). Similar experiments on tiger salamanders (Wood et al. 1982) which tested the effects of hypoxia over an intermediate period (8 days) showed that lowered concentrations of organic phosphate (ATP and DPG) did contribute to an increased oxygen affinity, whilst there was no haemopoietic response in these animals.

The developmental changes in haemoglobin which have been identified in R. catesbeiana are no doubt broadly representative of the Anura as a whole, although both the timing and number of components involved vary among (and sometimes within) species. In Xenopus, the first haemoglobin transition occurs in young tadpoles, and may coincide with the demise of cells from the ventral blood islands (Kobel and Wolff 1983); the second occurs under the partial influence of thyroid hormones at metamorphosis and entails a complete change of globin subunits (Maclean and Jurd 1972; Just et al. 1980). This sequence is very similar to that in R. catesbeiana. Earlier studies by Moss and Ingram (1968) on wild populations of R. catesbeiana which showed variations in the balance of haemoglobins among larvae of the same size and appearance were also confirmed in Xenopus. There is disagreement over the timing of the main (i.e. metamorphic) replacement of haemoglobins in Xenopus, nor is it clear whether a new or multipotential cell line is responsible for the adult haemoglobins (Maclean and Jurd 1972; Just et al. 1980). Investigations of more anurans (Bufo, Hyla, Bombina spp.) by Cardellini and Sala (1978, 1979, 1983) suggested no connection between the timing of the main haemoglobin shift and the animals' phylogenetic position. They also confirmed that the shift is total and contended that it is usually completed within three weeks after metamorphosis.

In anurans where the adult remains in water, functional similarities between larval and adult haemoglobins would be expected. Thus the purified proteins from adult aquatic pipids show a high oxygen affinity, no Bohr effect and a pronounced allosteric effect of ATP, as in larval bullfrogs (Vieira et al. 1982). The semi-aquatic toad, Leptodactylus labyrinthicus, is also interesting in this respect: the major adult haemoglobin (60% of total) shows a lower Bohr effect and greater allosteric response to organic phosphate than the second haemoglobin (30% of total). It was suggested that the first is responsible for oxygenation while the animal is in water, the other when it is in air (Lima et al. 1983). Structural differences were greater in the β-globins than in the α-chains.

In contrast with bullfrogs and other anurans, metamorphosis in urodeles is not always accompanied by a haemoglobin transition, a point best illustrated by the Ambystomidae. The axolotl (Ambystoma mexicanum) shows a changeover in haemoglobins (and erythrocytes) as the youngster grows, but when neotenous adults are induced to transform there is a significant decrease in oxygen affinity with (according to some authors) no change in haemoglobin type (see Gahlenbeck and Bartels 1970; Sullivan 1974). Studies on Dicamptodon ensatus (Wood 1971) suggested that where developmental changes in respiration do not coincide with functional changes in the haemoglobins themselves, organic phosphates have an important role: this ambystomid transfers spontaneously from aquatic to aerial respiration at metamorphosis, yet the oxygen affinity of purified (phosphate-free) larval and adult haemoglobin is similar; the reduced oxygen affinity of adult blood is achieved instead by increased levels of ATP in the erythrocytes. Unlike axolotls and Dicamptodon, the blood of tiger salamanders (A. tigrinum) shows no reduction in oxygen affinity at metamorphosis; haemoglobin profiles and total organic phosphate are unchanged (Wood et al. 1982). There is a loss of gills and thickening of skin at metamorphosis, but air breathing is established well before this and indeed becomes obligatory at 25°C or more.

In mammals, interspecific differences in the organic phosphates of erythrocytes have been related to respiratory strategies. A study by Hazard and Hutchison (1982) on a variety of amphibians has shown that no single organic phosphate is characteristic of terrestrial versus aquatic forms, and that phosphate responses to imposed stresses of temperature and altered photoperiod were not universal or substantial. Nevertheless, some interesting correlations emerged. Toads (e.g. Scaphiopus, bufonids) which are largely reliant on aerobic respiration showed not only elevated amounts of haemoglobin in their blood but also high concentrations of erythrocytic phosphates and low blood-oxygen affinity, thus providing a rapid delivery of oxygen in quantity. These animals tend to be sluggish, but they can sustain activity. The intermittent jumping activity of frogs (ranids and hylids) on the other hand is fuelled anaerobically, and the blood parameters (including lower concentrations of organic phosphates) show correspondingly less emphasis on oxygen delivery. This association between high phosphate levels and dependence on aerobic production of energy for activity does not hold for aquatic amphibians.

Apodans show some remarkable respiratory adaptations. Typhlonectes compressicauda, for example, makes underwater burrows (in mudbanks of the Amazon) and is viviparous. Transfer of oxygen from mother to larva is facilitated by close contacts between larval gills and the uterine surface and by higher levels of ATP in the maternal blood cells. Thus the oxygen affinity of larval blood is higher than the mother's and the Bohr effect lower, even though the respective haemoglobins appear to be identical (Garlick et al. 1979). Adult Typhlonectes, though aquatic, visit the surface for air. Their lungs occupy a large portion of the body cavity and reach from the buccal cavity to the cloaca. The burrowing habit is also assisted by a very large blood volume (a quarter of total body weight) and a high haemoglobin concentration which together provide a large oxygen capacity and a buffer reserve against short-term acid-base fluctuations (Toews and Macintyre 1978). The blood of Typhlonectes has a higher oxygen affinity than the terrestrial burrower Boulengerula taitanus (another apodan), but the purified haemoglobins do not. Thus some factor other than the haemoglobin itself is responsible for the differences in whole blood (Garlick et al. 1979). Similarly, haemoglobins from adults of three species of urodele (Triturus spp) did not show significant differences in functional behaviour, notwithstanding different habitats (Condo et al. 1983). Moreover, electrophoretic patterns of haemoglobin in T. cristatus were not affected by temperature acclimation at 4°C versus 20°C for 160 days (Condo et al. 1981).

Haematological responses to life at high altitudes are well illustrated by Andean frogs of the genus Telmatobius. These frogs are aquatic and respire mainly through the skin. Their erythrocytes are particularly small in size (see section on erythrocyte structure), large in number and high in haemoglobin content; oxygen affinity of the blood is also high (Hutchison et al. 1976; Ruiz et al. 1983). Thus oxygen uptake and capacity are maximized. The metabolic rate of Telmatobius is the lowest of any anuran, and only the large aquatic salamanders such as Amphiuma and Necturus have lower values. It is worth noting here that whilst the two urodeles, like Telmatobius, can survive hypoxic conditions their erythrocytes are large. There is, then, no consistent relationship between erythrocyte size and metabolic rate or oxygen availability in amphibians, although large red cells are usually a feature of aquatic species (see section on erythrocyte structure).

Aquatic amphibians in general have lower levels of CO_2 in the blood than their terrestrial counterparts. The gas is excreted readily through the skin into the surrounding water, but there are indications that this process is poorly controlled (see Foxon 1964; Boutilier and Toews 1981). The CO_2 buffering capacity of the blood is higher in animals living in dry conditions or in burrows, where the skin loses its value as an excretory site. Induced hypercapnia (excess CO_2)

in bufonid toads (Boutilier et al. 1979) and Typhlonectes (Toews and Macintyre 1978) leads to increased concentrations of bicarbonate; the work on toads suggests a conventional mechanism whereby the bicarbonate ions are produced in erythrocytes and then pass into the plasma via a chloride shift.

Bohr effects across the class are weak compared with most vertebrates (see Sullivan 1974; Boutilier and Toews 1981). This weakening has been achieved through amino-acid substitutions in haemoglobin molecules at a few key positions (Perutz and Brunori 1982). In extreme cases, e.g. Amphiuma, purified haemoglobin shows a large reversed Bohr effect (Bonaventura et al. 1977). Addition of organic phosphate to Amphiuma haemoglobin results in a "normal" Bohr effect and decreased oxygen affinity. Similar effects of organic phosphates on oxygen affinity and on the magnitude (if not the sign) of the Bohr effect have been noted in a variety of amphibians, and are unlikely to be of value to the animals only in a developmental context (see above): a drop in organic phosphate might occur when the environment deteriorates (i.e. becomes hypoxic and/or hypercapnic) so that oxygen extraction can continue.

Thrombocytes
These are the functional equivalents of mammalian platelets, as suggested by similarities in ultrastructure (notably the canalicular network - see earlier section on thrombocytes and Figs. 19-21). Their ability to promote the coagulation of blood has been noted in a number of amphibian genera, including Xenopus, Bufo, Rana and Batrachoceps (Fey 1966; Spurling 1981). Using phase contrast microscopy, Fey (1966) observed that when Xenopus thrombocytes spread themselves flat on the substratum and aggregated, fibrin threads were formed rapidly in the vicinity. The thrombocytes disintegrated progressively after thread formation, whilst other blood cells remained largely intact. Xenopus blood is said to coagulate rapidly over a wide temperature range (see Spurling 1981). However, Ahmad et al. (1979) showed that clotting of R. tigrina blood was retarded when the temperature was lowered, and they suggested that such a change is important in ectotherms since the sluggish blood circulation at low temperature might otherwise increase the risk of intravascular clotting.

The precise roles of amphibian thrombocytes in coagulating and inflammatory reactions have yet to be elucidated: there are obvious difficulties in isolating such cells for detailed physiological studies. It is thus unclear whether their modes of activation or the range and properties of their secretory products are as sophisticated as their counterparts in higher vertebrates.

Some mammalian-type procoagulant factors (e.g. Factor XIII) have been identified in amphibian plasma, others (e.g. Fitzgerald factor) are apparently missing (see Spurling 1981). Comparative studies of the chemical pathways involved have been handicapped by species or class specificity of procoagulant proteins. A fibrinolytic system (Srivastava et al. 1981) and a heparin-dependent anti-coagulant mechanism (Jordan 1983) have been described in some amphibian species, whilst heparin-like substances have been identified in amphibian basophils (see earlier section on granulocyte structure and cytochemistry).

Lymphocytes
The level of immunological sophistication in anuran amphibians approaches that seen in "higher" vertebrates (Table 8), and several distinct lymphocyte subsets with effector and regulatory roles can now be recognised. Recent work in this area has been intensive and has centred on a few favoured models, notably Xenopus. Immunologically, the urodeles remain a more enigmatic group, and the apodans virtually ignored. Most functional studies have been performed on lymphocytes from the spleen and thymus: unlike most other lymphoid tissues in the class they are compact and discrete and show a reasonably clear-cut division of labour (see section on haemopoietic and lymphoid tissues). With a few exceptions (e.g. McKnight et al. 1982), techniques have not been adapted for purifying amphibian blood lymphocytes.

Amphibian antibody-forming cells appear to be the product of stimulated lymphocytes. Studies on anurans have shown that humoral immunity is usually (but not invariably) preceded by intense prolif-erative activity and the appearance of large pyroninophilic cells in lymphocytic areas of the spleen and other secondary lymphoid organs, close to sites of antigen trapping (see Manning and Horton 1982). The end result may be the production of plasma cells, but antibody secretion is not exclusive to these highly differentiated cells. Kraft and Shortman (1972), for example, detected large dividing blast-like antibody-forming cells in the spleen of Bufo marinus early in the response to polymerized Salmonella flagellin, with non-dividing lymphocyte-like cells appearing later. Similarly, antibody-secreting lymphocytes of various sizes as well as plasma cells have been identified in anti-erythrocyte plaques (Ambrosius and Hanstein 1971; see also Turner and Manning 1973). The more mature cells of the antibody-forming series may be under-represented in organs such as spleen, however, due to emigration (Manning and Horton 1982).

In adult anurans, different antigens elicit different classes of antibody. Anti-bacterial and anti-erythrocyte agglutinins, for example, are exclusively high molecular weight IgMs which show plateau-like titres over a prolonged period, whereas viruses and foreign serum proteins elicit IgMs which are later accompanied (but not replaced) by distinct, low-molecular weight antibodies (IgRAA, also described as "IgG" or IgY). In Xenopus, the antibody responses divide along classical lines into thymus-dependent and independent types (Table 9), and experiments using early-thymectomized diploid animals reconstituted with triploid histocompatible T cells show that the role of the latter is an indirect one: plaque-forming cells obtained after immunization with foreign erythrocytes were diploid

(Katagiri et al. 1980). Thus, the antibody-forming cells are clearly
of non-thymic (B cell) origin. The nature of the help given by T
cells is not clear but, as Table 9 illustrates, the ability to secrete
IgG-like antibodies does not depend on their presence, nor are
antigens which induce only IgMs necessarily thymus-independent.

Further evidence for lymphocyte collaboration in anuran antibody
production has come from hapten-carrier experiments. Injection of the
appropriate, specific erythrocyte carrier into Xenopus toadlets two
days before immunizing with trinitrophenylated erythrocytes (TNP-RBC)
increased the numbers of hapten-specific secretory-type rosette-
forming cells in the spleen (Horton et al. 1980). Toadlets
thymectomized as larvae or as young adults (Gruenewald and Ruben 1979)
showed impaired anti-TNP reactivity. Moreover, Blomberg et al. (1980)
have shown that precursors of the anti-hapten producing cells from
Xenopus are adherent to nylon wool, X-ray sensitive and positive for
surface Ig (i.e. B cells), whilst the carrier-reactive (helper) cells
do not adhere to nylon wool, are X-ray resistant and negative for
surface Ig (i.e. T cells).

Table 8 : LYMPHOCYTE FUNCTIONS WITHIN THE AMPHIBIA

Function		Anura	Urodela
Allograft rejection		acute/subacute	chronic
Mixed lymphocyte reactivity		strong	weak
Graft-versus-host reactivity		strong	weak
Cytotoxic T cells		present	?
Lymphokine production (migration inhibition factor)		present	present
Antibodies:	IgM	present	present
	IgRAA	present	present?
	IgA	absent	absent
	IgE	absent	absent
Regulatory cells:	helpers	present	present?
	suppressors	present	present

Two more categories of anti-hapten response have been recognised in Xenopus: a completely thymus-independent response when TNP-LPS is used; and a response to TNP-Ficoll which is thymus-dependent but not amplified by priming with the carrier (Horton et al. 1980). In mammals, both LPS and Ficoll are classified as T-independent antigens, but belong to subgroups 1 and 2 respectively. The latter will only fire off the more mature B cells, and indeed there is some evidence for a T cell contribution (Mosier and Subbarao 1982). Thus in Xenopus, the role played by the thymus in TNP-Ficoll reactivity may be in bringing about the maturation of B cells, rather than provision of carrier-specific helper T cells.

Anurans become immunologically competent as larvae, once the lymphoid system has differentiated. Both IgM and IgRAA can be produced, although the latter is less evident in larvae than in adults (Du Pasquier 1976; Du Pasquier and Haimovich 1976). Experiments on Xenopus larvae using TNP-RBC have shown that carrier-dependent, specific helper function appears during larval development (Ruben et al. 1980). The level of this anti-TNP response increases during metamorphosis: unlike both larvae and adults, substantial numbers of rosette-forming cells can be generated without carrier priming, whilst priming boosts the response beyond larval and adult levels (Ruben et al. 1980; Jones and Ruben 1981). This is unexpected, since according to several reports there is a decline in immunological reactivity during metamorphosis, presumably to allow new "self" antigens to become established. Jones and Ruben (1981) found that the response of Xenopus to TNP-LPS was unaffected by metamorphosis, which made it unlikely that increases in the TNP-RBC response were due to changes in the sensitivity of B cells or in the relative numbers capable of a response. They suggest instead that the emerging lymphocytes with new, adult-type histocompatibility antigens (Du Pasquier et al. 1979) will interact with larval-type "allogeneic" lymphocytes to secrete amplifying signals which provide a substitute for signals from carrier-initiated T helper cells.

The evidence for suppressor lymphocytes acting on anuran B cells is limited. Responses to antigens such as LPS are independent not only of helper T but also suppressor T cells, since antibody titres are neither enhanced nor diminished by early thymectomy (Manning and Collie 1975). When T-dependent antigens are used, suppression tends to be masked by helper effects, but can be demonstrated in Xenopus cell culture if the helper lymphocytes are first depleted (Hsu et al. 1983). Thus if splenic T cells, which provide effective help, are separated from the B cells by passage through nylon wool and replaced by thymic T cells, antibody production is diminished compared with cultures of adherent spleen cells alone (direct help from the thymus is not forthcoming presumably because of the immaturity of the cells concerned). The small numbers of B cells found in amphibian thymuses could, then, be migratory cells coming under a local inhibitory influence in the normal, intact animal.

Ruben et al. (1983) have shown a modest thymic suppression of antibody production in Xenopus organ culture. Haemagglutinin titres produced

by spleen fragments alone after the first week of culture were higher
than when fragments of spleen and thymus from the same animal were co-
cultured. On further incubation, the balance shifted in favour of
helpers. The suppressors were induced by antigen, but their effects
were not always specific. This technique was also used to confirm in
vivo findings (see above) that responses to thymus-independent
antigens are not subject to T suppression.

It is widely believed that larval and adult urodeles, unlike the
anurans, produce antibody of a single class (IgM). However, one study
on hyperimmunized axolotls (Ambystoma mexicanum) does suggest that a
low molecular weight antibody with non-μ heavy chains (IgRAA?) can
also be secreted (Warr et al. 1982). This work needs to be
substantiated.

The question of helper T lymphocytes in the urodeles is also
unsettled. Axolotls thymectomized at 5 weeks of age actually produce
higher levels of circulating antibody (IgM) to horse erythrocytes

Table 9 : T-DEPENDENCE OF ANTIBODY PRODUCTION IN XENOPUS

	Antigen[a]	Intact animals		Thymectomized animals[b]	
		IgM	IgRAA	IgM	IgRAA
T-dependent	(Human gamma globulin	+	+	-	-
	(Sheep erythrocytes	+		-	
T-independent	(Haemocyanin (Limulus)	+	+	+	+
	(Aeromonas salmonicida	+		+	
	(Escherichia coli LPS	+		+	
	(Polyvinylpyrrolidone(PVP)	+		+	

[a]Animals immunized after metamorphosis

[b]Operation performed on young larvae (1 week old)

For references see Tochinai (1976b) and Manning and
Jurd (1981).

compared with intact animals (Charlemagne and Tournefier 1977a), whilst responses of thymectomized Pleurodeles waltlii to Salmonella typhimurium flagellar (H) antigen – another T-dependent antigen in mammals – are only slightly impaired (Charlemagne and Tournefier 1977b). From these reports it appears that if helper cells do exist in urodeles, they are less dependent on the thymus than anuran helper cells are, or they are effective on fewer occasions with respect to the nature and dose of antigen. The increased antibody response in thymectomized axolotls suggests that T suppressors are normally present. Incidentally, more recent work (Tournefier 1982) has distinguished the T suppressors from alloreactive T cells in the axolotl on the basis of differences in sensitivity to hydrocortisone.

Evidence for helper cells in urodeles has also been sought using hapten-carrier (TNP-RBC) techniques. Ruben's work on the newt Notophthalmus viridescens (see Ruben and Edwards 1980) has shown that low-dose priming with the appropriate erythrocyte carrier significantly enhances the number of splenic anti-TNP rosette-forming cells, an effect similar to that seen in anurans and other vertebrates and one which, prima facie, suggests a helper population of some description. Non-secretory, anti-RBC rosettes (helpers?) peaked quickly after immunization, whilst the anti-TNP rosettes appearing later were nearly all of the secretory type (B cells). However, these rosetting cells may not belong to different subsets: Ruben has found that some antigen-binding newt cells have dual (i.e. hapten and carrier) specificity. Using axolotls, Tahan and Jurd (1981) confirmed the need for carrier priming; they further concluded that help in this urodele is provided by T cells, since adult thymectomy severely reduced the numbers of anti-TNP rosette-forming cells. This contrasts with Charlemagne's (1982) report of elevated circulating anti-TNP in the axolotl due to thymectomy (using the same carrier but a longer immunization schedule). A "background" of anti-TNP rosettes and antibodies in unimmunized animals was also claimed, together with increased anti-TNP titres in axolotls immunized only with RBC (Charlemagne and Tournefier 1977a; Charlemagne 1982).

These discrepancies underline the problem of relating numbers of rosette-forming cells to the output of circulatory antibody, and the need for unequivocal identification of anti-hapten reactivity by means of blocking agents. The most telling need, however, is for cell transfer experiments on histocompatible urodeles in vivo and lymphocyte fractionations and recombinations in vitro along the lines of those already performed to such good effect in anurans.

The best evidence for T effector lymphocytes in amphibians has come from studies on alloimmune responses. Hildemann and Haas (1962) showed that the ability of anuran larvae (R. catesbeiana) to recognise and destroy skin allografts coincides with the appearance of small lymphocytes in the blood during development. In an immuno-competent Xenopus, blood capillaries under the allograft dilate and become packed with lymphocytes; these subsequently invade the graft in large numbers and proliferate at the site (Horton and Horton 1975).

The alloreactive cells are known to be of thymic origin since grafts persist on early-thymectomized animals (e.g. Kaye and Tompkins 1983), and triploid T cells used to restore histocompatible diploid thymectomized animals can be recovered from allografted sites (Katagiri et al. 1980). Grafts not invaded by lymphocytes do not degenerate, although the mode of killing in vivo is uncertain: cytotoxic T cells with specificity for alloantigens have been demonstrated in vitro (Bernard et al. 1979), but other leucocyte types are represented in the population invading a graft, albeit in relatively small numbers. The loss of reactivity to allografts which occurs in Xenopus during metamorphosis has been attributed to active suppression by T cells, since their transfer from metamorphosing animals into adults of the same strain prevents rejection of a skin graft differing from the adult host by minor histocompatibility antigens (Du Pasquier and Bernard 1980). By implication, the same mechanism dampens down responses to new "self" antigens, although as was discussed above, this would not extend to postulated secretory signals acting on B cells.

Different T cell populations seem to be responsible for allograft and xenograft rejection, since Xenopus injected with N-methyl-N-nitrosurea retain allografts but not xenografts (Balls et al. 1981). The injected animals sustain permanent damage to their thymic cortex but not the medulla (James et al. 1982). Alloreactive lymphocytes can also be distinguished from helper T cells: sequential thymectomy shows that the thymus is vital only for a short time during development before alloimmune cells manage to seed themselves into the periphery and become independent, whereas the gland retains its influence over antibody production into late larval and adult life (Table 10; see also Gruenewald and Ruben 1979).

Immune responses to grafts of foreign tissue have also been observed in urodeles and apodans. In urodeles, early thymectomy severely impairs the rejection of skin allografts, under circumstances which have minimal effects on antibody production (Charlemagne and Tournefier 1977b). The normal rejection process is, however, much slower than in anurans and this, coupled with poorer graft-versus-host and mixed lymphocyte reactions (Table 8), has led to the suggestion that anurans but not the urodeles (nor apodans?) possess a major histocompatibility complex (MHC) (see Cohen 1980). This poses some fascinating questions about the phylogeny of MHC restriction and the biological importance of the phenomenon: to what extent are T cells in lower vertebrates guided by MHC products? Judging by in vitro tests, both helper and cytotoxic T cells from Xenopus are functionally restricted by the MHC (Bernard et al. 1979;1981), but investigations in vivo have shown good helper-dependent antibody production in thymectomized Xenopus implanted with mismatched thymuses (thymectomy was performed on 4-7 day old larvae; restoration in late larval or early adult life). This may mean that there is no MHC restriction in vivo in the amphibian, that non-thymic elements are responsible for educating the donor T cells, or that T cells were educated very early in the host thymus and moved out of the gland at a stage before thymectomy was performed, needing only to complete their maturation under hormonal influence from the donor thymus (Du Pasquier and Horton 1982; Nagata and Cohen 1984). Recent work by Flajnik et al. (1985) suggests that the thymus does have a part to play in educating helper T cells, but not a critical one: chimaeric Xenopus derived by grafting an anterior embryonic portion (containing thymic anlagen) onto an MHC-mismatched posterior portion (delivering lymphoid stem cells) and then immunized as adults showed no impairment of IgM production, but "IgG" (IgRAA) titres were decreased or delayed. An intensification of such experiments on MHC-defined Xenopus is anticipated, and should help significantly our perception of T cell education during development: these amphibians have the advantage of simplicity over currently available mammalian models, and are much closer to the physiological "norm" (see Cohen and Warr 1983).

Table 10 : EFFECTS OF SEQUENTIAL THYMECTOMY ON ALLOGRAFT

REACTIVITY Vs. T-DEPENDENT ANTIBODY PRODUCTION

IN XENOPUS

Thymectomy at		Skin allografts[a]	Antibody response to sheep RBC[b]
Age (days)	Larval stage no.		
3	42	1°: long term acceptance 2°: long-term acceptance (usually)	not done
7/8	47/8	1°: long-term acceptance (usually) 2°: normal rejection (usually)	none
15/16	51	1°: normal rejection 2°: normal rejection	none
30	54/5		none

a)   see Kaye and Tompkins (1983)

b)   see Horton et al. (1977)

A number of other immune mechanisms which have been identified in amphibians are probably lymphocyte-dependent, but have not been explored in depth. These include immediate-type hypersensitivity reactions (see section on granulocyte functions) and the production of lymphokine-like factors (see functions of monocytes and macrophages). Antibody-dependent cell-mediated cytotoxicity (ADCC) has also been reported in urodeles and anurans (Jurd and Doritis 1977). Both orders are known to be susceptible to a wide range of tumours, but the immunological aspects of neoplasia have been examined in only a few cases, notably the lymphosarcomas of Xenopus and Ambystoma (see Balls and Ruben 1976) and the renal adenocarcinoma of R. pipiens (Rollins and Cohen 1981). The work on R. pipiens suggests that T-dependent immunity plays no major role in preventing development of the tumour. Susceptibility of Xenopus to a nematode infection of the skin (Capillaria xenopodia) does seem to be associated with a T cell deficiency (Nagata and Cohen 1984).

Finally, various immunosuppressive reagents have been administered to newts in an attempt to establish a link between lymphocyte activity and the ability of an animal to regenerate new limb tissues (e.g. Sicard and Laffond 1983). This is an intriguing variation on the lymphocyte theme; however, the most effective reagents used to date are of mammalian origin and specificity, and a convincing correlation between immunological and regenerative capacities has yet to be made.

Monocytes and macrophages
The ability of large mononuclear leucocytes to engulf foreign material has been recognised for many years. Jordan (1938) reported that India ink injected into the dorsal lymph sac of R. pipiens evoked a more active phagocytic response by monocytes than other blood cells. The monocytes are also the first leucocytes to appear in the blood during development (Hildemann and Haas 1962; Foxon 1964) and are known to migrate through capillary walls to ingest material in the tissues, i.e. become functioning macrophages (see Noble 1931). The phagocytes of the blood-filtering organs may be of this lineage.

Killed staphylococci injected into the dorsal lymph sac of R. pipiens were taken up by monocytes and polymorphs, and caused a dramatic increase in monocyte numbers (Everly and Hanson 1965; see also Noble 1931). Similarly, Clothier (1972) found that Xenopus infected with live mycobacteria accumulated large numbers of macrophages in the target tissues and lymph. As the disease progressed, the percentage of macrophages capable of ingesting the bacteria when challenged rose dramatically. Clothier's work correlates with the in vitro demonstration by Drössler and Ambrosius (1972) of macrophage migration inhibition factors in toads (Bufo bufo) following BCG sensitization. The source and molecular properties of these lymphokine-like factors are not known in detail, but studies on R. temporaria have shown that the inhibitory effect is specifically induced by antigen, is brought about by soluble (non-antibody) factors of molecular weight 27-50,000 which are released rapidly into the medium after antigenic challenge, and can be induced also by a T cell mitogen, Con A (Rimmer and Gearing 1980; Gearing and Rimmer 1981, 1985).

It seems likely that amphibian macrophages are not only acting as effector cells, but are involved also in the inductive events leading to antibody production. Turner (1970) showed that colloidal carbon injected 4-24 h before sheep RBC (a T-dependent antigen) caused increased levels of circulating haemagglutinin in Xenopus. In the newt Notophthalmus viridescens, on the other hand, the number of rosette-forming cells in the spleen following injection of horse RBC was diminished when carbon was administered 1-4 days before the antigen, and preliminary tests suggested that serum antibody titres too were diminished (Ruben et al. 1981). The different outcome of these two experiments probably reflects differences in the extent of "blockade" achieved. Further work on newts (Ruben and Stack 1982; see also Ruben 1984) has shown that responses to DNP-dextran (regarded as a group 2 T-independent antigen in mammals) were also affected by carbon, but those to TNP-LPS (a group 1 T-independent antigen) were not. Thus, only certain categories of antigen appear to be "processed" by macrophages during antibody induction. Moreover, the failure of newts to secrete antibodies to soluble (c.f. particulate) antigens of the "T-dependent" type seems to stem from a failure in uptake by phagocytes.

Granulocytes

Amphibian neutrophils, like the monocytes, are migratory cells with a capacity to phagocytose inert particles (Jordan 1938) and microbial cells (Everly and Hanson 1965), but they begin to function later in development. The metabolic studies of David and McMullen (1972) on Amphiuma have shown that only about two thirds of the neutrophils in the blood are potentially capable of amoeboid movement, diapedesis and phagocytosis; the rest form a reserve population of vegetative cells. This state of readiness is not related to the cells' degree of maturity as judged by morphological criteria.

Eosinophils, although closely related to the neutrophils, show a poorer ability to phagocytose small particulate material; instead, they appear to be implicated in responses to metazoan parasites. Mitchell (1982) fed ranid frogs with metacercarial cysts of Gorgoderina vitelliloba and found that juvenile flukes invading the kidneys of newly metamorphosed froglets were attacked by eosinophils which adhered to and dissolved the tegument; this presumably killed the flukes, because the infections did not persist. Similar experiments on mature hosts had a different outcome: here the response involved neutrophils and lymphocytes but very few eosinophils, and the parasites did not sustain any damage. Mitchell further suggested that the neutrophils were used extensively as food by the parasites! Elkan (1976), in his survey of amphibian diseases, was unimpressed by the anti-parasite role of eosinophils as well as neutrophils: larval trematodes were frequently found in diverse organs, with little evidence of effective infiltrations by defensive leucocytes; the most obvious responses were a rise in the number of eosinophils in the blood, and attempts by tissues around the parasites to wall them off. In nematode infections, large numbers of eosinophils were seen surrounding worms or filling lymphatic sheaths, without appearing to have much effect on the worms' survival.

The roles of amphibian mast cells and basophils have not been established. These cells are well placed around the body to monitor invasions by parasites and to recruit eosinophils to the defence, but there is, as yet, no evidence for such an interaction. Moreover, there have been very few successful demonstrations in amphibians of immediate type hypersensitivity, a phenomenon which shows close mechanistic similarities with anti-helminth reactions in higher vertebrates. Cohen et al. (1971) gave R. pipiens repeated injections of killed Salmonella typhi subcutaneously at 3-4 day intervals and later challenged with a soluble form of the antigen; in several cases this caused an immediate but usually non-fatal flaccid paralysis, but no particular cell type or organ could be implicated in the reaction. Cytochemical findings (see section on granulocyte structure and cytochemistry) suggest that the chemical armoury of mast cells and basophils is incomplete: heparin or heparin-like mucosubstances are present; serotonin has been detected in some species but not others (including R. pipiens); and histamine is absent. Reite (1972) reviewed the pharmacological effects of histamine on amphibians and concluded that it has a negligible influence on their cardiovascular system, although some changes in gastric secretion and extravascular smooth muscle were noted. The identity of the mediator(s) responsible for the histamine-like shock reported by Cohen et al. (1971) thus remains a mystery.

Studies on mammalian granulocytes have made enormous strides in recent years through techniques of cell separation, functional tests on purified cells in vitro, and chemical analyses of isolated granules. These techniques have not been adapted for amphibians, and we are forced to fall back on descriptive ultrastructural studies and traditional cytochemical tests. These have been a valuable source of comparative data, but at best provide only clues about the cells' functional capabilities. Do the known - sometimes marked - differences between species in respect of granule substructure or the presence of a particular enzyme really reflect differences in a cell's killing power, for example?

A detailed, systematic appraisal of membrane receptors on these amphibian leucocytes is also awaited. In mammals, the activities of all three granulocyte types are triggered by interaction with immunoglobulin and complement components. The key immunoglobulin classes in this respect are IgG and IgE, but these are both late evolutionary acquisitions, and both appear to be absent in all amphibians. This may mean increased versatility of IgM at the amphibian level (or substitution by IgRAA), but it seems more likely that the roles of granulocytes and lymphocyte products are intertwined to a smaller extent than in higher vertebrates, with perhaps a more prominent role for complement.

CONCLUDING REMARKS

Frogs may not spring immediately to one's mind as a key to medical progress nor, despite the efforts of certain canning companies and handbag manufacturers, can they claim the economic significance of, say, fishes or chickens. On the other hand, their peculiar ability to exploit both aquatic and terrestrial environments continues to intrigue biologists of many persuasions, whilst experimental studies on the sources and properties of their blood-forming stem cells and the heterogeneity of their lymphocytes have illustrated the advantages of free-living embryonic and larval stages.

The partial move to land is reflected in a complex and varied arrangement of the amphibian circulatory system to allow for oxygen uptake at different respiratory surfaces, and there are correspondingly wide differences among species and stages in the red blood cells and their contents. Although the structural and molecular features of amphibian erythrocytes can often be correlated with particular respiratory needs, the cells' ability to bind and release oxygen should ideally be linked also to conditions at key sites in vivo. This is a difficult goal in small animals and where there is variable mixing and distribution of blood.

The move to land also appears to have brought about changes in the siting of haemopoietic and lymphoid tissues. Bone marrow is seen for the first time in this group, and structures resembling lymph nodes have been identified in some species. As in "higher" vertebrates, blood cells undergoing maturation in the haemopoietic tissues of amphibians are derived from an immigrant population of precursors. In frogs, these precursors can be traced back to the embryonic kidneys (or adjacent tissue), but the limits of their developmental potential and the nature of controlling signals need to be examined further.

To immunologists, the toad Xenopus has become the laboratory mouse of the poikilotherm world, concerted efforts having been made in recent years to unravel the developmental, structural and functional properties of its lymphocytes. Immunoglobulin-producing lymphocytes of non-thymic origin (i.e. B cells) have been identified, there is evidence of functional diversity within the thymus-dependent (T cell) population - helpers, suppressors and different kinds of effector have all been demonstrated - and a major histocompatibility complex (MHC) has been described. There seems to be a strong case for intensifying studies on MHC-defined Xenopus (and, for that matter, on chickens) for comparison with the picture that is emerging in mammals, if we are to appreciate fully the physiological significance of the MHC.

The benefits of intensive immunological studies on readily available, genetically defined animals are obvious, but there are dangers: is Xenopus representative, or is it the odd toad out? Studies on other anurans (notably ranid frogs) show a degree of immunological sophistication similar to and perhaps surpassing that of Xenopus, but the urodeles present a very different, albeit blurred, picture. There is apparently no MHC, their immunoglobulin isotypes may be more limited, the existence of helper T cells has been questioned, and the blood-forming and lymphoid tissues have their own characteristic locations and organization.

In our efforts to unravel the functions and interrelationships of amphibian lymphocytes, the roles of other leucocytes have gone almost unnoticed. This is unfortunate, since there are significant periods in development when the lymphocytic system is not fully functional, yet the animals do not seem particularly susceptible to infection. This may apply to the metamorphic period, and it certainly applies to the period of development between the emergence of the larva from its protective jelly coat and the acquisition of a complete battery of lymphocytes, a time span of a few days or several weeks depending on species. It is also worth noting that thymectomized amphibians have been known to survive for long periods after the operation. The existing work on amphibian granulocytes is mainly of a descriptive nature, and more functional studies are needed: how do the observed cytochemical and ultrastructural variations translate into functional differences and limitations among cell types and species, and to what extent are the cells' functions limited by the absence of specialized immunoglobulin isotypes? Moreover, are macrophage functions bound up as closely with lymphocytes as in "higher" vertebrates?

A fuller elucidation of leucocyte function in amphibians is likely to depend on the successful adaptation of techniques used initially in "higher" vertebrates for obtaining and testing the different cell types. Some disappointment is inevitable: cell techniques do not always transfer easily from species to species, let alone from class to class (cell separation based on density gradient centrifugation, for example). However, other approaches, notably the use of monoclonal antibodies for preparative and analytical work, hold great promise for cell studies in the "lower" vertebrates. Finally, special technical advantages offered by amphibian cells in particular (the sheer size of urodele leucocytes, for example, or the abundance of basophils in the blood of certain species) should not be overlooked.

ACKNOWLEDGEMENTS
I am indebted to Sue Davies for her patient and painstaking secretarial assistance, to Bob Moore for help with illustrations, and to Raimundo Santos Barros, Pietro Cenini, Robert Havard and Len Jones for help with translations. The various colleagues who were approached for photomicrographs (see captions) are also thanked for their support and generosity.

REFERENCES

Ahmad, N., Dube, B., Agarwal, G.P. & Dube, R.K. (1979). Comparative studies of blood coagulation in hibernating and non-hibernating frogs (Rana tigrina). Thrombos. Haemostas. (Stuttg.),42, 959-64.

Aleporou, V. & Maclean, N. (1978). Feedback inhibition of erythro-poiesis induced in anaemic Xenopus. J. Embryol. Exp. Morphol.,43, 221-31.

Ambrosius, H. & Hanstein, R. (1971). Beiträge zur Immunbiologie poikilothermer Wirbeltiere VI. Die Dynamik Antikörper produzierender Zellen in den lymphoiden Organen des Wasserfrosches Rana esculenta L. Acta Biol. Med. Germ.,27, 771-82.

Araki, T., Okazaki, T., Kajita, A. & Shukuya, R. (1974). Polymerization of oxygenated and deoxygenated bullfrog hemoglobins. Biochem. Biophys. Acta,351, 427-36.

Baldwin, W.M. & Cohen, N. (1981). A primitive dendritic splenocyte in Xenopus laevis with morphological similarities to Reed-Sternberg cells. In Aspects of Developmental and Comparative Immunology I, ed. J.B. Solomon, pp. 179-82. Oxford: Pergamon.

Balls, M., Clothier, R.H., Foster, N.J. & Knowles, K.R. (1981). Skin allograft and xenograft responses in NMU-treated Xenopus laevis. In Aspects of Developmental and Comparative Immunology I, ed. J.B. Solomon, pp. 499-500. Oxford: Pergamon.

Balls, M. & Ruben, L.N. (1976). Phylogeny of neoplasia and immune reactions to tumours. in Comparative Immunology, ed. J.J. Marchalonis, pp. 167-208. Oxford: Blackwell.

Barber, D.L. & Westermann, J.E.M. (1978). Occurrence of the periodic acid-Schiff positive granular leucocyte (PAS-GL) in some fishes and its significance. J. Fish Biol.,12, 35-43.

Benbassat, J. (1970). Erythroid cell development during natural amphibian metamorphosis. Dev. Biol.,21, 557-83.

Benbassat, J. (1974). The transition from tadpole to frog haemoglobulin during natural amphibian metamorphosis. II Immunofluorescence studies. J. Cell. Sci.,16, 143-56.

Bernard, C.C.A., Bordmann, G., Blomberg, B. & Du Pasquier, L. (1979). Immunogenetic studies on the cell-mediated cytotoxicity in the clawed toad Xenopus laevis. Immunogenetics,9, 443-54.

Bernard, C.C.A., Bordmann, G., Blomberg, B. & Du Pasquier, L. (1981). Genetic control of T helper cell function in the clawed toad Xenopus laevis. Eur. J. Immunol.,11, 151-5.

Bertolini, B. & Monaco, G. (1976). The microtubule marginal band of the newt erythrocyte. J. Ultrastructr. Res.,54, 59-67.

Bleicher, P.A. & Cohen, N. (1981). Monoclonal anti-immunoglobulin M can separate T cell from B cell proliferative responses in the frog, Xenopus laevis. J. Immunol.,127, 1549-55.

Bleicher, P.A., Rollins-Smith, L.A., Jacobs, D.M. & Cohen, N. (1983). Mitogenic responses of frog lymphocytes to crude and purified preparations of bacterial lipopolysaccharide (LPS). Dev. Comp. Immunol., 7, 483-97.

Blomberg, B., Bernard, C.C.A. & Du Pasquier, L. (1980). In vitro
     evidence for T-B lymphocyte collaboration in the clawed toad,
     Xenopus laevis. Eur. J. Immunol., 10, 869-76.
Bonaventura, C., Sullivan, B., Bonaventura, J. & Bourne, S. (1977).
     Anion modulation of the negative Bohr effect of haemoglobin
     from a primitive amphibian. Nature, 265, 474-6.
Boschini Filho, J. (1979). Elementos figurados do sangue periferico
     de Typhlonectes compressicaudus. Descricao das
     formas eritrocitarias aos niveis optico e eletronico. Bol.
     Fisiol. Animal, Univ. S. Paulo, 3, 33-8.
Boschini Filho, J. (1980). Caracterizacao do trombocito do sangue
     periferico de Typhlonectes compressicaudus (Amphibia-Apoda)
     pela reacao citoquimica do P.A.S. Bol. Fisiol. Animal,
     Univ. S. Paulo, 5, 75-80.
Boutilier, R.G., Randall, D.J., Shelton, G. & Toews, D.P. (1979).
     Acid-base relationships in the blood of the toad Bufo
     marinus. I. The effects of environmental CO_2. J. Exp.
     Biol., 82, 331-44.
Boutilier, R.G. & Toews, D.P. (1981). Respiratory properties of
     blood in a strictly aquatic and predominantly skin-breathing
     urodele, Cryptobranchus alleganiensis. Respir.
     Physiol., 46, 161-76.
Brown, G.L., Rutledge, R.G. & Neelin, J.M. (1981). Anuran
     erythrocytes and liver both contain satellite histone HI[s].
     Life Sci., 28, 2993-9.
Brown, R.D. (1981). Identification and properties of terminal
     deoxynucleotidyl transferase from Xenopus laevis embryos
     prior to the development of the thymus. J. Biol.
     Chem., 256, 3627-9.
Broyles, R.H., Dorn, A.R., Maples, P.B., Johnson, G.M., Kindell, G.R.
     & Parkinson, A.M. (1981). Choice of hemoglobin type in
     erythroid cells of Rana catesbeiana. In Hemoglobins in
     Development and Differentiation, eds. G. Stamatoyannopoulos
     and A.W. Nienhuis, pp. 179-91. New York: A.R. Liss.
Bruning, H., Cochran, M.D., Suhar, T., Brown, R.D. & Armentrout, R.W.
     (1981). Early events in the development of the immune
     system of Xenopus laevis: characterization of lymphocytes in
     2-day-old embryos. Dev. Comp. Immunol., 5, 607-16.
Campbell, F.R. (1969). Electron microscopic studies on
     granulopoiesis in the slender salamander. Anat. Rec., 163,
     427-42.
Campbell, F.R. (1970). Ultrastructure of the bone marrow of the
     frog. Amer. J. Anat., 129, 329-56.
Cannon, M.S. & Cannon, A.M. (1979). The blood leucocytes of Bufo
     alvarius: a light, phase-contrast and histochemical study.
     Can. J. Zool., 57, 314-22.
Cardellini, P. & Sala, M. (1978). Metamorphic variations in the
     hemoglobins of Hyla arborea L. Comp. Biochem. Physiol., 61B,
     21-4.
Cardellini, P. & Sala, M. (1979). Hemoglobin transition in the toad,
     Bufo viridis, during development. Boll. Zool., 46, 51-6.

Cardellini, P. & Sala, M. (1983). Developmental time of the
    hemoglobin transition in the anuran Bombina orientalis.
    Comp. Biochem. Physiol.,75B,, 259-62.
Carver, F.J. & Meints, R.H. (1977). Studies of the development of
    frog hemopoietic tissue in vitro. I. Spleen culture assay
    of an erythropoietic factor in anemic frog blood. J. Exp.
    Zool.,201, 37-46.
Charlemagne, J. (1982). Helper function in Ambystoma mexicanum - a
    comment. Dev. Comp. Immunol.,6, 181-3.
Charlemagne, J. & Tournefier, A. (1975). Cell surface
    immunoglobulins of thymus and spleen lymphocytes in urodele
    amphibian Pleurodeles waltlii (Salamandridae). In
    Immunologic Phylogeny, eds. W.H. Hildemann and A.A.
    Benedict, Adv. Exp. Med. Biol.,64, pp. 251-6. New York:
    Plenum.
Charlemagne, J. & Tournefier, A. (1977a). Anti-horse red blood cells
    antibody synthesis in the Mexican axolotl (Ambystoma
    mexicanum). Effect of thymectomy. In Developmental
    Immunobiology, eds. J.B. Solomon and J.D. Horton, pp. 267-
    76. Amsterdam: Elsevier.
Charlemagne, J. & Tournefier, A. (1977b). Humoral response to
    Salmonella typhimurium antigens in normal and thymectomized
    urodele amphibian Pleurodeles waltlii Michah. Eur. J.
    Immunol.,7, 500-2.
Chegini, N., Aleporou, V., Bell, G., Hilder, V.A. & Maclean, N.
    (1979). Production and fate of erythroid cells in anaemic
    Xenopus laevis. J. Cell Sci.,35, 403-16.
Chin, K.N. & Wong, W.C. (1977). Some ultrastructural observations on
    the intestinal mucosa of the toad (Bufo melanostictus). J.
    Anat.,123, 331-40.
Chiu, H. & Lagunoff, D. (1972). Histochemical comparison of
    vertebrate mast cells. Histochem. J.,4, 135-44,
Clothier, R.H. (1972). The histopathology of a lymphoreticular
    disease in Xenopus laevis. Ph.D. thesis, University of East
    Anglia.
Cohen, N. (1980). Salamanders and the evolution of the major
    histocompatibility complex. In Contemporary Topics in
    Immunobiology, Vol. 9, eds. J.J. Marchalonis and N. Cohen,
    pp. 109-40. New York: Plenum.
Cohen, N. & Warr, G.W. (1983). A cold-blooded look at T-cell
    education. Immunol. Today,4, 70-1.
Cohen, S.G., Sapp, T.M. & Shaskas, J.R. (1971). Phylogeny of
    hypersensitivity I. Anaphylactic responsiveness of the
    frog, Rana pipiens.   J. Allergy 47, 121-30.
Cohen, W.D. (1982). The cytomorphic system of anucleate nonmammalian
    erythrocytes. Protoplasma,113, 23-32.
Condo, S.G., Giardina, B., Belleli, A., Lunadei, M., Ferracin, A. &
    Brunori, M. (1983). Comparative studies of hemoglobins
    fron newts (Triturus cristatus, T. vulgaris, T. alpestris).
    A kinetic approach. Comp. Biochem. Physiol., 74A, 545-8.
Condo, S.G., Giardina, B., Lunadei, M., Ferracin, A. & Brunori, M.
    (1981). Functional properties of hemoglobins from Triturus
    cristatus. Eur. J. Biochem.,120, 323-7.

Cooper, E.L. (1973). The thymus and lymphomyeloid system in poikilothermic vertebrates. Contemp. Topics in Immunobiol.,2, 13-38.

Cooper, E.L. (1976). Immunity mechanisms. In Physiology of the Amphibia Vol.III, ed. B. Lofts, pp. 164-272. New York: Academic Press.

Cooper, E.L., Klempau, A.E., Ramirez, J.A. & Zapata, A.G. (1980). Sources of stem cells in evolution. In Development and Differentiation of Vertebrate Lymphocytes, ed. J.D. Horton, pp. 3-14. Amsterdam: Elsevier.

Cooper, E.L. & Schafer, D.W. (1970). Bone marrow restoration of transplantation immunity in the leopard frog Rana pipiens. Proc. Soc. Exp. Biol. Med.,135, 406-11.

Cowden, R.R. (1965). Quantitative and qualitative cytochemical studies on the Amphiuma basophil leucocyte. Z. Zellforsch.,67, 219-33.

Cowden, R.R. & Dyer, R.F. (1971). Lymphopoietic tissue and plasma cells in amphibians. Amer. Zool.,11, 183-92.

Cowden, R.R., Narain, A. & Beveridge, G.C. (1964). Basophil leukocytes and tissue mast cells in newt, Diemyctylus viridescens. Acta Haematol.,32, 250-5.

Csaba, G., Olah, I. & Kapa, E. (1970). Phylogenesis of mast cells. II. Ultrastructure of the mast cells in the frog. Acta Biol. Acad. Sci. Hung.,21, 255-64.

Curtis, S.K., Cowden, R.R. & Nagel, J.W. (1979a). Ultrastructure of the bone marrow of the salamander Plethodon glutinosus (Caudata : Plethodontidae). J. Morphol.,159, 151-84.

Curtis, S.K., Cowden, R.R. & Nagel, J.W. (1979b). Ultrastructural and histochemical features of the thymus glands of the adult lungless salamander, Plethodon glutinosus (Caudata : Plethodontidae). J. Morphol.,160, 241-74.

Daimon, T., Mizuhira, V. & Uchida, K. (1979). Fine structural distribution of the surface-connected canalicular system in frog thrombocytes. Cell Tissue Res.,201, 431-40.

Dasgupta, S. (1962). Induction of triploidy by hydrostatic pressure in the leopard frog, Rana pipiens. J. Exp. Zool.,151, 105-21.

David, G.B. & McMullen, J.M. (1972). Quantitative cytochemical observations on the control of respiration in polymorphonuclear neutrophil leucocytes of Amphiuma tridactylum. J. Cell Sci.,10, 719-47.

Deparis, P. (1968). Recherches experimentales sur l'origine des erythrocytes de l'individu adulte par la methode des greffes embryonnaires d'ilots sanguins diploides et triploides chez Pleurodeles waltlii (Amphibien Urodele). Ann. Embryol. Morphol.,1, 351-9.

Deparis, P. & Jaylet, A. (1976). Thymic lymphocyte origin in the newt, Pleurodeles waltlii studied by embryonic grafts between diploid and tetraploid embryos. Ann. Immunol.,127C, 827-31.

Deparis, P. & Jaylet, A. (1984). The role of endoderm in blood cell ontogeny in the newt Pleurodeles waltlii. J. Embryol. Exp. Morphol.,81, 37-47.

Dickinson, K.E.J. & Nahorski, S.R. (1981). Atypical characteristics of frog (Rana pipiens) and chick erythrocyte beta-adrenoceptors. Eur. J. Pharmacol.,74, 43-52.

Dieterlen-Lièvre, F., Beaupain, D., Lassila, O., Toivonen, P. & Martin, C. (1980). Embryonic origin of lymphoid stem cells investigated in avian inter- and intra-specific chimeras. In Development and Differentiation of Vertebrate Lymphocytes, ed. J.D. Horton, pp. 35-43. Amsterdam: Elsevier.

Dodd, M.H.I. & Dodd, J.M. (1976). The biology of metamorphosis. In Physiology of the Amphibia Vol.III,ed. B. Lofts, pp. 467-600. New York: Academic Press.

Donnelly, N., Manning, M.J. & Cohen, N. (1976). Thymus dependency of lymphocyte subpopulations in Xenopus laevis. In Phylogeny of Thymus and Bone Marrow-Bursa Cells, eds. R.K. Wright and E.L. Cooper, pp. 133-41. Amsterdam: Elsevier.

Dorn, A.R. & Broyles, R.H. (1982). Erythrocyte differentiation during the metamorphic hemoglobin switch of Rana catesbeiana. Proc. Nat. Acad. Sci. USA,79, 5592-6.

Drössler, K. & Ambrosius, H. (1972). Spezifische zellvermittelte Immunität bei Froschlurchen. II Verzögerte Überempfindlichkeit bei der Erdkröte (Bufo bufo L.) gegen BCG. Acta Biol. Med. Germ.,29, 441-5.

Du Pasquier, L. (1976). Amphibian models for study of the ontogeny of immunity. In Comparative Immunology, ed. J.J. Marchalonis, pp. 390-418. Oxford: Blackwell.

Du Pasquier, L. & Bernard, C.C.A. (1980). Active suppression of the allogeneic histocompatibility reactions during the metamorphosis of the clawed toad Xenopus. Differentiation,16, 1-8.

Du Pasquier, L., Blomberg, B. & Bernard, C.C.A. (1979). Ontogeny of immunity in amphibians : changes in antibody repertoires and appearance of adult major histocompatibility antigens in Xenopus laevis. Eur. J. Immunol.,9, 900-6.

Du Pasquier, L. & Haimovich, J. (1976). The antibody response during amphibian ontogeny. Immunogenetics,3, 381-1.

Du Pasquier, L. & Horton J.D. (1982). Restoration of antibody responsiveness in early thymectomized Xenopus by implantation of major histocompatibility complex - mismatched larval thymus. Eur. J. Immunol.,12, 546-51.

Du Pasquier, L., Weiss, N. & Loor, F. (1972). Direct evidence for immunoglobulins on the surface of thymus lymphocytes of amphibian larvae. Eur. J. Immunol.,2, 366-70.

Edwards, B.F. & Ruben, L.N. (1982). Aspects of amphibian immunity. In Animal Models of Immunological Processes, ed. J.B. Hay, pp. 255-86. London: Academic Press.

Eipert, E.F., Klempau, A.E., Lallone, R.L. & Cooper, E.L. (1979). Bone marrow as a major lymphoid organ in Rana. Cell Immunol.,46, 275-80.

Elkan, E. (1976). Pathology in the Amphibia. In Physiology of the Amphibia Vol III, ed. B. Lofts, pp. 273-314. New York: Academic Press.

Elli, R., Giuliani, A., Tentori, L., Chiancone, E. & Antonini, E. (1970). The hemoglobin of amphibia X. Sedimentation behaviour of frog, triton and axolotl hemoglobins. Comp. Biochem. Physiol.,36, 163-71.

Etkin, W. (1964). Metamorphosis. In Physiology of the Amphibia, ed. J.A. Moore, pp. 427-68. New York: Academic Press.

Everly, M.E. & Hanson, R.J. (1965). Bacterial phagocytosis in the dorsal lymph sac of Rana pipiens. Proc. Indiana Acad. Sci.,75, 56.

Fey, F. (1962). Hamatologische Untersuchungen an Xenopus laevis Daudin. I. Die morphologie des Blutes mit einigen vergleichenden Betrachtungen bei Rana esculenta und R. temporaria. Morph. Jb.,103, 9-20.

Fey, F. (1963). Untersuchungen zur vergleichenden Hamatologie niederer Wirbeltiere. Folia Haematol.,81, 21-9.

Fey, F. (1965a). Hamozytologische Untersuchungen nach Splenektomie an Xenopus - Froschen. Acta Biol. Med. Germ.,14, 417-22.

Fey, F. (1965b). Hamatologische Untersuchungen der blutbildenen Gewebe niederer Wirbeltiere. Folia Haematol.,84, 122-46.

Fey, F. (1966). Vergleichende Hamozytologie niederer Vertebraten. II Thrombozyten. Folia Haematol.,85, 205-17.

Fey, F. (1967a). Vergleichende Hamozytologie niederer Vertebraten. III Granulozyten. Folia Haematol.,86, 1-20.

Fey, F. (1967b). Vergleichende Hamozytologie niederer Vertebraten, IV Monozyten, Plasmozyten, Lymphozyten. Folia Haematol.,86, 133-47.

Flajnik, M.F., Du Pasquier, L. & Cohen, N. (1985). Immune responses of thymus/lymphocyte embryonic chimeras : studies on tolerance and major histocompatibility complex restriction in Xenopus. Eur. J. Immunol.,15, 540-47.

Forman, L.J. & Just, J.J. (1981). Cellular quantitation of hemoglobin transition during natural and thyroid-induced metamorphosis of the bullfrog Rana catesbeiana. Gen. Comp. Endocrinol.,44, 1-12.

Foxon, G.E.H. (1964). Blood and respiration. In Physiology of the Amphibia, ed. J.A. Moore, pp. 151-209. London: Academic Press.

Futamura, M., Terashi, Y., Okazaki, T. & Shukuya, R. (1982). Haemoglobin transition in early developmental stages of the tadpole, Rana catesbeiana. Biochem. Biophys. Acta.,704, 37-42.

Gahlenbeck, H. & Bartels, H. (1970). Blood gas transport properties in gill and lung forms of the axolotyl (Ambystoma mexicanum). Respir. Physiol.,9, 175-82.

Gambino, J., Gavin, R.H. & Eckhardt, R.A. (1981). Studies on Xenopus erythrocyte cytoskeletons and cytonuclear skeletons. J. Cell Biol.,91, 331A.

Gangwar, P.C. & Untawala, G.G. (1971). Peroxidase activity of leucocytes in different animals and man. Ind. J. Physiol. Pharmacol.,15, 15-9.

Garavini, C. (1981). Nonspecific acid-esterase activity in lymphoid cells of Bufo bufo. Experientia, 37, 516-7.

Garlick, R.L., Davis, B.J., Farmer, M., Fyhn, H.J., Fyhn, U.E.H., Noble, R.W., Powers, D.A., Riggs, A. & Weber, E. (1979). A fetal-maternal shift in the oxygen equilibrium of hemoglobin from the viviparous caecilian, Typhlonectes compressicauda. Comp. Biochem. Physiol.,62A, 239-44.

Gearing, A.J.H. & Rimmer, J.J. (1981). Leucocyte migration inhibition in Rana temporaria following challenge with thymus-dependent antigens. In Aspects of Developmental and Comparative Immunology I, ed. J.B. Solomon, pp. 509-10. Oxford: Pergamon.

Gearing, A.J.H. & Rimmer, J.J. (1985). Amphibian lymphokines II. Migration inhibition factor produced by antigenic and mitogenic stimulation of amphibian leukocytes. Dev. Comp. Immunol.,9, 291-300.

Gibbs, E.L., Nace, G.W. & Emmons, M.B. (1971). The live frog is almost dead. Bioscience,21, 1027-34.

Goldschneider, I., Metcalf, D., Battye, F. & Mandel, T. (1980). Analysis of rat hemopoietic cells on the flourescence-activated cell sorter. I. Isolation of pluripotent hemopoietic stem cells and granulocyte-macrophage progenitor cells. J. Exp. Med.,152, 419-37.

Goldstine, S.N., Manickavel, V. & Cohen, N. (1975). Phylogeny of gut-associated lymphoid tissue. Amer. Zool.,15, 107-18.

Gouchi, H. (1982). Ultrastructure of eosinophil granules of bullfrogs, Rana catesbeiana. J. Med. Soc. Toho Univ.,29, 9-16.

Grasso, J.A. (1973a). Erythropoiesis in the newt, Triturus cristatus Laur. I. Identification of the erythroid precursor cell. J. Cell Sci.,12, 463-89.

Grasso, J.A. (1973b). Erythropoiesis in the newt, Triturus cristatus Laur. II. Characteristics of the erythropoietic process. J. Cell Sci.,12, 491-523.

Gruenewald, D.A. & Ruben, L.N. (1979). The effect of adult thymectomy upon helper function in Xenopus laevis, the South African clawed toad. Immunology,38, 191-4.

Guillet, F. & Tournefier, A. (1981). Lactate dehydrogenase isoenzyme pattern as a marker of lymphocyte populations in the axolotl (Ambystoma mexicanum). Dev. Comp. Immunol.,5, 617-28.

Hadji-Azimi, I. & Schwager, J. (1982). Biochemical studies of Xenopus laevis lymphocyte surface immunoglobulins. Dev. Comp. Immunol.,6, 703-17.

Hazard, E.S. & Hutchison, V.H. (1982). Distribution of acid-soluble phosphates in the erythrocytes of selected species of amphibians. Comp. Biochem. Physiol.,73A, 111-24.

Hightower, J.A. & Haar, J.L. (1975). A light and electron microscopic study of myelopoietic cells in the perihepatic subcapsular region of the liver in the adult aquatic newt, Notophthalmus viridescens. Cell Tissue Res.,159, 63-71.

Hildemann, W.H. & Haas, R. (1962). Developmental changes in leukocytes in relation to immunological maturity. In Mechanisms of Immunological Tolerance, eds. M. Hasek, A. Langerova and M. Vojtiskova, pp. 35-49. Prague: Publishing House of the Czechoslovak Academy of Science.

Hollyfield, J.G. (1966). The origin of erythroblasts in Rana pipiens
    tadpoles. Dev. Biol.,14, 461-79.
Horton, J.D. (1971). Histogenesis of the lymphomyeloid complex in
    the larval leopard frog, Rana pipiens. J. Morphol.,134, 1-
    20.
Horton, J.D. & Horton, T.L. (1975). Development of transplantation
    immunity and restoration experiments in the thymectomized
    amphibian. Amer. Zool.,15, 73-84.
Horton, J.D. & Manning, M.J. (1972). Response to skin allografts in
    Xenopus laevis following thymectomy at early stages of
    lymphoid organ maturation. Transplantation,14, 141-54.
Horton, J.D., Rimmer, J.J. & Horton, T.L. (1977). Critical role of
    the thymus in establishing humoral immunity in amphibians:
    studies on Xenopus thymectomized in larval and adult life.
    Dev. Comp. Immunol.,1, 119-32.
Horton, J.D., Smith, A.R., Williams, N.H., Edwards, B.F. & Ruben, L.N.
    (1980). B-equivalent lymphocyte development in the
    amphibian thymus? In Development and Differentiation of
    Vertebrate Lymphocytes, ed. J.D. Horton, pp. 173-82.
    Amsterdam: Elsevier.
Hsu, E., Julius, M.H. & Du Pasquier, L. (1983). Effector and
    regulator functions of splenic and thymic lymphocytes in the
    clawed toad Xenopus. Ann. Immunol.,134D, 277-92.
Hutchison, V., Haines, H. & Engbretson, G. (1976). Aquatic life at
    high altitude : respiratory adaptations in the Lake Titicaca
    frog, Telmatobius culeus. Respir. Physiol.,27, 115-29.
Ishizeki, K. (1980). Differentiation of granulocytes in the newt
    Triturus pyrrhogaster with special reference to the granule
    formation. Acta Anat. Nippon.,55, 305-19.
James, H., Clothier, R., Ferrer, I. & Balls, M. (1982). Effects of
    N-methyl-N-nitrosourea on carrier primed anti-hapten
    responses in Xenopus laevis. Dev. Comp. Immunol.,6, 499-
    508.
Jones, S.E. & Ruben, L.N. (1981). Internal histoincompatibility
    during amphibian metamorphosis? Immunology,43, 741-5.
Jordan, H.E. (1938). Comparative hematology. In Handbook of
    Hematology Vol.II, ed. H. Downey, pp. 704-862. New York:
    Harper.
Jordan, R.E. (1983). Antithrombin in vertebrate species :
    conservation of the heparine-dependent anticoagulant
    mechanism. Arch. Biochem. Biophys.,227, 587-95.
Jurd, R.D. (1977). Secretory immunoglobulins and gut-associated
    lymphoid tissue in Xenopus laevis. In Developmental
    Immunobiology, ed. J.B. Solomon and J.D. Horton, pp. 307-
    14. Amsterdam: Elsevier.
Jurd, R.D. & Doritis, A. (1977). Antibody-dependent cellular
    cytotoxicity in poikilotherms. Dev. Comp. Immunol.,1, 341-
    52.
Jurd, R.D. & Stevenson, G.T. (1976). Surface immunoglobulins on
    Xenopus laevis lymphocytes. Comp. Biochem. Physiol.,53A,
    381-7.

Just, J.J. & Atkinson, B.G. (1972). Hemoglobin transitions in the bullfrog, Rana catesbeiana, during spontaneous and induced metamorphosis. J. Exp. Zool.,182, 271-80.

Just, J.J., Schwager, J., Weber, R., Fey, R. & Pfister, H. (1980). Immunological analysis of hemoglobin transition during metamorphosis of normal and isogenic Xenopus laevis. Roux Arch. Dev. Biol.,188, 75-80.

Kaloustian, K.V. & Dulac, R.W. (1982). Relationships between red blood cell indices and the effects of thyroxine in three species of amphibians. Comp. Biochem. Physiol.,73A, 427-30.

Kampmeier, O.F. (1969). Evolution and comparative morphology of the lymphatic system, pp. 266-336. Springfield, Illinois: C.C. Thomas.

Kapa, E. & Csaba, G. (1972). Phylogenesis of mast cells. III Effect of hormonal induction on the maturation of mast cells in the frog. Acta Biol. Acad. Sci. Hung.,23, 47-54.

Kapa, E., Szigeti, M., Juhasz, A. & Csaba, G. (1970). Phylogenesis of mast cells. I Mast cells of the frog, Rana esculenta. Acta Biol. Acad. Sci. Hung.,21, 141-7.

Katagiri, C., Kawahara, H., Nagata, S. & Tochinai, S. (1980). The mode of participation of T-cells in immune reactions as studied by transfer of triploid lymphocytes into early-thymectomized diploid Xenopus. In Development and Differentiation of Vertebrate Lymphocytes, ed. J.D. Horton, pp. 163-71. Amsterdam: Elsevier.

Kau, C.L. and Turpen, J.B. (1983). Dual contribution of embryonic ventral blood island and dorsal lateral plate mesoderm during ontogeny of hemopoietic cells in Xenopus laevis. J. Immunol., 131, 2262-6.

Kaye, C. & Tompkins, R. (1983). Allograft rejection in Xenopus laevis following larval thymectomy. Dev. Comp. Immunol.,7, 287-94.

Kelényi, G. & Nemeth, A. (1969). Comparative histochemistry and electron microscopy of the eosinophil leucocytes of vertebrates. 1 A study of avian, reptile, amphibian and fish leucocytes. Acta Biol. Acad. Sci. Hung.,20, 405-22.

Kent, R. (1966). Uptake of carbon particles by the RES of the fowl, the frog and the chick embryo. J. Reticuloendothel. Soc.,3, 271-93.

Klempau, A.E. & Cooper, E.L. (1983). T-lymphocyte and B-lymphocyte dichotomy in anuran amphibians. I T-lymphocyte proportions, distribution and ontogeny, as measured by E-rosetting, nylon wool adherence, postmetamorphic thymectomy, and non-specific esterase staining. Dev. Comp. Immunol.,7, 99-110.

Klempau, A.E. & Cooper, E.L. (1984). T-lymphocyte and B-lymphocyte dichotomy in anuran amphibians. II Further investigations on the E-rosetting lymphocyte by using monoclonal antibody azathioprine inhibition and mitogen-induced polyclonal expansion. Dev. Comp. Immunol.,8, 323-38.

Kobel, H.R. & Du Pasquier, L. (1977). Strains and species of Xenopus
    for immunological research. In Developmental Immunobiology,
    eds. J.B. Solomon and J.D. Horton, pp. 299-306. Amsterdam:
    Elsevier.
Kobel, H.R. & Wolff, J. (1983). Two transitions of hemoglobin
    expression in Xenopus : from embryonic to larval and from
    larval to adult. Differentiation,24, 24-6.
Koppenheffer, T.L. & Stadig, B.K. (1981). Receptors for antigen-
    bound immunoglobulin on Xenopus laevis spleen cells. Amer.
    Zool.,21, 974.
Kraft, N. & Shortman, K. (1972). Differentiation of antibody-forming
    cells in toad spleen : a study using density and
    sedimentation velocity cell separation. J. Cell Biol.,52,
    438-52.
Kubai, L. & Auerbach, R. (1983). A new source of embryonic
    lymphocytes in the mouse. Nature 301, 154-6.
Kuramoto, M. (1981). Relationships between number, size and shape of
    red blood cells in amphibians. Comp. Biochem. Physiol.,69A,
    771-5.
Lassen, U.V., Pape, L. & Vestergaard-Bogind, B. (1978). Chloride
    conductance of the Amphiuma red cell membrane. J. Membr.
    Biol.,39, 27-48.
Le Douarin, N. (1966). L'hematopoïèse dans les formes embryonnaires
    et jeunes des vertebres. Annee Biol.,5, 105-71.
Lima, A.A.B., Airoldi, L.P.S., Meirelles, N.C. & Focesi, A. (1983).
    Separation and functional characterization of Leptodactylus
    labyrinthicus hemoglobin components. Comp. Biochem.
    Physiol.,76A, 123-5.
Lykkeboe, G. & Johansen, K. (1978). An O2-Hb "paradox" in frog
    blood? (n - values exceeding 4.0). Respir. Physiol.,35,
    119-28.
Maclean, N. & Jurd, R.D. (1972). The control of haemoglobin
    synthesis. Biol.Rev.,47, 393-437.
Manning, M.J. (1978). The amphibian immune system and emerging
    adaptations to life on land. In Proc. Zodiac Symp. on
    Adaptation, pp. 88-91. Wageningen, Netherlands: Pudoc
    Agricultural Publications.
Manning, M.J. & Collie, M.H. (1975). Thymic function in amphibians.
    In Immunologic Phylogeny, eds. W.H. Hildemann and A.A.
    Benedict, Adv. Exp. Med. Biol.,64, 353-62. New York: Plenum.
Manning, M.J., Donnelly, N. & Cohen, N. (1976). Thymus-dependent and
    thymus-independent components of the amphibian immune
    system. In Phylogeny of Thymus and Bone Marrow - Bursa
    Cells, eds. R.K. Wright and E.L. Cooper, pp. 123-32.
    Amsterdam: Elsevier.
Manning, M.J. & Horton, J.D. (1982). RES structure and function of
    the Amphibia. In The Reticuloendothelial System 3, eds. N.
    Cohen and M.M. Sigel, pp. 423-59. New York: Plenum.
Manning, M.J. & Jurd, R.D. Antibody production in thymectomized
    Xenopus. In Aspects of Developmental and Comparative
    Immunology I, ed. J.B. Solomon, pp. 495-6. Oxford:
    Pergamon.

Manning, M.J. & Turner, R.J. (1976). Comparative Immunobiology, 184pp. Glasgow: Blackie.

Mattes, M.J. & Steiner, L.A. (1978). Antisera to frog immunoglobulins cross-react with a periodate-sensitive cell surface determinant. Nature,273, 761-3.

McCutcheon, F.H. (1936). Hemoglobin function during the life history of the bullfrog. J. Cell. Comp. Physiol.,8, 63-81.

McKnight, B.J., Ford, T.C., Rickwood, D. (1982). An improved method for the isolation of leukocytes from Xenopus laevis blood. Dev. Comp. Immunol.,6, 381-4.

Mitchell, J.B. (1982). The effect of host age on Rana temporaria and Gorgoderina vitelliloba interactions. Int. J. Parasitol.,12, 601-4.

Mitsui, T. (1965). Light microscope and electron microscope study of the peroxidase reaction of the eosinophil leucocytes in cold-blooded animals. Okajimas Folia Anat. Jpn.,40, 893-909.

Mitsui, T. & Fukui, A. (1964). Gigantic peroxidase granules in the leukocytes of the giant salamander Megalobatrachus japonicus. Okajimas Folia Anat. Jpn.,40, 301-10.

Monaco, G., Salustri, A. & Bertolini, B. (1982). Observations on the molecular components stabilizing the microtubular system of the marginal band in the newt (Triturus cristatus) erythrocyte. J. Cell Sci.,58, 149-64.

Moore, M.A.S. & Owen, J.J.T. (1967). Stem cell migration in developing myeloid and lymphoid systems. Lancet ii, 658-9.

Mosier, D.E. & Subbarao, B. (1982). Thymus-independent antigens : complexity of B-lymphocyte activation revealed. Immunol. Today,3, 217-22.

Moss, B. & Ingram, V.M. (1968). Hemoglobin synthesis during amphibian metamorphosis. I Chemical studies on the hemoglobins from the larval and adult stages of Rana catesbeiana. J. Mol. Biol.,32, 481-92.

Nagata, S. & Cohen, N. (1984). Induction of T cell differentiation in early-thymectomized Xenopus by grafting adult thymuses from either MHC-matched or from partially or totally MHC-mismatched donors. Thymus,6, 89-103.

Nelson, R.D. & Yunis, J.J. (1969). Species and tissue specificity of very lysine rich and serine rich histones. Exp. Cell Res.,57, 311-8.

Noble, G.K. (1931). The Biology of the Amphibia, 577pp. New York: McGraw Hill.

Okazaki, T., Ishihara, H. & Shukuya, R. (1982). Changes in the density of circulating erythrocytes of the bullfrog tadpole, Rana catesbeiana, in relation to the transition of hemoglobin during metamorphosis. Comp. Biochem. Physiol.,73B, 309-12.

Orenstein, N.S., Galli, S.J., Hammond, M.E., Smith, G.N., Silbert, J.E. & Dvorak, H.F. (1977). Mucopolysaccharides synthesized by guinea pig basophilic leucocytes. Fed. Proc.,36, 1329.

Pantelouris, E.M., Knox, B. & Wallace, H. (1963). Iron in amphibian oocytes and embryos. Exp. Cell Res.,32, 469-75.

Perutz, M.F. & Brunori, M. (1982). Stereochemistry of cooperative effects in fish and amphibian hemoglobins. Nature,299, 421-6.

Pinder, A. & Burggren, W. (1983). Respiration during chronic hypoxia and hyperoxia in larval and adult bullfrogs (Rana catesbeiana). II Changes in respiratory properties of whole blood. J. Exp. Biol.,105, 205-13.

Reite, O.B. (1972). Comparative physiology of histamine. Physiol. Rev.,52, 778-819.

Rimmer, J.J. & Gearing, A.J.H. (1980). Antigen specific migration inhibition of peritoneal exudate cells in an anuran (Rana temporaria). In Development and Differentiation of Vertebrate Lymphocytes, ed. J.D. Horton, pp. 195-200. Amsterdam: Elsevier.

Rogovin, V.V., Fomina, V.A. & Piruzyan, L.A. (1978a). Electron cytochemistry of peroxidase activity in neutrophils of the bone marrow of the frog Rana temporaria. Biol. Bull. Acad. Sci. USSR,4, 492-6.

Rogovin, V.V., Fomina, V.A. & Piruzyan, L.A. (1978b). Three types of bone-marrow eosinophils of the frog Rana temporaria (an electron-cytochemical study). Biol. Bull. Acad. Sci. USSR,5, 363-6.

Rollins, L.A. & Cohen, N. (1981). Effect of early thymectomy on development of renal tumors in Lucke tumor herpesvirus-infected leopard frogs. In Aspects of Developmental and Comparative Immunology I, ed. J.B. Solomon, pp. 505-7. Oxford: Pergamon.

Rollins-Smith, L. & Cohen, N. (1982). Effects of early larval thymectomy on mitogen responses in leopard frog (Rana pipiens) tadpoles. Dev. Comp. Immunol.,6, 303-10.

Ruben, L.N. (1984). Some aspects of the phylogeny of macrophage-lymphocyte immune regulation. Dev. Comp. Immunol.,8, 247-56.

Ruben, L.N., Buenafe, A. & Seivert, D. (1983). Some characteristics of thymus suppresion of antibody production in vitro in Xenopus laevis, the South African clawed toad. Thymus,5, 13-8.

Ruben, L.N. & Edwards, B.F. (1980). Phylogeny of the emergence of T-B collaboration in humoral immunity. Contemp. Topics in Immunobiol.,9, 55-90.

Ruben, L.N., Jones, H. & Stack, J. (1981). Immunoregulation by phagocytic cells in the common American newt, Notophthalmus viridescens. In Aspects of Developmental and Comparative Immunology I, ed. J.B. Solomon, pp. 171-8. Oxford: Pergamon.

Ruben, L.N. & Stack, J. (1982). Limitations in response capacity of the newt, Notophthalmus viridescens, to soluble and particulate antigens. Dev. Comp. Immunol.,6, 491-8.

Ruben, L.N., Welch, J.M. & Jones, R.E. (1980). Carrier primed anti-hapten responses in larval and metamorphosing Xenopus laevis, the South African clawed toad. In Development and Differentiation of Vertebrate Lymphocytes, ed, J.D. Horton, pp. 227-37. Amsterdam: Elsevier.

Ruiz, G., Rosenmann, M. & Veloso, A. (1983). Respiratory and hematological adaptations to high altitude in Telmatobius frogs from the Chilean Andes. Comp. Biochem. Physiol.,76A, 109-13.

Schwager, J. & Hadji-Azimi, I. (1982). Separation of splenic T and B lymphocytes of Xenopus laevis. Experientia, 38, 748.

Sean, K.E., Lassalle, B. & Boilly, B. (1980). Ultrastructure de l'erythrocyte du triton Pleurodeles waltlii Michah. Can. J. Zool.,58, 1193-9.

Setoguti, T., Fujii, H. & Isono, H. (1970). An electron microscopic study on neutrophil leukocytes of the toad, Bufo vulgaris japonicus. Arch. Histol. Jpn.,32, 87-94.

Shelton, G. (1976). Gas exchange, pulmonary blood supply, and the partially divided amphibian heart. In Perspectives in Experimental Biology. Vol.1. Zoology, ed. P. Spencer Davies, pp. 247-59. Oxford: Pergamon.

Sicard, R.E. & Laffond, W.T. (1983). Putative immunological influence on amphibian forelimb regeneration. 2. Effect of several immunoactive agents on regeneration rate and gross morphology. Exp. Cell. Biol.,51, 337-44.

Sinclair, G.D. & Brasch, K. (1975). The nucleated erythrocyte : a model of cell differentiation. Rev. Can. Biol.,34, 287-303.

Small, J.V. & Davies, H.G. (1970). The haemoglobin in the condensed chromatin of mature amphibian erythrocytes - a further study. J. Cell Sci.,7, 15-33.

Spurling, N.W. (1981). Comparative physiology of blood clotting. Comp. Biochem. Physiol.,68A, 541-8.

Srivastava, V.M., Dube, B., Dube, R.K., Agarwal, G.P. & Ahmad, N. (1981). Blood fibrinolytic system in Rana tigrina. Thrombos. Haemostas. (Stuttg.).,45, 252-4.

Stearner, S.P. (1950). The effects of X-irradiation on Rana pipiens (leopard frog), with special reference to survival and to the response of the peripheral blood. J. Exp. Zool.,115, 251-62.

Sullivan, B. (1974). Amphibian haemoglobins. In Chemical Zoology, 9, ed. M. Florkin and B.T. Scheer, pp. 77-122. London: Academic Press.

Surbis, A.Y. (1978). Ultrastructural study of granulocytes of Bufo marinus. Florida Sci.,41, 45-52.

Tahan, A.M. & Jurd, R.D. (1981). Thymus dependency in anti-trinitrophenyl binding responses in the spleen of Ambystoma mexicanum : effects of thymectomy and anti-thymocyte serum treatments. Dev. Comp. Immunol.,5, 85-94.

Takaya, K. (1968). Mast cells and histamine in a newt, Triturus pyrrhogaster Boie. Experientia,24, 1053-4.

Takaya, K., Fujita, T. & Endo, K. (1967). Mast cells free of histamine in Rana catesbeiana. Nature,215, 776-7.

Tam, L-T. & Riggs, A.F. (1984). Oxygen binding and aggregation of bullfrog (Rana catesbeiana) hemoglobin. J. Biol. Chem.,259, 2610-6.

Theil, E.C. (1981). Red cell ferritin and iron storage during the early hemoglobin switch. In Hemoglobins in Development and Differentiation, eds. G. Stamatoyannopoulos and A.W. Nienhuis, pp. 423-31. New York: A.R. Liss.

Thomas, N. & Maclean, N. (1975). The erythroid cells of anaemic Xenopus laevis. 1. Studies on cellular morphology and protein and nucleic acid synthesis during differentiation. J. Cell Sci.,19, 509-20.

Tochinai, S. (1976a). Lymphoid changes in Xenopus laevis following thymectomy at the initial stage of its histogenesis. J. Fac. Sci. Hokkaido Univ. Ser.6, 175-82.

Tochinai, S. (1976b). Demonstration of thymus-independent immune system in Xenopus laevis. Response to polyvinylpyrrolidone. Immunology,31, 125-8.

Toews, D.P. (1971). A mechanism for the selective distribution of blood in the amphibia. Can. J. Zool.,49, 957-9.

Toews, D.P. & Macintyre, D. (1978). Respiration and circulation in an apodan amphibian. Can. J. Zool.,56, 998-1004.

Tompkins, R. (1978). Triploid and gynogenetic diploid Xenopus laevis. J. Exp. Zool.,203, 251-6.

Tompkins, R., Volpe, E.P. & Reinschmidt, D.C. (1980). Origin of hemopoietic stem cells in amphibian ontogeny. In Development and Differentiation of Vertebrate Lymphocytes, ed. J.D. Horton, pp. 25-34. Amsterdam: Elsevier.

Tooze, J. & Davies, H.G. (1965). Cytolysosomes in amphibian erythrocytes. J. Cell Biol.,24, 146-50.

Tooze, J. & Davies, H.G. (1968). Light and electron microscopic observations on the spleen and the splenic leukocytes of the newt Triturus cristatus. Amer. J. Anat.,123, 521-56.

Tournefier, A. (1982). Corticosteroid action on lymphocyte subpopulations and humoral immune response of axolotl (urodele amphibian). Immunology,46, 155-62.

Turner, R.J. (1969). The functional development of the reticulo-endothelial system in the toad, Xenopus laevis (Daudin). J. Exp. Zool.,170, 467-80.

Turner, R.J. (1970). The influence of colloidal carbon on hemagglutinin production in the toad, Xenopus laevis. J. Reticuloendothel. Soc.,8, 434-45.

Turner, R.J. (1973). Response of the toad, Xenopus laevis to circulating antigens. II Responses after splenectomy. J. Exp. Zool.,183, 35-46.

Turner, R.J. & Manning, M.J. (1973). Response of the toad, Xenopus laevis to circulating antigens. 1. Cellular changes in the spleen. J. Exp. Zool.,183, 21-34.

Turpen, J.B. (1980). Early embryogenesis of hematopoietic cells in Rana pipiens. In Development and Differentiation of Vertebrate Lymphocytes, ed. J.D. Horton, pp. 15-24. Amsterdam: Elsevier.

Turpen, J.B. & Knudson, C.M. (1982). Ontogeny of hematopoietic cells in Rana pipiens : precursor cell migration during embryogenesis. Dev. Biol.,89, 138-51.

Turpen, J.B., Knudson, C.M. & Hoefen, P.S. (1981). The early ontogeny of hematopoietic cells studied by grafting cytogenetically labelled tissue anlagen: localization of a prospective stem cell compartment. Dev. Biol.,85, 99-112.

Turpen, J.B. & Smith, P.B. (1985). Dorsal lateral plate mesoderm influences proliferation and differentiation of hemopoietic stem cells derived from ventral lateral plate mesoderm during early development of Xenopus laevis embryos. J. Leukocyte Biol., 38, 415-27.

Turpen, J.B., Turpen, C.J. & Flajnik, M. (1979). Experimental analysis of hematopoietic cell development in the liver of larval Rana pipiens. Dev. Biol.,69, 466-79.

Turpen, J.B., Volpe, E.P. & Cohen, N. (1973). Ontogeny and peripheralization of thymic lymphocytes. Science,182, 931-3.

Tyler, L.W., Piotrowski, D.C. & Kaltenbach, J.C. (1982). Tadpole erythrocytes : optical properties with dark field microscopy. Amer. Zool.,22, 946.

Van Rooijen, N. (1980). Immune complex trapping in lymphoid follicles : a discussion on possible functional implications. In The Phylogeny of Immunological Memory, ed. M.J. Manning, pp. 281-90. Amsterdam: Elsevier.

Vieira, H.F., Vieira, M.L.C., Meirelles, N.C. & Focesi, A. (1982). Some functional and structural properties of Bufo paracnemis and Pipa pipa hemoglobins. Comp. Biochem. Physiol.,73A, 197-200.

Villena, A. & Zapata, A. (1981). Ultrastructure of the jugular body of Rana pipiens. In Aspects of Developmental and Comparative Immunology I, ed. J.B. Solomon. pp. 491-2. Oxford: Pergamon.

Volpe, E.P., & McKinnell, R.G. (1966). Successful tissue transplantation in frogs produced by nuclear transfer. J. Hered.,57, 167-74.

Volpe, E.P., Tompkins, R. & Reinschmidt, D.C. (1979). Clarification of studies on the origin of thymic lymphocytes. J. Exp. Zool.,208, 57-66.

Volpe, E.P., Tompkins, R. & Reinschmidt, D.C. (1981). Evolutionary modifications of nephrogenic mesoderm to establish the embryonic centers of hemopoiesis. In Aspects of Developmental and Comparative Immunology I, ed. J.B. Solomon, pp. 193-201. Oxford: Pergamon.

Warr, G.W., Ruben, L.N. & Edwards, B.F. (1982). Evidence for low-molecular weight antibodies in the serum of a urodele amphibian, Ambystoma mexicanum. Immunol. Letters,4, 99-102.

Watt, K.W.K. & Riggs, A. (1975). Hemoglobins of the tadpole of the bullfrog Rana catesbeiana. Structure and function of isolated components. J. Biol. Chem.,250, 5934-44.

Welsch, U. & Storch, V. (1982). Light microscopic and electron microscopic observations on the caecilian spleen : a contribution to the evolution of lymphatic organs. Dev. Comp. Immunol.,6, 293-302.

West, N.H. & Burggren, W.W. (1982). Gill and lung ventilation responses to steady-state aquatic hypoxia and hyperoxia in the bullfrog tadpole (Rana catesbeiana). Respir. Physiol.,47, 165-76.

Wood, S.C. (1971). Effects of metamorphosis on blood respiratory properties and erythrocyte ATP level of the salamander Dicamptodon ensatus. Respir. Physiol.,12, 53-65.

Wood, S.C., Hoyt, R.W. & Burggren, W.W. (1982). Control of hemoglobin function in the salamander, Ambystoma tigrinum. Mol. Physiol.,2, 263-72.

Zapata, A., Gomariz, R.P., Garrido, E. & Cooper, E.L. (1982). Lymphoid organs and blood cells of the caecilian Ichthyophis kohtaoensis. Acta Zool.(Stockh.).,63, 11-6.

Zettergren, L.D. (1982). Ontogeny and distribution of cells in B lineage in the American leopard frog, Rana pipiens. Dev. Comp. Immunol.,6, 311-20.

Zettergren, L.D., Lydyard, P.M. & Parkhouse, R.M.E. (1977). Liver as a site of B cell generation in Xenopus laevis. Fed. Proc.,36, 1239.

4 Reptiles

J. Sypek & M. Borysenko
Tufts University School of Medicine, Boston, Massachusetts, USA

INTRODUCTION

The pioneering investigations on the nature of
reptilian blood described primarily the morphology of its various
cellular elements. It is rather unfortunate, however, that many
later studies have based their haematological terminology on these
early studies which in many instances are now outdated. Pienaar
(1962), in his treatise on the haematology of South African
reptiles, provided the first major attempt to integrate and
standardize the nomenclature used by earlier workers. This
terminology has been superceded in part by recent advances in
immunology, which have elucidated more clearly the definitive
nature of the blood cell elements. Furthermore, the terminology
applied to certain reptilian blood types has been adopted from
mammalian haematology, even though there are little more than vague
morphological similarities between the cells in question. The
application of such nomenclature should only be applied if the
cells possess in addition to the morphological criteria, functional
and ontogenetic similarities.

Because of the lack of functional knowledge the nomenclature
applied to certain reptilian leucocytes should be regarded as
tentative. In addition, morphological studies on reptilian
leucocytes have typically utilized only a limited number of
staining techniques to delineate such cells and this has led to
both tenuous interpretations and contradictory reports. This is
exemplified by the problems presented with reptilian "heterophils"
and "azurophils" (Ryerson 1943; Pienaar 1962). Environmental and
physiological factors are also known to affect many parameters of
blood measurements (Duguy 1970) and should be considered in any
haematological study. However, observations often lack such
information so that comparison with other studies is made more
difficult.

In this review, we have attempted to compare the information
existing in the literature relating to the morphological,
functional and ontogenetic nature of reptilian blood cells, and to
discuss this in the light of modern immunological knowledge. It is
apparent that the application of some of the haematological
terminology to reptilian cells rests on tenuous grounds and it is
hoped that this review will aid in providing guidelines to assess
the justification of applying mammalian terminology to reptilian
blood cells.

CIRCULATORY AND LYMPHOID SYSTEMS

The heart is almost completely four-chambered in reptiles, since a partial or, as in crocodiles and alligators, a complete interventricular septum is present. There is also an unusual tripartite division of the truncus arteriosus leaving the heart. As a result, venous blood returning to the heart from the body and going to the lungs is nearly completely separated from arterial blood returning to the heart from the lungs and going to the rest of the body. Thus, nearly all the blood entering the systemic circulation is oxygenated. The elimination of cutaneous respiration in reptiles, places greater importance on pulmonary circulation which is reflected in larger pulmonary veins and smaller cutaneous veins. As in other vertebrates, a hepatic portal system and a renal portal system are present.

Diffuse lymphoid aggregates are commonly found in members of all the reptilian classes. These lymphoid tissues are widely distributed throughout the organ systems, however, they are not as extensive or well-developed as those of mammals and birds. These aggregations and large accumulations of densely packed lymphocytes are found in association with the pharyngeal region, the oesophageal folds, throughout the lamina propria of the alimentary tract, and to a lesser extent in the inguinal and axillary regions, the lungs, kidneys and urinary bladder (McCauley 1956; Borysenko & Cooper 1972; Hussein et al. 1978, 1979a,b; Zapata & Solas 1979; El Ridi et al. 1980; Jacobson & Collins 1980). The number and size of these lymphoid tissues may vary tremendously with the season of year and become particularly sparse or poorly developed, especially in those species which hibernate (Hussein et al. 1978, 1979a,b,c; El Ridi et al. 1980).

The most prominent and well-developed lymphoid organs are the thymus and the spleen. The reptilian thymus is usually lobulated, although the numbers of lobules varies greatly. The thymic tissue is located in the cervical region and may be contiguous with the parathyroid glands and adjacent to the ultimobranchial bodies and great vessels. The thymus is clearly separated into cortical and medullary regions (Fig. 1). Arteries and veins are prominent in the connective tissue septa and in the medulla while only capillaries are encountered in the cortex. The cortex consists primarily of small densely packed lymphocytes in a framework of stellate reticular-epithelial cells. Fewer lymphocytes are found in the medulla and the predominant cells in this region are reticular-epithelial cells. Large myoid cells are also constituents of the medulla, but occur in smaller numbers in the cortex as well (Fig. 2). Variable numbers of macrophages and eosinophils occur in the medulla and connective tissue septa (Borysenko & Cooper 1972).

Thymic involution occurs with age in reptiles as in other vertebrates. The distinction between cortex and medulla becomes reduced and much of the lymphoid tissue is replaced by connective tissue elements. Seasonal fluctuations are transitory and reversible, while long term starvation or chronic disease may lead to permanent involution, as with ageing (Bockman 1970; Borysenko & Lewis 1979; Hussein et al. 1978, 1979a,b,c).

The spleen of most reptiles is spherical or elongated and located near the transverse colon. It is encapsulated by fibrous connective tissue from which a number of trabeculae extend into the parenchyma. The parenchyma of the spleen shows a definite demarcation into a red and white pulp in most species (Figs. 3 & 4). The red pulp is composed of a system of cords and sinuses which contain all the elements of the peripheral blood and plasma cells (Marchalonis et al. 1969; Kanakambika & Muthukkaruppan 1973; Borysenko 1976a; Zapata et al. 1981; Kroese & Van Rooijen 1982). In several species of lizards, however, the red and white pulps are not distinctly delineated as in other reptiles. In these species, the red pulp consists of discontinuous narrow strands of tissue surrounded by blood sinuses between confluent areas of white pulp (Kanakambika & Mathukkaruppan 1973). More typically in other reptiles, the red pulp cord and sinuses are separated from the white pulp by a marginal zone of flat reticular cells. The white pulp is comprised of two lymphoid compartments. Lymphoid tissue surrounds both central arterioles and ellipsoids, forming periarteriolar and periellipsoidal lymphocyte sheaths (Fig. 4). The ellipsoids consist of a layer of reticular tissue, enclosing a capillary and cuboidal epithelium. In contrast with periarteriolar lymphocyte sheaths, reticular fibres are rarely observed in periellipsoidal lymphocyte sheaths (Borysenko 1976a; Zapata etal. 1981; Kroese & Van Rooijen 1982). A variety of leucocytes can be seen within the venous drainage of the spleen (Fig. 5).

Immune complexes appear to be trapped by dendritic cells located in the periellipsoidal lymphocyte sheaths of the white pulp, where follicles and germinal centres are not yet present unlike mammals (Kroese & Van Rooijen 1983). Upon antigenic stimulation lymphoblast proliferation occurs in the periarteriolar lymphocyte sheaths followed by migration and differentiation of blast cells into plasma cells in the red pulp and production of specific antibody (Borysenko 1976b). Although it is apparent that the spleen possesses lymphocytes which elicit both cell-mediated immunity and humoral immune responses, T and B cell zones, have not yet been defined immunohistochemically.

BLOOD CELL FORMATION

Erythropoiesis
The normal site for production of red cells in reptiles is the bone marrow, which utilizes the progressive maturation of early red cell stem lines. Efrati et al. (1970) in their study on the haemopoietic system observed pronormoblasts in the bone marrow of *Agama stellio* which they described as being morphologically similar to those observed in mammals. Pienaar (1962) in his observations on erythropoiesis in *Cordylus vittifer* describes the site of erythrocyte proliferation as occurring within the vascular spaces of the reticular stroma of bone marrow. It appears, however, that red cell production may also occur in parenchymatous organs such as liver and spleen (extramedullary erythropoiesis), particularly during haemoprotozoan infections where there is peripheral destruction of red cells (Pienaar 1962; Frye 1981; Schall 1983).

Lymphopoiesis
The thymus in reptiles is the first lymphoid organ to develop. Stem cells observed in the yolk sac islets are later seen infiltrating the epithelial rudiments of the thymus. Subsequently, large and small lymphocytes appear in the thymus followed by differentiation into the medullary and cortical zones. These events suggest that reptilian lymphocytes are derived from blood-borne stem cells, whose origin remains to be determined. Similarly, no direct evidence is available indicating the origin of immunoglobulin producing cells. The yolk sac islets or, embryonic liver and bone marrow may be likely candidates, (Pitchappan & Mathukkaruppan 1977a) however, definitive studies have yet to be conducted. Although lymphoid aggregates commonly occur in the cloacal mucosa, a true phylogenetic precursor of the avian bursa of Fabricius has not been found (Borysenko & Cooper 1972).

Granulopoiesis
The spleen is functionally the most important secondary lymphoid organ, containing large numbers of immunocompetent cells of diverse origin. The spleen develops into a lymphopoietic organ much later than the thymus. In the early stages of spleen development it contains large numbers of granulocytes. Eosinophils are restricted to the subcapsular region while basophils are scattered throughout the parenchyma. It appears that the eosinophils proliferate in the subcapsular regions, while basophils enter the spleen as blood elements during this phase of development. Later, the subcapsular eosinophils disappear and the spleen becomes primarily lymphopoietic (Kanakambika & Muthukkaruppan 1973; Borysenko 1978). Efrati et al. (1970) reported that promyelocytes and myelocytes morphologically similar to their mammalian counterparts occurred mainly in the bone marrow and rarely in the thymus. Pienaar (1962) observed that

differentiation of these cells occurred in the extravascular environment of the reticular stroma of the bone marrow and that the maturing granulocytes subsequently migrated through the endothelial barrier of the sinusoids into the blood stream.

According to many earlier investigators, hypothetically all of the cells in the peripheral blood derive from a multipotent type of haemocytoblast which is capable of differentiating in the blood marrow or spleen to produce erythrocytes, granulocytes, monocytes and lymphocytes (see Pienaar 1962 for extensive literature review). The evidence regarding the foci of haemopoietic activity, particularly granulopoietic and lymphopoietic mechanisms (including thrombopoiesis, myelopoiesis, etc.) and cell lineage inherent in reptiles, however, still remains an open question, although it is likely that most early haemopoiesis occurs in the bone marrow.

BLOOD CELL STRUCTURE AND FUNCTION

Erythrocytes

The reptilian erythrocyte is an ellipsoidal cell containing a centrally situated oval nucleus (Figs. 5,6 a-e). In blood smears stained with Romanowsky stains, the yellowish cytoplasm most often appears translucent and homogeneous and the nucleus heterochromatic. Although seemingly contradictory by definition, true reticulocytes as well as polychromatophilic juvenile erythrocytes, can be demonstrated in reptiles with new methylene blue, brilliant cresyl blue and other supravital stains. Similarly, Howell-Jolly bodies can be demonstrated in the erythrocytes of some reptiles (Pienaar 1962; Efrati et al. 1970). As the erythrocyte ages to the point of senescence, the nucleus becomes pycnotic. The cell membrane eventually lyses and the haemoglobin is lost to the circulating plasma or the cell is phagocytized by various phagocytic cells (Frye 1981). The circulating blood of reptiles may also contain immature cells of the erythrocyte series characterized by a rounded form, basophilic cytoplasm and a large nucleus which is more euchromatic than that of a mature cell. These cells are especially common in young or moulting animals or those heavily infected by certain parasitic Protozoa (Pienaar 1962; Schall 1983).

Reptilian erythrocytes display a wide variation in size as shown in Table 1. The data on erythrocytes suggests that the size of cells may be related to the systematic position of the species from which they are derived. Wintrobe (1933) suggested that the size of erythrocytes may reflect the position of a species in the evolutionary scale. St. Girons (1970) also postulated that size may be related to the metabolic rate of a given species. The number of circulating erythrocytes is lower in reptiles than in mammals or birds. Lizards in general have more erythrocytes than snakes and turtles have the fewest (Pienaar 1962; Duguy 1970).

TABLE 1

SUMMARY OF REPTILIAN ERYTHROCYTE DIMENSIONS[1]

Group	Erythrocytes		Nuclei		N/C^2 Ratio
	Length	Width	Length	Width	
Testudines	18.5-20μm	10-12μm	5-6.5μm	4-5μm	0.08-0.15
Rhynchocephalia	19-25μm	13-16μm	8-9μm	5-6μm	0.15
Sauria	13-22μm	5-13.5μm	5.5-8μm	2.5-4.5μm	0.11-0.215
Ophidia	15-19μm	8-11μm	5-8μm	3-4μm	0.09-0.22
Crocodilia	16-17μm	9-10μm	5-6μm	3.5-4μm	0.13-0.135

[1]References: Pienaar 1962; St. Girons 1970; Frye 1981.

[2]N/C = nucleus to cytoplasm

Since lizards have among the smallest erythrocytes and the turtles among the largest, there may be an inverse correlation between the number of erythrocytes and their size (Ryerson 1949). The number of red blood cells of reptiles is reported to vary according to season, with an increase before winter and a decrease thereafter in those species which hibernate; with sex, being higher in males of certain species (Duguy 1970); with nutritional status and climatic conditions; and with parasitic infections - showing declines in cell numbers (Pienaar 1962).

At the ultrastructural level, reptilian erythrocytes appear ellipsoidal, with nuclei possessing chromatin in dense clumps and nucleoli which are difficult to discern (Fig. 7). A finely granular cytoplasm is present which usually possesses sparse endoplasmic reticulum, a Golgi complex near the nucleus, and a scattering of mitochondria. Beneath the plasmalemma and encircling the cell in a plane parallel to their flat surface are prominent "marginal bands" of microtubules (Taylor et al. 1963; Desser & Weller 1979a).

Reptilian red blood cells appear to have exceptionally long life spans (600-800 days) and the red blood cell turnover appears to be directly related to the low metabolic rate of reptiles. Haemoglobin makes up the greater share of the solid content of the erythrocytes with most estimates falling between 25-32% of the net weight. Reptilian blood contains 6-12g% of haemoglobin (Dessauer

1970; Frye 1981). The concentration is directly proportional to the packed cell volume (20-45%) unless large numbers of immature cells are present in which case methaemoglobin is frequently present in significant quantities. The blood of individual reptiles may contain two or more haemoglobins, distinguishable by molecular weight, surface charge and chemical properties (Dessauer, 1970). Variability in these haemoglobins lies in the structure of the globin and iron poryphyrin present (Dessauer 1970). Oxygen transport, the primary function of haemoglobins, depends on the ability of the respiratory pigment to combine reversibly with oxygen. Oxygen equilibrium curves describing this property display the sigmoid shape typical of mammalian haemoglobins (Dessauer 1970). The shape of the curve is attributed to interactions between oxygen combining sites. A rise in temperature or fall in pH (the Bohr effect) within the physiological range leads to a decrease in oxygen affinity. Considering the broad ranges of temperature and pH characteristic of the reptilian inner environment under normal physiological conditions, oxygen equilibrium curves vary considerably from one reptilian species to another. These curves, however, are remarkably similar at the activity temperature of a given species (Dessauer 1970). Red blood cells of reptiles probably possess broader and more typical metabolic potentials than the highly specialized erythrocytes of mammals since DNA is present. Presumably nucleic acid metabolism and protein synthesis occur throughout the long life span of the reptilian erythrocyte. Nucleotides occur in high concentrations and nucleic acid phosphatases are active as well (Dessauer, 1970). All amino acids that are common constituents of proteins are present within the cell. Carbonic anhydrase, which is important in facilitating shifts in ions between plasma and red blood cells, is present in all the major reptilian groups. Energy rich phosphate compounds which are a result of metabolic reactions are present in higher concentrations in reptiles compared with mammals. The glycolytic pathway, the tricarboxylic acid cycle and the glucose-6-phosphate pathway are probably all involved in energy production in the erythrocytes of reptiles (Dessauer 1970).

A summary of histochemical data on reptilian erythrocytes is presented in Table 2.

Eosinophilic granulocytes
    The reptilian eosinophil has received much speculative attention with regard to its definitive characteristics and subsequent classification (Ryerson 1943; Pienaar 1962; Elkan & Zwart 1967; Kelényi & Nemeth 1969; Efrati et al. 1970. Desser & Weller 1979b). In all reptiles, these granulocytes are large, generally rounded cells in which the nuclei are variable in shape and usually located eccentrically. In general, two types of "eosinophils" may be distinguished on the basis of the following characteristics; those with spherical intracytoplasmic granules

TABLE 2

SUMMARY OF REPTILIAN BLOOD CELL HISTOCHEMISTRY

BLOOD CELL TYPE	ENZYMES						POLY-SACCHARIDES				Lipids			NUCLEIC ACIDS			BASIC PROTEINS			REFERENCES
	Acid phosphatase	Alkaline phosphatase	β-glucuronidase	Esterase	Peroxidase	Lactic dehydrogenase	Periodic acid-Schiff	Alcian blue	Toluidine blue	Astrablau	Sudan black B	Nile blue sulphate (Acidic)	Sudan III (Neutral)	Methyl-green pyronin	Acridine orange	Feulgen reagent	Arginine	Bierbrich scarlet	Ninhydrin	
Erthrocyte	+	-	-	-	+	+	±	ND	ND	ND	-	ND	ND	ND	ND	ND	ND	ND	ND	1,2,3,4
Eosinophilic Granulocyte	+	+	+	±	+	ND	+	+	-	-	±	-	-	+	+	ND	ND	ND	ND	1,5,6,7
Heterophilic Granulocyte	+	±	+	+	±	ND	+	-	-	-	+	+	-	+	+	ND	ND	ND	ND	1,5,6,7
Basophilic Granulocyte	±	±	+	±	-	ND	+	+	+	+	-	ND	ND	+	+	+	-	-	-	1,5,6,7
Monocyte	+	-	+	ND	±	ND	+	ND	ND	ND	+	ND	ND	ND	ND	ND	ND	ND	ND	1,2,5
Thrombocyte	+	-	-	ND	-	ND	±	ND	ND	ND	-	ND	ND	ND	ND	ND	ND	ND	ND	2,5,8
Lymphocyte	±	±	-	ND	-	ND	±	ND	ND	ND	-	ND	ND	ND	ND	ND	ND	ND	ND	1,2,5

References: 1. Caxton-Martins 1977; 2. Caxton-Martins & Nganwuchu 1978; 3. Dessauer 1970; 4. Gerzeli 1954; 5. Pienaar 1962; 6. Efrati et al. 1970; 7. Desser 1978; 8. Desser & Weller 1979a.

(Fig. 5b) which react positively to both neutral and alkaline benzidine peroxidase reactions and those with fusiform granules (Fig. 5d) which are peroxidase negative at alkaline pH (Kelényi & Nemeth 1969). The latter category of cells, commonly known as "heterophils", will be discussed separately. (Heterophil granules in reptiles, as in birds and in certain mammals such as the rabbit, are stainable with either acidic or basic dyes. However, they may also show a predilection for eosin. In these species, therefore, heterophils have been also termed "pseudoeosinophils" (Kelényi & Nemeth 1969)).

Studies on the haematology of numerous reptilian species have shown that 0-20% of leucocytes in a normal differential cell count may be eosinophils (Charipper & Davis 1932; Bernstein 1938; Ryerson 1949; Taylor & Kaplan 1961; Pienaar 1962; Heady & Rogers 1962; Duguy 1970; Desser 1978; Kumar De & Marti 1981; Wood & Ebanks 1984). This cell type, however, appears to be more prevalent in turtles (15-20%) than in any other reptilian group. In contrast, in lizards eosinophil numbers may be very low. The interspecific variation in size of the eosinophilic leucocytes ranges from 9-20 µm with the smallest cells found in lizards, and the largest in snakes, while turtles, crocodiles and the tuatara possess eosinophils intermediate in size (St. Girons 1970). Eosinophils appear to be greatly influenced by seasonal factors; the lowest numbers occur during the period of summer activity while the maximum numbers are found during hibernation (Duguy 1970).

At the ultrastructural level, reptilian eosinophils possess the following characteristics: an elongated or lobulated nucleus, and round or slightly oval intracytoplasmic granules (Fig. 7). The granules are surrounded by single membranes and originate from the Golgi transport vesicles as small electron-dense droplets. The nucleus is regular in appearance and the chromatin occurs in dense clumps. The nucleoplasm is densely granulated. Oval mitochondria, β-glycogen and assorted smaller granular inclusions are present among the cytoplasmic granules. Endoplasmic reticulum, Golgi complexes and ribosomes are scarce (Taylor et al. 1963; Kelényi & Nemeth 1969; Efrati et al. 1970; Desser & Weller 1979b).

Histochemical observations on the granular contents of reptilian eosinophils have shown that with methanol fixation the granules appear orange-yellow in colour and bluish black with acetone fixation upon Giemsa staining (Table 3). The high isoelectric point (alkaline erythrosin) of the specific granules is due to the presence of a major basic protein which binds acidic dyes in alkaline media by ionic-type linkages e.g. eosin, erythrosin, acid fuchsin, phloxine, aniline blue, Biebrich scarlet and Cochineal red. Selective anisotropic staining and fluorochroming is observed with substantive dyes and fluorochroming by non-ionic type linkages e.g. Congo red, Sirius red F3B, and thioflavine S (Kelényi & Nemeth 1969). Positive benzidine peroxidase reactions are noted at an alkaline (pH 11) and neutral (pH 7-8) conditions (Kelényi & Nemeth 1969). In supravital preparations after treatment with neutral red

TABLE 3

SUMMARY OF REPTILIAN GRANULOCYTES AND MONOCYTES

BLOOD CELL TYPE	Size	Shape	Benzidine peroxidase		Neutral red	Methanol fixation	Acetone fixation	High Iso-electric pt.	Anisotropic staining	E.M. crystalloids	Phagocytic	Pseudopodia	Cytophilic antibody	References
			GRANULE CHARACTERISTICS											
			Neutral	Alkaline										
Eosinophilic Granulocyte	9-20 μm	round-oval	+	+	+	orange-red	bluish-black	+	+	-	Ab Needed	±	+	1,2,3,4,5,6,7
Heterophilic Granulocyte	10-23 μm	oval-fusiform	+	-	-	muddy brown or orange-pink	bluish-black	+	+	-	+	+	?	1,3,4,5,6,8
Basophilic Granulocyte	7-20 μm	ovoid	ND	ND	+	deep blue, metachromatic	ND	ND	ND	-	-	±	+	2,4,5,6,7
Monocyte	8-20 μm	various shapes	ND	ND	+	light blue-grey or azure	ND	ND	ND	-	+	+	+	2,4,5,7,9

References: 1. Ryerson 1943; 2. Pienaar 1962; 3. Kelényi & Nemeth 1969; 4. Efrati et al. 1970; 5. St. Girons 1970; 6. Desser & Weller 1979b; 7. Frye 1981; 8. Mead & Borysenko 1984a; 9. Borysenko 1976b.

the granules stain progressively with a deep-rust colour which persists and does not fade indicating their acidic nature (Ryerson 1943; Pienaar 1962). A summary of histochemical data on reptilian eosinophilic granulocytes is presented in Table 2.

The presence of cytophilic immunoglobulin on reptilian eosinophils has been demonstrated by antigen-specific interactions with sheep red blood cells (SRBC) in SRBC-immunized turtles. Turtle eosinophils phagocytize immune complexes, although the specific antibody is bound to the cells' surface and immune complexes form when eosinophils encounter SRBC (Mead & Borysenko 1984a). Three classes of serum immunoglobulin have been reported in turtles IgM, IgY and IgN (Chartrand et al. 1971; Benedict & Pollard 1972; Leslie & Clem 1972). It is postulated that the cytophilic antibody may be one or more of these immunoglobulin classes.

The numerical values of reptilian eosinophils vary widely in differential white cell counts as eosinophils respond to a variety of parasitic, environmental and non-specific stimuli (Wood 1935; Duguy 1970; Leibovitz et al. 1978; Glazebrook et al. 1981; Burke & Rodgers 1982; Wolke et al. 1982). Their numbers reflect cell mobilization at a particular point in time. Since some investigators have reported heterophils as eosinophils and vice versa, it is difficult to compare values for these latter cells on a valid basis with other published data.

Heterophilic granulocytes
        The reptilian heterophil is frequently counted with other acidophilic granulocytes, making it difficult to compare data on this cell type with those of other workers. Heterophils are distinct from eosinophils in that they usually have fusiform granules (Fig. 6d) which tend to stain muddy brown. In other species, however, the cytoplasm appears finely granular (Fig. 6c), staining orange-pink with most Romanowsky stains. In contrast, eosinophils contain large spherical granules (Fig. 6b) and stain bright orange-red. The heterophil nucleus stains intensely and is usually lobed or polymorphonuclear (Fig. 6 c&d)(Frye 1981).

Studies on reptilian haematology have reported that 0-40% of leucocytes (species variation) in a normal differential cell count may be heterophils (Charipper & Davis 1932; Bernstein 1938; Ryerson 1949; Taylor & Kaplan 1961; Pienaar 1962; Heady & Rogers 1962; Duguy 1970; Desser 1978; Kumar De & Marti 1981; Wood & Ebanks 1984). The size of these cells is quite variable even in individual specimens and ranges from 10-23 µm with a similar size distribution in the various reptilian groups as observed with the eosinophils (St. Girons 1970). Heterophils have been observed to be influenced by seasonal factors with greater numbers occurring in the summer months and decreasing numbers present during hibernation (Duguy 1970).

At the ultrastructural level, heterophils have a number of characteristic features. Immature heterophils (myelocytes) possess granules which are ovoid and electron-dense and with cellular maturation become progressively elongated. Heterophilic myelocytes are regular in outline with a compact nucleus. The most apparent feature of these cells is the presence of numerous fine vesicles throughout the cytoplasm (Kelényi & Nemeth 1969; Desser & Weller 1979b). Mature heterophils are irregular in outline, the nucleus is segmented and chromatin occurs in discrete clumps. The large granules are rod-like and the number of fine vesicles is greatly reduced (Fig. 9). A second smaller type of granular spheroid inclusion is also present in mature heterophils. Both types of granules are homogeneous with no crystalloid cores apparent. The nucleoplasm is finely granulated. The centrosomal region is adjacent to the nucleus with paired centrioles, and a Golgi complex. Ellipsoid mitochondria with parallel cristae, lipid droplets and numerous microfilaments are arranged randomly or in bundles in the cytoplasm. Endoplasmic reticulum is abundant but scattered. Pseudopods are present and the cytoplasm is clear and homogeneous (Taylor et al. 1963; Kelényi & Nemeth 1969; Efrati et al. 1970; Desser & Weller 1979b).

Histochemical observations on the intracytoplasmic granules of reptilian heterophils show that as with eosinophils, upon methanol fixation intracytoplasmic granules appear orange-yellow in colour and bluish-black with acetone fixation upon Giemsa staining. The high isoelectric point of the specific granules is due to the presence of an arginine-rich major basic protein which binds acidic dyes in alkaline media by ionic-type linkages. Selective anisotropic staining and fluorochroming is observed with substantive dyes and fluorochroming by non-ionic type linkages (Kelényi & Nemeth 1969). In supravital staining neither neutral red or Janus green B appears to be taken up very readily (Ryerson 1943). Further histochemical data on reptilian heterophilic granulocytes is presented in Table 2.

Reptilian heterophils, like mammalian neutrophils, appear to be phagocytic (Efrati et al. 1970) and are evident in inflammatory responses to a variety of infectious, parasitic and non-specific agents (Wood 1935; Ryerson 1943; Duguy 1970; Hiradhar et al. 1979; Jacobson 1980; Jacobson et al. 1980a, 1983). In addition, they have been observed to be responsive to various chemotactic stimuli, such as turpentine and RNA (Ryerson 1943). As with other granulocytes occurring in reptiles, more studies on the functional aspects of this cell are necessary.

Basophilic granulocytes
    In all reptile species studied, basophils are small, nearly spherical compact cells with a cytoplasm usually filled with intensely basophilic granules which are metachromatic (Figs. 5 & 6a).

The granulation may be so dense that the small slightly eccentric nucleus may be almost completely obscured. Morphological studies of blood cells in numerous reptilian species have shown that 0-40% of the leucocytes in a normal differential cell count are basophils (Charipper & Davis 1932; Bernstein 1938; Ryerson 1949; Taylor & Kaplan 1961; Pienaar 1962; Heady & Rogers 1962; Duguy 1970; Desser 1978; Kumar De & Maiti 1981; Wood & Ebanks 1984). In certain species of turtles, *Chrysemys picta* and *Chelydra serpentina*, basophils comprise greater than 50% of the total leucocytes (Michels 1923; Mead et al. 1983). Cells with similar structural properties (mast cells) have also been observed in the subcutaneous connective tissue, mesenteries, gut mucosa, lung parenchyma and the spleen (Michels 1923; Pienaar 1962). Interspecies variation in the size of basophilic granulocytes ranges from 7-20 μm, with the smallest occurring in lizards, larger cells in snakes, still larger in turtles and crocodiles, and the largest in the tuatara (St. Girons 1970). Little seasonal variation has been observed with these cells, although it appears that their numbers are lower during hibernation and higher during periods of greater activity (St. Girons 1970). In this regard, Reite (1973a) reported redistribution of basophil leucocytes of turtles in response to cold exposure from the blood to the liver.

At the ultrastructural level, reptilian basophils possess a regular to irregular cell periphery. The nucleus is round and possesses a diffuse chromatin structure. A dense ovoid nucleolus is present and the nucleoplasm is finely granulated. Small vacuoles and numerous homogeneous ovoid granules, which are electron-dense and devoid of any crystalline matrix, occur in the cytoplasm (Fig. 7). In addition, oval mitochondria with parallel cristae, β -glycogen particles and some microfilaments are present. A sparse granular endoplasmic reticulum, ribosomes, mitochondria and glycogen particles are also observed along with a small centrosomal region which includes a small Golgi complex surrounding the centrioles. Actin-like filaments and "fibre bundles" also occur (Taylor et al. 1963; Efrati et al. 1970; Desser & Weller 1979b).

In supravital preparations basophil granules have an affinity for neutral red and Janus green B indicating an acidic nature of the granular contents (Pienaar 1962). Water is found to have a deleterious effect on preservation or subsequent staining of the granular contents. Fixation in absolute alcohol and treatment with Romanowsky stains provides optimum staining of basophil granules (Michels 1923; Pienaar 1962). A summary of histochemical data on reptilian basophilic granulocytes is presented in Table 2.

Morphological and histochemical observations indicate that reptilian basophilic leucocytes resemble their mammalian counterparts (Mead et al. 1983). Basophils of the snapping turtle have been shown to possess surface immunoglobulins. Upon treatment

with rabbit antiserum to turtle immunoglobulins they undergo
degranulation and release histamine in vitro. Degranulation can
also be induced in turtles with compound 48/80 and calcium
ionophore A23187 (Reite 1973b; Mead et al. 1983). The level of
histamine release induced by rabbit anti-turtle immunoglobulin sera
has been found to be dependent on the concentration of the inducing
reagent, the length of exposure to this reagent and the level of
calcium. In addition, release is temperature dependent, increasing
over a range of 10°C to maximal release at 27°C (Sypek et al.
1984). In addition, turtle basophils have been found to express
antigen-specific cytophilic antibody with reaginic function after
immunization with sheep red blood cells (SRBC) and form
basophil-SRBC rosettes in vitro. The formation of these rosettes
can be induced by passive sensitization in vitro of non-immune
turtle basophils with SRBC immune turtle sera (Mead et al. 1983).
The class of cytophilic antibody present on the turtle basophil has
been postulated as being either one of the two low molecular weight
antibodies occurring in turtles. These turtle antibodies are IgY,
a 7.5S immunoglobulin and IgN, a 5.7S immunoglobulin (Mead et al.
1983). The reason for high basophil cell numbers in reptiles
remains unclear, but perhaps future studies on the immune function
of this cell will demonstrate some selective advantage for the
prevalence of this cell type. Basophil levels have been reported
to be elevated in certain haematozoan infections such as
haemogregarines, trypanosomes, and Pirahemacyton (Wood 1935,
Pienaar 1962). Studies of basophil involvement in the
immunopathologic response to these and other infections may help
clarify the role of the basophil in host resistance.

   Monocytes
   Monocytes may account for 0-10% of the leucocytes in a normal
differential count of reptiles (Charipper & Davis 1932; Bernstein
1938; Ryerson 1949; Taylor & Kaplan 1961; Pienaar 1962; Heady &
Rogers 1962; Duguy 1970; Otis 1973; Desser 1978; Kumar De & Maiti
1981), with values higher than 20% reported for certain species of
snakes (Pienaar 1962). The size of these cells is quite variable
ranging from 8-20 um and is variable even in individual animals.
In view of the number of cytological variations of these cells
there is considerable ambiguity in regard to the identification of
this group of leucocytes, and as such they have also been referred
to as "azurophils" by St. Girons (1970) and Pienaar (1962); and as
"neutrophils" by others (Alder & Huber 1923; Bernstein 1938).
Monocytes appear to undergo little seasonal variation in their
numbers, as their percentage remains relatively constant throughout
the year under normal conditions (Duguy 1970).

Monocytes generally possess a single, indented nucleus and a finely
granular or vacuolated cytoplasm that stains a light blue-grey or
azure with Romanowsky stains. At the ultrastructural level the
nucleus contains moderate amounts of heterochromatin distributed

along the nuclear envelope and the dense ovoid nucleolus. Within
the cytoplasm are ellipsoid mitochondria, Golgi, short segments of
rough endoplasmic reticulum, lysosomes and vacuoles of various
sizes and densities. The plasma membrane is irregular with
pseudopodia prominent even in immature forms. Both the nucleoplasm
and cytoplasm are coarsely granulated (Fig. 8) (Taylor et al. 1963;
Borysenko 1976b).

The cytoplasmic constituents of reptilian monocytes and macrophages
stain selectively with pure azur dyes and pyronin or methyl-green
pyronin mixtures (Pienaar 1962). In supravital preparations
neutral red is taken up, particularly in the lysosomes and vacuoles
(Pienaar 1962). A summary of histochemical data on reptilian
monocytes is presented in Table 2.

Monocytes and macrophages can be observed in substantial numbers
following antigenic challenge and in the spleen they occur most
frequently along the sinus walls and in the red pulp cords
(Borysenko 1976a). These cells also possess a high degree of
phagocytic and locomotory ability (Pienaar 1962; Efrati et al.
1970) and are evident in a wide variety of inflammatory responses.
They have an active role in granuloma and giant cell formation and
have been associated in particular with the granulomatous response
of turtles to bacterial infections (Evans 1983) and to the ova of
various spirorchid trematodes (Greiner et al. 1980; Glazebrook et
al. 1981; Wolke et al. 1982); as well as to a variety of other
infectious agents (Cox et al. 1980; Jacobson 1980; Jacobson et al.
1980b, 1981, 1983). The presence of cytophilic immunoglobulin on
monocytes has been confirmed by antigen-specific interactions in
SRBC-immunized turtles (Mead & Borysenko 1984a). This cytophilic
antibody has been postulated to be one or more of the
immunoglobulin classes reported from turtles, and in the anuran
amphibian *Xenopus* these cytophilic antibodies on monocytes have
been demonstrated to be IgM and IgY (Sekizawa et al. 1984).

Thrombocytes
Thrombocytes have been reported in reptilian blood in cell
numbers ranging from 25 to 350 cells per 100 leucocytes. (Pienaar
1962). These cells vary in size from 8-16 μm in length and 5-9 μm
in width (Bernstein 1938; Ryerson 1949; Taylor & Kaplan 1961;
Pienaar 1962; Desser 1978; Wood & Ebanks 1984). Thrombocytes are
rarely observed in their natural shape as their cytoplasm is of a
particular delicate and viscous nature, thus easily prone to
rupture. As such, these cells are often observed devoid of
cytoplasm or with an irregular ragged cytoplasm. When undisrupted,
thrombocytes are ellipsoidal or fusiform in shape (Fig. 6e).
Thrombocytes frequently are found clumped or aggregated together in
blood smears.

The centrally placed deeply staining nuclei in thrombocytes tend to be rather large and smooth-edged and the cytoplasm is usually neutral in appearance with occasional azurophilic granulation. Electron microscopy reveals a lobulated nucleus with dense heterochromatin attached to the nuclear membrane. In "activated" cells (aggregated together to form a clot), pseudopodia are observed which contain a fine granular material. The perinuclear cytoplasm is encircled by closely spaced microtubules which are frequently disorganized in appearance in cells that are aggregated together. Ribosomes, ovoid mitochondria and many actin-like filaments are observed in the perinuclear cytoplasm. Smooth endoplasmic reticulum is sparsely distributed throughout the peripheral cytoplasm. Vacuoles and spheroidal electron-dense inclusions are also observed in the cytoplasm. The nucleolus is present but usually indistinct and centrioles are present adjacent to the nucleus. The plasmalemma of thrombocytes is coated with an electron-dense material and in aggregates of "activated" cells numerous dense fibrin-like filaments may be observed between and around the cells (Taylor et al. 1963; Efrati et al. 1970; Desser & Weller 1979a).

A summary of histochemical data on reptilian thrombocytes is presented in Table 2. β-glycogen particles are scattered throughout the cytoplasm. Thrombocytes are probably involved in thrombus formation, blood clotting and wound healing in a manner somewhat analagous to that of mammalian blood platelets, however, the exact nature and extent of their participation has not yet been thoroughly investigated (Pienaar 1962).

Lymphocytes
    The reptilian lymphocyte varies in size, and both large (~ 14.5 µm) and small (5.5-10 µm) mononuclear cells are found in the peripheral blood (Fig. 5) (Ryerson 1949; Pienaar 1962; St. Girons 1970). The cytoplasm of reptilian lymphocytes is finely granular and lightly basophilic with Romanowsky stains, and may contain azurophilic or hyaline granules. The lymphocyte is the most prevalent leucocyte in the peripheral blood and haemopoietic tissues of most reptiles and its numbers may comprise greater than 80% of the total white blood cell count (Ryerson 1949; Pienaar 1962; St. Girons 1970).

As with other cellular constituents of reptilian blood, lymphocyte numbers are highly variable and may be influenced by a variety of factors. These include variations due to sex, with females of certain species having higher numbers than males of the same species (Duguy 1970); age, with juvenile animals possessing higher numbers than adults of the same species (Pienaar 1962); nutritional state, with malnutrition causing a depletion in lymphocyte numbers (Borysenko & Lewis 1979); and concomitant pathology due to wound healing causing an increase in lymphocyte numbers (Hiradhar et al.

1979); and a variety of parasitic infections, such as
spirochidiasis (Glazebrook et al. 1981; Wolke et al. 1982),
anasakiasis (Burke & Rodgers, 1982), haematozoa (Wood 1935; Pienaar
1962) and viruses (Jacobson et al. 1981). The highest lymphocyte
numbers within a given species appear to occur during the spring
and summer and the lowest during the winter in both splenic and
circulating blood populations (Duguy 1970; Hussein et al. 1978,
1979a,b,c) Desser 1979; In tropical reptiles, this winter decline
in lymphocyte numbers appears to be associated only with a decrease
in splenic and circulating cells, while in temperate species this
decline may be due to an outright decrease or absence of
lymphocytes in the spleen, thymus and bone marrow along with a
reduction in circulating numbers. With the onset of spring or the
end of hibernation, lymphocyte proliferation is apparent in the
haemopoietic organs, particularly the spleen, as circulating
lymphocytes begin to increase in numbers to their summer maxima
(Hussein et al. 1978, 1979a,b,c; Wright & Cooper 1981). This
fluctuation in lymphocyte numbers appears to be reflected in the
ability of temperate species, in particular, to mount primary
immune responses which may be suppressed or inhibited during
hibernation or low temperatures (Wright & Cooper 1981).

A summary of histochemical data on reptilian lymphocytes is
presented in Table 2.

At the ultrastructural level, reptilian lymphocytes possess nearly
round nuclei with a thin rim of cytoplasm surrounding the nucleus
(Fig . 7). Heterochromatin is clustered along the nuclear envelope
and a prominent central nucleolus comprises most of the nuclear
material. Cytoplasmic organelles consist of free ribosomes and a
number of mitochondria which are usually located at one pole of the
cell. Vacuoles and granular inclusions may be apparent along with
the presence of pseudopodia (Taylor et al. 1963; Ambrosius &
Hoheisel 1973a,b; Borysenko 1976b). Lymphoblasts possess large
nuclei, with prominent nucleoli and very little dense chromatin is
apparent. Proportionally more cytoplasm is present than in small
lymphocytes. Polyribosomes occur in small clusters scattered
evenly throughout the cytoplasm and mitochondria again are located
at one pole of the cell. A few short segments of rough endoplasmic
reticulum may also be apparent (Ambrosius & Hoheisel 1973a,b;
Borysenko 1976b).

Plasma cells are superficially similar to the lymphocyte in size
and shape and are generally quite rare in peripheral blood. Their
cytoplasm is intensely basophilic. The distinguishing
characteristic of this cell, other than its basophilic cytoplasm,
is a perinuclear halo (Golgi complex) which encompasses about a
third of the nucleus. The reptilian plasma cell has an eccentric
nucleus with a dense perinuclear chromatic pattern. Rough
endoplasmic reticulum fills most of the cytoplasm which is lamellar

or saccular in nature (Fig. 10). A well developed Golgi complex is
apparent and polyribosomes are few in number (Borysenko 1976b).

   Lymphocyte Heterogeneity
   The major classes of lymphocytes, T and B cells and their
subsets, are known to be involved in a diversity of lymphocyte
functions. While functional heterogeneity has been suggested in
the lymphocytes of reptiles, specific membrane markers to
distinguish different cell populations have not been well
characterized. Attempts have been made to delineate major
categories of lymphocytes by means of heterologous antisera raised
in rabbits against thymocytes and immunoglobulins of a few
reptilian species. The properties of a reptilian heterologous
anti-thymocyte serum were first studied by Pitchappan and
Muthukkaruppan (1977b) using the lizard, *Calotes versicolor*.
Rabbit anti-thymocyte serum was characterized *in vitro* by using
cytotoxicity assays and quantitative absorption analysis. The
cytolytic activity of such antiserum appears to be greatest towards
thymocytes and to a much lesser degree against lymphocyte
constituents of the spleen, bone marrow and peripheral blood.
Subsequently, similar results have been reported in the lizards
Chalcides ocellatus (El Ridi & Kandil 1981) and *Agama stellio* (Negm
& Mansour 1983), the snake *Spalerosophis diadema* (Mansour et al.
1980), and the snapping turtle, *Chelydra serpentina* (Mead &
Borysenko 1984b). From these studies, T cells appear to be the
predominant lymphocyte constituent in the thymus, as well as in the
spleen and peripheral blood. The studies by Pitchappan and
Muthukkaruppan (1977b) and Mead and Borysenko (1984b) suggest the
existence of two T lymphocyte antigens, one which is thymus
specific and another which is lymphocyte specific for both thymic
and splenic lymphocyte populations.

Attempts have also been made to distinguish lymphocyte
subpopulations on the basis of their surface immunoglobulins.
Heterologous rabbit antiserum directed against surface
immunoglobulin has been characterized in the lizards *Chalcides
ocellatus* (El Ridi & Kandil 1981) and *Agama stellio* (Negm & Mansour
1982); in the snakes, *Spalerosophis diadema* (Mansour et al. 1980)
and *Elaphe quadrivirgata* (Kawaguchi et al. 1980); and in the
snapping turtle, *Chelydra serpentina* (Mead & Borysenko 1984b). In
each of these studies, an immunoglobulin positive subset was
identified in lymphoid organs and blood. Very low reactivity was
observed with thymocytes in these studies.

The nature of lymphocyte heterogeneity has also been studied
examining mitogen responsiveness in the Florida alligator,
Alligator mississippensis (Cuchens & Clem 1979 a,b). Peripheral
blood lymphocytes were isolated by Ficoll-hypaque centrifugation,
fractionated in glass wool columns and subjected to classical T and
B cell mitogens *in vitro*. The studies showed that the non-adherent

populations (T cell-like) responded to phytohaemagglutinin (PHA) and Concanavalin A (Con A), but not to lipopolysaccharide (LPS); while the adherent population (B cell-like) was stimulated by LPS. Subsequent treatment with anti-surface immunoglobulin antisera ablated the response to LPS but not to PHA. In similar studies, lymphocyte heterogeneity was examined employing nylon wool adherent and non-adherent splenocytes from the lizard *Calotes versicolor* and assessing the capacity of these treated cells to participate in cell-mediated immunity (Manickasundari et al. 1984). It was observed that the *in vitro* capacity to migrate out of capillary tubes and to mediate migration inhibition resided in the non-adherent splenocyte population (T cell-like) and that these properties could be abrogated by treatment with anti-thymocyte antisera. These studies taken together strongly suggest the existence of considerable lymphocyte heterogeneity in reptiles, beyond the level of mere T and B cell diversity.

Humoral Immune Responses
Characteristics of antibody response to various antigens. Several studies have revealed that all extant groups of reptiles, turtles, tortoises, snakes, lizards, alligators and tuataras are capable of responding to a variety of antigens including proteins (Evans 1963; Ambrosius & Lehmann 1965; Grey 1966; Lerch et al. 1967; Lykakis 1968), bacteria (Metchnikoff 1901; Evans 1963, Maung 1963) and heterologous erythrocytes (Rothe & Ambrosius 1968). These earlier studies have been reviewed by Evans (1963), Grey (1966), Ambrosius et al. (1970) and Cohen (1971). Although naturally-occurring antibodies have been detected in a variety of reptiles (see review by Kawaguchi et al. 1978), this area is beyond the scope of our discussion, which we will limit to adaptive immune responses.

More recently, detailed analysis has been made on the kinetics of the antibody response in several species of lizards and snakes to soluble and particulate antigens (Kanakambika & Muthukkaruppan, 1972; Wetherall & Turner 1972; Coe et al. 1976; El Kes 1978; El Rouby 1978; Hussein et al. 1979a,b; El Deeb et al. 1980). A more comprehensive study on the development, morphology and the diverse functions of the immune system has been made in the garden lizard, *Calotes versicolor* (Muthukkaruppan et al. 1970; Kanakambika 1971; Manickavel 1972; Pitchappan 1975; Jayaraman 1976; Pillai 1977; Baskar 1978; Ramila 1978). Using this species, the sequence of early cellular events and the kinetics of the serum antibody response to sheep red blood cells (SRBC) have also been investigated.

Antigen-specific rosette forming cells and antibody producing cells were enumerated in the spleen after immunization with SRBC. Within three days post-immunization there is a significant increase in the number of specific antigen-binding cells in the spleen over background, reaching a peak level by the 7th day (Pillai &

Muthukkaruppan 1977). This is followed by the appearance of
plaque-forming cells attaining a peak level by the 14th day. The
serum antibody response peaks by the 21st day (Kanakambika &
Muthukkaruppan 1972). This illustrates the sequence of early
cellular events leading to antibody formation. However, the survey
of literature on reptilian immune response reveals that there is a
wide range of variability in the kinetics, level and type of
antibody generated, depending on the nature of antigen, dose and
route of antigen injected, temperature at which the animals are
held, the type of assay used, species specific characteristics, as
well as reproductive and seasonal rhythms (Cohen 1971).

Specificity and memory, the two important characteristics of
immunity have been demonstrated in some turtles, lizards and
alligators. A secondary response is characterized by early
appearance of antibody, higher peak titres and a change in
molecular species (Lerch et al. 1967; Rothe & Ambrosius 1968;
Ambrosius et al. 1970; Wetherall & Turner 1972). However, this
pattern of anamnestic response has not always been observed in
other reptilian species studied (Maung 1963; Grey 1966; Rothe &
Ambrosius 1968; Kanakambika & Muthukkaruppan 1972).

Immunoglobulin classes. Reptiles have been shown to synthesize at
least two immunoglobulin classes (Grey 1963; 1966; Ambrosius 1966;
Ambrosius et al. 1970; Lykakis 1968; Coe 1972).

Salient findings are that during the primary response "heavy" 18S
and "light" 7S antibodies are synthesized, both of which are
sensitive to mild treatment with 2-mercaptoethanol (2-ME). With
multiple immunizations, 7S and subsequently 4.5S molecular species,
both resistant to 2-ME, appear in the serum. The existence of
three different sizes of immunoglobulins has been confirmed by
Benedict & Pollard (1972) and Leslie & Clem (1972) from their
studies on different species of turtles. The sedimentation
coefficients of these immunoglobulins are   17S, 7.5S and 5.7S.
Antigenic analysis suggests that the 5.7S molecule may be a
fragment of the 7.5S molecule, and that the heavy chain of 7.5S is
distinct from that of 17S molecule.

With reference to the molecular species of antibodies, again there
is considerable variation among reptiles. For example, Wetherall &
Turner (1972) have shown that both 19S and 7S antibodies are
produced against rat erythrocytes and bovine serum albumin in the
lizard, *Tiliqua rugosa*, but *Salmonella typhimurium* could induce the
formation of only 19S antibodies which are susceptible to mild
reduction by 2-ME. An interesting finding among reptiles is that
the 19S response is quite prolonged even in the presence of high
titres of 7S antibody. This, in fact, is a common feature in
several other reptilian species studied (Maung 1963; Ambrosius &
Lehmann 1965; Grey 1966; Lerch et al. 1967; Lykakis 1968;

Marchalonis, et al. 1969). Furthermore, the humoral response to
certain antigens (eg. *S. typhimurium*) is characterized by the
formation of only 19S antibody even after repeated immunization
(Wetherall & Turner 1972), as already reported for other bacterial
antigens in tortoises (Maung 1963) and tuataras (Marchalonis et al.
1969). Thus, bacterial antigens appear to induce only IgM
antibodies in reptiles.

A more detailed analysis has been made to characterize the
antibodies produced by the tuatara, *Sphenodon punctatum*
(Marchalonis et al. 1969). The serum of immunized animals consists
of 18S and 7S immunoglobulins. Anti-flagellin (bacterial) antibody
activity is confined to the high molecular weight (18S)
immunoglobulin type, which resembles the IgM of higher vertebrates
in size and polypeptide chain structure (Marchalonis et al. 1969).
The 7S molecule possesses light chains resembling those of the 18S
protein, but the heavy chains of the 7S molecules differ from those
of the 18S, thereby indicating the presence of distinct
immunoglobulin classes. Furthermore, immunoelectrophoretic analysis
revealed the presence of slow moving proteins corresponding to IgM
and IgG of mammals (Marchalonis et al. 1969). There is very little
information with reference to other immunoglobulin classes in
reptiles. The presence of reaginic antibody may be inferred from
the demonstration of cytophilic antibody on turtle basophils (Mead
et al. 1983). Anti-surface immunoglobulin induces degranulation of
these cells and the subsequent release of histamine suggesting the
presence of cytophilic antibody which may be responsible for
immediate hypersensitivity-type reactions in these reptiles.
Because IgE has not been reported in reptiles, two classes of low
molecular weight antibodies are postulated as candidates for the
cytophilic antibody-IgY and IgN.

Rosette forming cells (RFC) & plaque-forming cells (PFC). The
types of cells involved in antigen recognition and binding have
been identified in the lizard, *Calotes versicolor*. Small, medium
and large lymphocytes exhibit the ability of specific antigen
binding, as shown by the formation of rosettes with SRBC. These
studies show that RFC in normal spleen are predominantly small and
medium lymphocytes (76%), whereas seven days after immunization
with 0.1 ml of 25% SRBC more than 55% of rosettes are formed by
large lymphoblast cells (Pillai 1977).

Utilizing the Jerne plaque assay, antibody-forming cells have been
identified in the spleen but not kidney, thymus or intestine of
chelonians (Rothe & Ambrosius 1968; Sidky & Auerbach 1968; Kassin &
Pevnitskii 1969). More detailed analysis in the lizard, *C.
versicolor* shows the presence of PFC in the spleen, blood and bone
marrow, but not in the thymus, lung, liver, kidney, or cloacal
complex, after immunization with SRBC (Kanakambika & Muthukkaruppan
1972; Pillai, 1977).

The distribution and kinetics of RFC and PFC in various lymphoid
tissues following immunization with SRBC has also been studied in
Calotes versicolor (Pillai 1977). A rapid increase in the number
of specific antigen-binding cells was observed both in the spleen
and bone marrow, attaining a peak level on the 7th day. However,
in peripheral blood, the RFC remained at the base level except for
a slight increase on the 21st day. The PFC peak response occurred
on day 14 in the spleen, bone marrow and peripheral blood. In
general, the spleen contains more RFC and PFC than bone marrow and
peripheral blood. Peritoneal exudate cells contain very few PFC
and no immune RFC. These results suggest that cellular events
leading to the differentiation of antibody-producing cells occur in
the spleen of intact lizards and that only a proportion of PFC are
recirculated. This is further evidenced by the finding that
splenectomy in the lizard completely supresses the appearance of
RFC and PFC in bone marrow and blood (Pillai 1977). Separation of
immune rosettes on fetal calf serum gradients or immune spleen
cells on a Ficoll-isopycnic gradient demonstrated that RFC and PFC
reside in different gradient fractions. Analysis of antigen
binding cells demonstrated the absence of antibody secretion at the
peak of the immune response (Pillai & Muthukkaruppan 1982).

Helper function. The procedure of low dose priming with
heterologous erythrocytes is known to stimulate the helper activity
maximally in mice (Playfair 1971; Grantham & Fitch 1975). This
model, known to selectively prime T cells, has been employed to
understand the nature of helper function in the lizard, *Calotes
versicolor*, with reference to the priming dose, and the duration
and specificity of helper activity (Muthukkaruppan et al. 1976;
Ramila 1978).

A significant increase in the number of PFC is generated in the
spleen, when lizards are primed with 2.5% SRBC or formaldehyde
treated SRBC (F-SRBC) before challenge. The enhancement is as much
as four times that of controls which were immunized with a single
dose of 2.5% SRBC. Furthermore, 25% F-SRBC is more effective in
evoking an accelerated anti-SRBC response than the other doses
tested, even though F-SRBC by itself is unable to induce the
anti-SRBC PFC response in lizards (Muthukkaruppan et al. 1976) as
in mice (Dennert & Tucker 1972). These studies also show the
specificity of the helper activity generated by SRBC and F-SRBC
priming (Ramila 1978). Furthermore, the helper function is maximal
from 5 to 7 days after priming with 2.5% SRBC and from 5 to 10 days
after priming with F-SRBC, thus demonstrating the involvement of
short lived (helper) memory cells in the accelerated humoral immune
response in reptiles.

Immune response to haptens. It is well known that the co-operative
interaction between carrier-specific and hapten-specific lymphoid
cells are essential for the development of anti-hapten antibody

response (Katz & Benacerraf 1972). Hapten-carrier conjugates are widely used in determining the function of the various cells in the immune response and to understand helper and effector cell interactions in antibody formation. This model has been successfully used to study the mechanism of cellular interaction to antibody formation in tortoises (Ambrosius & Frenzel 1972). Recently, the mechanism and types of lymphocytes involved in cell collaboration have been studied in the garden lizard using trinitrophenol (TNP) as a hapten and SRBC and ovalbumin (ova) as carriers (Ramila 1978). The peak anti-TNP PFC response occurs 10 days after immunization with TNP-SRBC (0.1 ml 25%) or TNP-ova. Lower concentrations of SRBC (2.5%) or F-SRBC effectively prime the lizard for an accelerated anti-TNP PFC response. Maximum carrier effect was observed when the hapten-carrier complex was injected 10 days after priming for both TNP-SRBC or TNP-protein conjugates. An accelerated anti-TNP response is elicited only when the same carrier is used for both priming and challenge, thereby indicating the carrier specificity in anti-hapten responses in the lizard.

These results are in general agreement with the findings in mice, with reference to effective priming by F-SRBC (Dennert & Tucker 1972), requirement for a critical time interval between priming and challenge (Mitchison 1971) and carrier specificity (Cheers & Breitner 1971). Thus, in analogy it may be suggested that the carrier primed enhancement in the anti-TNP antibody response is the result of cell collaboration between carrier-specific helper T cells and hapten-specific antibody producing precursor B cells in the lizard. This contention is confirmed by adult thymectomy experiments in lizards (Ramila 1978).

Cell-mediated immune responses

Transplantation reactions. The first account of allograft rejection in reptiles is the classic study by May (1923), using the chameleon, *Anolis carolinensis*. He reported that while autografts survived permanently, allografts appeared healthy for a time, but were totally destroyed by 60-90 days post-transplantation at 22.5°C. He provided a vivid and highly accurate description of the inflammatory reaction, round cell infiltration and graft destruction, long before these events were established as hallmarks of cell-mediated immunity. With the renewed interest in transplantation immunity in the past two decades, a number of other reptilian species have been examined. Although one sees much variation from species to species, the general pattern is one of chronic rejection of first-set skin allografts and accelerated rejection of the second-set, at moderate temperatures (see reviews, Borysenko 1970; Worley & Jurd 1979). The relative slowness of allograft rejection has been attributed to the absence of a major histocompatibility locus, rather than to a primitive or deficient immune system (Cohen & Borysenko 1970).

Studies on several species of reptiles, the snapping turtle,
Chelydra serpentina (Borysenko 1970), the teiid lizard,
Cnemidophorus sexlineatus (Maslin 1967), the iguana, *Ctenosaura
pectinata* (Cooper & Aponte 1968), the garden lizard, *Calotes
versicolor* (Manickavel & Muthukkaruppan 1969; Manickavel 1972), the
European green lizard, *Lacerta viridis* (Worley & Jurd 1979) and the
garter snake, *Thamnophis sirtalis* (Terebey 1972) have documented a
number of characteristic features of transplantation immunity in
the Reptilia. Firstly, skin xenografts are rejected faster than
allografts, Secondly, second-set allografts and xenografts are
rejected more rapidly than the first-set, and finally the rate of
graft rejection is markedly affected by environmental temperature.

With regard to the last point, in turtles the relationship between
allograft survival time and temperature is not linear. A break
occurs at about 25°C, implying that there are at least two
temperature sensitive phases in allograft rejection. Although
allograft rejection is completely suppressed at 10°C, antigen
recognition apparently is not inhibited, since transfer of turtles
to higher temperatures resulted in accelerated rejection (day of
transfer considered day 1). Alternatively, transfer of turtles to
low temperatures after graft rejection had once initiated slowed
the rejection process. These findings led to the hypothesis that
at low temperatures an early, but post-recognition event, such as
proliferation was primarily affected, while at higher temperature
ranges, effector functions such as antibody and lymphokine
production were primarily affected (Borysenko 1979).

In vitro correlates of cell-mediated immunity (CMI). The *in vitro*
capillary leucocyte migration inhibition (MI) technique has been
successfully adapted to assess the CMI response to skin allografts
in the garden lizard, *Calotes versicolor* (Jayaraman &
Muthukkaruppan 1977). After grafting allogeneic skin tissue in a
unidirectional fashion (Manickavel & Muthukkaruppan 1969),
sensitized spleen cells were cultured in capillary tubes in the
presence or absence of antigen prepared from the donor spleen. For
the sequential study of the onset of CMI to skin allografts, spleen
cells were tested at specified intervals using the MI assay and
compared with morphological criteria of rejection. It is evident
that the lizards respond to skin allografts, as indicated by
melanophore degranulation, and that a significant inhibition of
migration of sensitized spleen cells occurs by the 7th day, in the
presence of the respective donor antigen. This response is
maintained as long as one month after grafting, the time at which
the clinical manifestation of allograft rejection is observed. The
MI of allograft sensitized spleen cells is an antigen specific
event (Jayaraman & Muthukkaruppan 1977). Thus, the *in vitro* MI
technique is a sensitive and specific method of measuring the CMI
response to skin allografts in reptiles. This is a highly useful
assay mainly because the assessment of allograft reaction in

reptiles is problematic due to its chronic nature and due to the use of multiple morphological criteria as the end point. Recently, the ability of peripheral blood lymphocytes to participate in two-way mixed lymphocyte reactions has been demonstrated in alligators (Cuchens & Clem 1979b). In a recent study on *Calotes versicolor*, the property of mediating antigen-specific migration inhibition was shown to reside in a non-adherent (on nylon wool) cell population whose capacity to do so was abrogated by treatment with anti-thymocyte antiserum (Manickasundari et al. 1984).

Graft-vs-host (GVH) reactions. The GVH reaction has been studied in only one reptilian species, *Chelydra serpentina*. *In vitro* experiments have shown that splenomegaly can be induced in immature spleen explants upon exposure to adult allogeneic spleen cells (Sidky & Auerbach 1968). Another study, using the same turtle species, showed a wide spectrum of *in vivo* GVH reactions, from acute to chronic, which again suggests the presence of multiple weak histocompatibility differences in outbred populations of turtles. The incidence of acute reactivity, as evidenced by splenomegaly and early mortality, was highest among those turtles that received spleen cells from donors of distant geographic locations. The incidence of lethal GVH reactions and the rapidity of these reactions were also greater at elevated temperatures (Borysenko & Tulipan 1973) and completely suppressed at 10°C (Borysenko 1972). The parallels between GVH reactions and graft rejection are quite striking, with the single exception that xenogeneic spleen cells were ineffective in inducing GVH reactivity, while skin xenografts were rejected more rapidly than allografts.

Delayed hypersensitivity reaction. Lizards *(Tarentola annularis)* stimulated in summer with 1 mg Bacillus Calmette-Guerin, displayed cutaneous hypersensitivity following intradermal challenge with purified protein derivative (PPD) at 21 to 34 days post-sensitization. Lesions involving desquamation, erythema, induration and dryness began to appear 3 to 5 days post-challenge, persisted 5 to 6 days and then subsided. Histologically, degeneration of horny layer, degranulation of melanophores and infiltration of lymphoid cells were observed. Thus, the lizard's cutaneous response to tuberculin has many of the features characteristic of the mammalian delayed hypersensitivity reaction (Badir et al. 1981).

CMI response to sheep erythrocytes and hapten-carrier complex. The capillary leucocyte migration inhibition (MI) technique has also been applied to detect CMI directed against SRBC in the lizard, *Calotes versicolor* (Jayaraman & Muthukkaruppan 1978a,b). The migration of spleen cells obtained from SRBC injected lizards is inhibited in the presence of specific antigen. The MI response is mediated by sensitized lymphocytes and is an antigen-specific

phenomenon. Sensitization of spleen cells against SRBC is evident
as early as four days after immunization by the inhibition of their
migration *in vitro* in the presence of specific antigen. In this
system, MI reaches a maximum on day 7 and is maintained steadily at
least up to 30 days. Administration of a very low dose of SRBC
(10,000) results in a high degree of MI, without the production of
PFC. Furthermore, formalinized SRBC (F-SRBC) which are unable to
stimulate the generation of PFC response in mice (Dennert & Tucker
1972) and in lizards (Ramila 1978) are shown to induce a good MI
response. These results and the finding that MI is a
thymus-dependent phenomenon (Jayaraman 1976; Manickasundari et al.
1984) indicate that the MI response of spleen cells is an *in vitro*
manifestation of CMI function in the lizard.

The nature of CMI response to hapten-carrier complex has been
investigated in the garden lizard, using the *in vitro* capillary MI
assay (Ramila 1978). A significant percentage of MI was observed
in spleen cells obtained from lizards immunized with 500 ug of
TNP-ova. It was also shown that the migration of cells is not
inhibited in the presence of TNP-BGG (TNP-conjugated bovine gamma
globulin), thereby indicating the specificity of MI response to
hapten-carrier complex as in mice (David et al. 1964; Snippe et al.
1975).

Normal lymphocyte transfer reaction. Chronic allograft destruction
in mammals results when skin and other tissue grafts are
transplanted across minor histocompatibility differences. By
analogy, it was suggested that chronic rejection in reptiles
reflects the activities of antigenic products of minor
histocompatibility loci acting in the absence of a major
histocompatibility system (Cohen & Borysenko 1970; Cohen 1971;
Cohen & Collins 1977). This problem has been approached by
utilizing the normal lymphocyte transfer (NLT) reaction. In
mammals, NLT reaction develops at the injection site, following
intradermal inoculation of immunocompetent lymphocytes into
unimmunized, allogeneic hosts differing from the donor at the major
histocompatibility locus. The reaction is a manifestation of a
local graft-vs-host reaction, as well as a host cellular immune
response against strong histocompatibility antigens on the
inoculated cells (Streilein & Billingham 1970; Zaakarian &
Billingham 1972; Sidky & Auerbach 1975). The appearance and tempo
of NLT reactions in irradiated or non-irradiated lizards, *Tarentola
annularis*, following intradermal injection of $2\text{-}3 \times 10^6$ viable
allogeneic spleen lymphocytes are similar to those observed in
guinea pigs, hamsters, dogs, rats and mice (Badir et al. 1981).
The data thus suggest that lizards are capable of eliciting
vigorous CMI responses to alloantigens. This implies that strong
antigenic disparity occurs in *T. annularis*. Furthermore, recent
studies clearly demonstrate the ability of the sensitized
lymphocytes of the lizard against alloantigens to manifest the

specific CMI response *in vitro*, as early as seven days after
grafting (Jayaraman & Muthukkaruppan 1977).

In light of the findings that lizards show a strong delayed
hypersensitivity reaction to tuberculin, a strong lymphocyte
transfer reaction (Badir et al. 1981), a specific MI response *in
vitro* to alloantigens and a pronounced and rapid MI response to
SRBC (Jayaraman & Muthukkaruppan 1977, 1978a,b), chronic allograft
rejection in these reptiles can not be ascribed to a deficiency in
cell-mediated immunity or to the absence of strong
histocompatibility antigens. It is quite possible that the chronic
response might be due to the characteristic organization of
reptilian integument rather than to the sluggish nature of the
immune response induced by transplantation antigens. For this
reason, it would be of interest to investigate the nature of the
allograft reaction in reptiles, grafting other tissues in
appropriate sites.

CONCLUDING REMARKS

        It is evident that all reptiles are capable of mounting
both nonspecific and highly specific immune responses mediated by
heterogeneous populations of blood cells. Reptilian lymphocytes,
in particular, produce several immunoglobulin types and in some
instances different subclasses of lymphocytes co-operate to amplify
or suppress immune respones. Overwhelming data demonstrate that
the highly diverse orders of reptiles are in no way primitive or
deficient with respect to their immune systems. Instead, among the
reptiles one finds some variation in the structure of lymphoid
organs that is reflected in the ontogeny of their haematopoietic
systems. Furthermore, because reptiles are distributed so widely
geographically and occupy so many ecological niches, environmental
influences on blood cell function are quite complex and not simply
related to environmental temperature and nutritional factors alone.
The immune responses of the endotherms, both birds and mammals, are
much more predictable in this sense. For this reason, comparative
immunologists and haematologists must take special precautions to
simulate natural environments when studying the systems of reptiles
and not overlook either environmental or internal factors which may
enhance or hinder their ability to resist disease.

The future prospects for the use of reptiles as immunological and
haematological models are many. Several species have been
successfully maintained in laboratory environments for years and an
abundance of baseline data is presently available. The current
applications of modern immunological and biotechnological
techniques to reptilian blood cells greatly enhances the prospects
of acquiring more detailed knowledge of cellular interactions and
functions.

Acknowledgements
The authors would like to express their gratitude to the
following individuals for providing haematologic material and data
in addition to many helpful comments and suggestions, George Benz,
University of Connecticut; Anita George, University of Rhode
Island; Molly Lutcavage, University of Miami; Frederic L. Frye,
D.V.M., Davis, CA; Joseph J. Schall, Ph.D., University of Vermont
and Emil Dolensek, D.V.M., The Bronx Zoo/New York Zoological
Society for the use of the Rand Collection and zoo's extensive
haematologic slide collection. The authors also thank Bela Bodey
M.D., Ph.D. for constructive criticism and suggestions and Diana
Lomber for preparation of the manuscript.

REFERENCES

Alder, A. & Huber, E. (1923). Untersuchungen uber Blutzellen und
    Zellbildung bei Amphibien und Reptilien. Folia Hemat., *29*,
    1-22.
Ambrosius, H. (1966). Comparative investigation of immune
    globulins of various vertebrate classes. Nature, *209*, 524.
Ambrosius, H. & Frenzel, E.M. (1972). Anti-DNP antibodies in
    carps and tortoises. Immunochemistry, *9*, 65-71.
Ambrosius, H.J., Hammerling, R., Richter, R. & Schimke, R. (1970).
    Immunoglobulins and the dynamics of antibody formation in
    poikilothermic vertebrates. *In* Developmental Aspects of
    Antibody Formation and Structure, eds. Sterzl, J. and Riha,
    I., Vol. II, pp. 727-44. New York, NY: Academic Press.
Ambrosius, H. & Hoheisel, G. (1973a). Ultrastructure of antibody
    producing cells of reptiles. I. Spleen cells of the tortoise
    (*Agrionemys horsfieldi* Gray). Acta Biol. Med. Ger., *31*,
    733-40.
Ambrosius, H. & Hoheisel, G. (1973b). Ultrastructure of antibody
    producing cells of reptiles. II. Blood cells of the tortoise
    (*Agrionemys horsfieldi* Gray). Acta Biol. Med. Ger., *31*,
    741-48.
Ambrosius, H. & Lehmann, R. (1965). Beitrage zur Immunobiologie
    Poikilothermer Wirbeltiere. II. Immunologische Untersuchungen
    an Schilokroeten (*Testudo hermanni* Gmelin). Z. Immun.
    Allergieforsch., *128*, 81-104.
Badir, N., Afifi, A. & El Ridi, R. (1981). Cell-mediated immunity
    in the gecko, *Tarentola annularis*. Folia Biol., *27*, 28-36.
Baskar, S. (1978). Mechanism of low-zone tolerance to sheep
    erythrocytes in the lizard, *Calotes versicolor*. Ph.D. Thesis,
    Madurai University, Tamilnadu, India.
Benedict, A.A. & Pollard, L.W. (1972). Three classes of
    immunoglobulins in the sea turtle, *Chelonia mydas*. Folia
    Microbiol., *17*, 75-78.
Bernstein, R.E. (1938). Blood cytology of the tortoise, *Testudo
    geometrica*. S. African J. Sci., *35*, 327-31.
Bockman, D.E. (1970). The thymus. *In* Biology of the Reptilia,
    eds. Gans, G. and Parsons, T.C., Vol. 3, Morphology, pp.
    111-33. New York, NY: Academic Press.
Borysenko, M. (1970). Transplantation immunity in Reptilia.
    Transplant. Proc., *2*, 299-306.
Borysenko, M. (1972). Immunosuppressive effects of low
    temperature on the graft-vs-host reaction in the snapping
    turtle. Anat. Rec., *175*, 275-76.
Borysenko, M. (1976a). Changes in spleen histology in response to
    antigenic stimulation in the snapping turtle, *Chelydra
    serpentina*. J. Morphol., *149*, 223-42.
Borysenko, M. (1976b). Ultrastructural analysis of normal and
    immunized spleen of the snapping turtle, *Chelydra serpentina*.
    J. Morphol., *149*, 243-64.

Borysenko, M. (1978). Lymphoid tissues and cellular components of
    the reptilian immune system. *In* Animal Models of Comparative
    and Developmental Aspects of Immunity and Disease, eds.
    Gershwin, M.E. and Cooper, E.L., pp. 63-79. New York, NY:
    Pergamon Press.
Borysenko, M. (1979). Evolution of lymphocytes and vertebrate
    alloimmune reactivity. Transplant. Proc., *11*, 1123-130.
Borysenko, M. & Cooper, E.L. (1972). Lymphoid tissue in the
    snapping turtle, *Chelydra serpentina*. J. Morphol., *138*,
    487-97.
Borysenko, M. & Lewis, S. (1979). The effect of malnutrition on
    immunocompetence and whole body resistance to infection in
    Chelydra serpentina. Dev. Comp. Immunol., *3*, 89-100.
Borysenko, M. & Tulipan, P. (1973). The graft-vs-host reaction in
    the snapping turtle, *Chelydra serpentina*. Transplantation, *16*,
    496-504.
Burke, J.B. & Rodgers, L.J. (1982). Gastric ulceration associated
    with larval nematodes (*Anasakis sp.* Type I) in pen reared
    green turtles *(Chelonia mydas)* from Torres Strait. J. Wildl.
    Dis., *18*, 41-46.
Caxton-Martins, A.E. (1977). Cytochemistry of blood cells in
    peripheral smears of some West African reptiles. J. Anat.,
    124, 393-400.
Caxton-Martins, A.E. & Nganwuchu, A.M. (1978). A cytochemical
    study of the blood of the rainbow lizard *(Agama agama)*. J.
    Anat., *125*, 477-80.
Charipper, H.A. & Davis, D. (1932). Studies on the Arneth count
    XX. A study of the blood cells of *Pseudemys elegans* with
    special reference to the polymorphonuclear leucocytes. Quart.
    J. exp. Physiol., *21*, 371-82.
Chartrand, S.L., Litman, G.W., LaPointe, N., Good, R.A. & Frommel,
    D. (1971). The evolution of the immune response. XII. The
    immunoglobulins of the turtle. Molecular requirements for
    biologic activity of 5.7S immunoglobulin. J. Immunol., *107*,
    1-11.
Cheers, C. & Breitner, J.C.S. (1971). Co-operation between
    carrier reactive and hapten sensitive cells *in vitro*. Nature,
    232, 248-50.
Coe, J.E. (1972). Immune response in the turtle *(Chrysemys
    picta)*. Immunology, *23*, 45-52.
Coe, J.E., Leong, D., Portis, J.L. & Thomas, L.A. (1976). Immune
    response in the garter snake *(Thamnophis ordinoides)*.
    Immunology, *31*, 417-24.
Cohen, N. (1971). Reptiles as models for the study of immunity
    and its phylogenesis. J. Amer. Vet. Med. Assoc., *159*,
    1662-671.
Cohen, N. & Borysenko, M. (1970). Acute and chronic graft
    rejection. Possible phylogeny of transplantation antigens.
    Transplant. Proc. *2*, 333-36.

Cohen, N. & Collins, N.H. (1977). Major and minor histocompatibility systems of ectothermic vertebrates. *In* The Major Histocompatibility Systems in Man and Animals, ed. Gotze, D., pp. 313-37. Berlin: Springer-Verlag.

Cooper, E.L. & Aponte, A. (1968). Chronic allograft rejection in the iguana, *Ctenasaura pectinata*. Proc. Soc. Exp. Biol. Med., *128*, 150-54.

Cox, W.R., Rapley, W.A. & Barker, I.K. (1980). Herpes virus-like infection in a painted turtle. *(Chrysemys picta)*. J. Wildl. Dis., *16*, 445-49.

Cuchens, M.A. & Clem, L.W. (1979a). Phylogeny of lymphocyte diversity. III. Mitogenic responses of reptilian lymphocytes. Dev. Comp. Immunol., *3*, 287-97.

Cuchens, M.A. & Clem, L.W. (1979b). Phylogeny of lymphocyte heterogeneity. IV. Evidence for T-like and B-like cells in reptiles. Dev. Comp. Immunol., *3*, 465-75.

David, J.R., Lawrence, H.S. & Thomas, L. (1964). Delayed hypersensitivity *in vitro*. III. The specificity of the hapten protein conjugates in the inhibition of cell migration. J. Immunol., *93*, 279-82.

Dennert, G. & Tucker, D.F. (1972). Selective priming of T-cells by chemically altered cell antigens. J. exp. Med., *136*, 656-61.

Dessauer, H.C. (1970). Blood chemistry of reptiles: Physiological and evolutionary aspects. *In* Biology of the Reptilia, eds. Gans, C. & Parsons, T.S., Vol. 3, Morphology, pp. 1-72. New York, NY: Academic Press.

Desser, S.S. (1978). Morphological, cytochemical and biochemical observations on the blood of the tuatara, *Sphenodon punctatus* N. Zealand J. Zool., *5*, 503-8.

Desser, S.S. (1979). Haematological observations on a hibernating tuatara, *Sphenodon punctatus*. N. Zealand J. Zool., *6*, 77-8.

Desser, S.S. & Weller, I. (1979a). Ultrastructural observations on the erythrocytes and thrombocytes of the tuatara. *Sphenodon punctatus* (Gray). Tissue & Cell., *11*, 717-26.

Desser, S.S. & Weller, I. (1979b). Ultrastructural observations on the granular leucocytes of the tuatara, *Sphenodon punctatus* (Gray). Tissue & Cell., *11*, 703-15.

Duguy, R. (1970). Numbers of blood cells and their variation. *In* Biology of the Reptilia, eds. Gans, C. & Parsons, T.C., Vol. 3, Morphology, pp. 93-109. New York, N.Y.: Academic Press.

Efrati, P., Nir, E. & Yaari, A. (1970). Morphological and cytochemical observations on cells of the hemopoietic system of *Agama stellio* (Linnaeus). Israel J. Med. Sci., *6*, 23-31.

El Deeb, S., El Ridi, R. & Badir, N. (1980). Effect of seasonal and temperature changes on humoral response of *Eumeces schneideri*. Dev. Comp. Immunol., *4*, 753-58.

Elkan, E. & Zwart, P. (1967). The ocular disease of young terrapins caused by Vitamin A deficiency. Path. Vet., *4*, 201-22.

El Kes, N. (1978). Experimental studies on immunologic performance of some Egyptian reptiles. M.Sc. Thesis, Faculty of Science, Cairo University.

El Ridi, R., El Deeb, S. & Zada, S. (1980). The gut-associated lympho-epithelial tissue (GALT) of lizards and snakes. *In* Aspects of Developmental and Comparative Immunology, ed. Solomon, J.B., Vol. I, pp. 233-39, New York, N.Y.: Pergamon Press.

El Ridi, R. & Kandil, O. (1981). Membrane markers of reptilian lymphocytes. Dev. Comp. Immunol., *5*, Suppl. 1, 143-50.

El Rouby, S. (1978). Studies of immunological competence in some Egyptian snakes, M.Sc. Thesis, Faculty of Science, Cairo University.

Evans, E.E. (1963). Antibody response in amphibia and reptilia. Fed. Proc., *22*, 1132-37.

Evans, R.H. (1983). Chronic bacterial pneumonia in free-ranging eastern box turtles (*Terrapene carolina carolina*). J. Wildl. Dis., *19*, 349-52.

Frye, F.L. (1981). Biomedical and Surgical Aspects of Captive Reptile Husbandry. Chapter 4, Hematology, pp. 61-111. Edwardsville, KS: Veterinary Medicine Publishing Co.

Gerzeli, G. (1954). Osservazioni d'istochimica comparata: I polisaccardi negli elementi ematici circolanti dei vertebrati inferiori. Arch. Zool. Ital., *39*, 1-14.

Glazebrook, J.S., Campbell, R.S.F. & Blair, D. (1981). Pathological changes associated with cardiovascular trematodes (Digenea: Spirorchidae) in a green sea turtle *Chelonia mydas* (L). J. Comp. Path, *91*, 361-68.

Grantham, W.G. & Fitch, F.W. (1975). The role of antibody feedback inhibition in the regulation of the secondary antibody response after high and low dose priming. J. Immunol., *114*, 394-98.

Greiner, E.C., Forrester, D.J., & Jacobson, E.R. (1980). Helminths of mariculture-reared green turtles (*Chelonia mydas mydas*) from Grand Cayman, British West Indies. Proc. Helm. Soc. Wash., *97*, 142-44.

Grey, H.M. (1963). Phylogeny of the immune response. Studies on some physical, chemical, and serologic characteristics of antibody produced in the turtle, J. Immunol., *91*, 819-25.

Grey, H.M. (1966). Structure and kinetics of formation of antibody in the turtle. *In* Phylogeny of Immunity, eds. Smith, R.T., Miescher, P.S.A. & Good, R.A., pp. 227-235. Gainesville, FL: University of Florida Press.

Heady, J. & M. Rogers. (1962). Turtle blood cell morphology. Proc. Iowa Acad. Sci., *69*, 587-90.

Hiradhar, P.K., Kothari, J.S. & Shah, R.V. (1979). Studies on changes in blood-cell populations during tail regeneration in the gekkonid lizard, *Hemidactylus flaviviridis*. Neth. J. Zool., *29*, 129-36.

Hussein, M.F., Badir, N., El Ridi, R. & Akef, M. (1978).
    Differential effect of seasonal variation on lymphoid tissue
    of the lizard, *Chalcides ocellatus*. Dev. Comp. Immunol., *2*,
    297-310.
Hussein, M.F., Badir, N., El Ridi, R. & Akef, M. (1979a).
    Lymphoid tissues of the snake, *Spalerosophis diadema*, in the
    different seasons. Dev. Comp. Immunol., *3*, 77-88.
Hussein, M.F., Badir, N., El Ridi, R. & Charmy, R. (1979b).
    Natural heterohemagglutinins in the serum of the lizard, *Agama
    stellio*. Dev. Comp. Immunol., *3*, 643-52.
Hussein, M.F., Badir, N., El Ridi, R. & El Deeb, S. (1979c).
    Effect of seasonal variation on the immune system of the
    lizard, *Scincus scincus*. J. exp. Zool., *209*, 91-6.
Jacobson, E.R. (1980). Necrotizing mycotic dermatitis in snakes:
    Clinical and pathologic features. J. Amer. Vet. Med. Assoc.,
    177, 838-41.
Jacobson, E., Clubb, S. & Greiner, E. (1983). Amebiasis in
    red-footed tortoises. J. Amer. Vet. Med Assoc., *183*, 1192-94.
Jacobson, E.R. & Collins, B.R. (1980). Tonsil-like esophageal
    lymphoid structures of Boid snakes. Dev. Comp. Immunol., *4*,
    703-11.
Jacobson, E.R., Gaskin, J., Iverson, W.D., Harvey, J. & Nelson, G.
    (1980a). Spirochetemia in a rhinoceros iguana. J. Amer. Vet.
    Med. Assoc., *177*, 918-21.
Jacobson, E., Gaskin, J.M., Simpson, C.F. & Terell, T.G. (1980b).
    Paramyxo-like virus infection in a rattlesnake. J. Amer. Vet.
    Med. Assoc., *177*, 796-99.
Jacobson, E., Gaskin, J.M., Page, D., Iverson, W.O. & Johnson, J.W.
    (1981). Illness associated with paramyxo-like virus infection
    in zoologic collection of snakes. J. Amer. Vet. Med. Assoc.,
    179, 1227-30.
Jayaraman, S. (1976). Modulation of humoral and cell-mediated
    immune response to sheep erythrocytes in the lizard, *Calotes
    versicolor*. Ph.D. Thesis, Madurai University, Tamilnadu,
    India.
Jayaraman, S. & Muthukkaruppan, V.R. (1977). *In vitro* correlate
    of transplantation immunity: Spleen cell migration inhibition
    in the lizard, *Calotes versicolor*. Dev. Comp. Immunol., *1*,
    133-44.
Jayaraman, S. & Muthukkaruppan, V.R. (1978a). The detection of
    cell-mediated immunity to sheep erythrocytes by the capillary
    migration inhibition technique in the lizard, *Calotes
    versicolor*. Immunology, *34*, 231-40.
Jayaraman, S. & Muthukkaruppan, V.R. (1978b). Influence of route
    and dose of antigen on the migration inhibition and plaque
    forming cell responses to sheep erythrocytes in the lizard,
    Calotes versicolor. Immunology, *34*, 241-46.

Kanakambika, P. (1971). Studies on the morphology, development and immunological function of the spleen in the lizard, *Calotes versicolor*. Ph.D. Thesis, Annamalai University, Tamilnadu, India.

Kanakambika, P. & Muthukkaruppan, V.R. (1972). Immune response to sheep erythrocytes in the lizard, *Calotes versicolor*, J. Immunol., *109*, 415-19.

Kanakambika, P. & Muthukkaruppan, V.R. (1973). Lymphoid differentiation and organization of the spleen in the lizard, *Calotes versicolor*. Proc. Ind. Acad. Sci. B, *78*, 37-44.

Kassin, L.F. & Pevnitskii, L.A. (1969). Detection of antibody-forming cells in turtles' spleen by means of a modified method of local hemolysis in gel. Bull. Exp. Biol. Med., *67*, 287-90.

Katz, D.H. & Benacerraf, B. (1972). Regulatory influence of activated T-cells on B-cell response to antigens. Adv. Immunol., *15*, 1-94.

Kawaguchi, S., Hiruki, T., Harada, T. & Morikawa, S. (1980). Frequencies of cell-surface or cytoplasmic IgM-bearing cells in the spleen, thymus and peripheral blood of the snake, *Elaphe quadrivirgata*. Dev. Comp. Immunol., *4*, 559-64.

Kawaguchi, S., Muramatsu, F. & Mitsuhashi, S. (1978). Natural hemolytic activity of snake serum. I. Natural antibody and complement. Dev. Comp. Immunol., *2*, 287-96.

Kelényi, G. & Nemeth, A. (1969). Comparative histochemistry and electron microscopy of the eosinophil leucocytes of vertebrates. I. Study of avian, reptile, amphibian and fish leucocytes. Acta Biol. Acad. Sci. Hung, *20*, 405-22.

Kroese, F.G.M. & van Rooijen, N. (1982). The architecture of the spleen of the red-eared slider, *Chrysemys scripta elegans* (Reptilia: Testudines). J. Morphol., *173*, 279-84.

Kroese, F.G.M. & van Rooijen, N. (1983). Antigen trapping in the spleen of the turtle, *Chrysemys scripta elegans*. Immunology, *49*, 61-68.

Kumar De, T. & Maiti, B.R. (1981). Differential leukocyte count in both sexes of an Indian soft-shelled turtle (*Lissemys punctata punctata*) Z. Mikrosk.-anat. Forsch., *95*, 1965-69.

Leibovitz, L., Rebell, G. & Boucher, G.C. (1978). *Caryospora cheloniae* sp.n.: A coccidial pathogen of mariculture-reared green sea turtles (*Chelonia mydas mydas*). J. Wildl. Dis., *14*, 269-75.

Lerch, E.G., Huggins, S.E. & Bartel, A.H. (1967). Comparative immunology. Active immunization of young alligators with haemocyanin. Proc. Soc. Exp. Biol. Med., *124*, 448-51.

Leslie, G.A. & Clem, L.W. (1972). Phylogeny of immunoglobulins structure and function. VI. 17S, 7.5S and 5.7S anti-DNP of the turtle, *Pseudamys scripta*. J. Immunol., *108*, 1656-64.

Lykakis, J.J. (1968). Immunoglobulin production in the European pond tortoise, *Emys orbicularis*, immunized with serum protein antigens. Immunology, *14*, 799-808.

Manickasundari, M., Selvaraj, P. & Pitchappan, R.M. (1984).
    Studies on the T-cells of the lizard, *Calotes versicolor*. Dev.
    Comp. Immunol., *8*, 367-74.
Manickavel, V. (1972). Studies on skin transplantation immunity
    in the lizard, *Calotes versicolor*. Ph.D. Thesis, Annamalai
    University, Tamilnadu, India.
Manickavel, V. & Muthukkaruppan, V.R. (1969). Allograft rejection
    in the lizard, *Calotes versicolor*. Transplantation, *8*, 307-11.
Mansour, M.H., El Ridi, R. & Badir, N. (1980). Surface markers of
    lymphocytes in the snake, *Spalerosophis diadema*. I.
    Investigation of lymphocyte surface markers. Immunology, *40*,
    605-13.
Marchalonis, J.J., Ealey, E.H.M. & Diener, E. (1969). Immune
    response of the tuatara, *Sphenodon punctatum*. Aust. J. exp.
    Biol. Med. Sci., *47*, 367-80.
Maslin, T.P. (1967). Skin grafting in the bisexual Teiid lizard,
    Cnemidophorus sexlineatus, and in the unisexual, *C.
    tesselatus*. J. exp. Zool., *166*, 137-50.
Maung, R.T. (1963). Immunity in the tortoise, *Testudo ibera*. J.
    Path. Bact., *85*, 51-66.
May, R.M. (1923). Skin grafts in the lizard, *Anolis carolinensis*.
    Brit. J. exp. Biol., *1*, 539-55.
McCauley, W.J. (1956). The gross anatomy of the lymphatic system
    of *Alligator mississippiensis*. Am. J. Anat., *99*, 189-209.
Mead, K.F. & Borysenko, M. (1984a). Surface immunoglobulin on
    granular and agranular leukocytes in the thymus and spleen of
    the snapping turtle, *Chelydra serpentina*. Dev. Comp.
    Immunol., *8*, 109-20.
Mead, K.F. & Borysenko, M. (1984b). Turtle lymphocyte surface
    antigens in *Chelydra serpentina* as characterized by rabbit
    anti-turtle thymocyte sera. Dev. Comp. Immunol., *8*, 351-58.
Mead, K.F., Borysenko, M. & Findlay, S.R. (1983). Naturally
    abundant basophils in the snapping turtle, *Chelydra
    serpentina*, possess cytophilic surface antibody with reaginic
    function. J. Immunol., *130*, 334-40.
Metchinikoff, E.L. (1901). L'immunite dans les maladies
    infectieuses. Paris: Masson.
Michels, N.A. (1923). The mast cell in lower vertebrates. La
    Cellule, *33*, 337-462.
Mitchison, N.A. (1971). The carrier effect in the secondary
    response to hapten-protein conjugates. II. Cellular
    co-operation. Eur. J. Immunol., *1*, 18-27.
Muthukkaruppan, V.R., Kanakambika, P., Manickavel, V. &
    Veeraraghavan, K. (1970). Analysis of the development of the
    lizard, *Calotes versicolor*. I. A series of normal stages in
    the embryonic development. J. Morphol., *130*, 479-90.
Muthukkaruppan, V.R., Pillai, P.S. & Jayaraman, S. (1976). Immune
    functions of the spleen in the lizard. *In* Immunoaspects of the
    spleen, eds. Battisto, J.R. & Streilein, J.W., pp. 61-73.
    Amsterdam: Elsevier/North-Holland Biomedical Press.

Negm, H. & Mansour, M.H. (1982). Phylogenesis of lymphocyte diversity. I. Immunoglobulin determinants on the lymphocyte surface of the lizard. *Agama stellio*. Dev. Comp. Immunol., *6*, 519-32.

Negm, H. & Mansour, M.H. (1983). Phylogenesis of lymphocyte diversity. II. Characterization of *Agama stellio* Ig-negative lymphocytes by a heterologous anti-thymocyte serum. Dev. Comp. Immunol., *7*, 507-16.

Otis, V.S. (1973). Hemocytological and serum chemistry parameters of the African puff adder, *Bitis arietans*. Herpetologica, *29*, 110-16.

Pienaar, U. de V. (1962). Haematology of some South African Reptiles. Johannesburg: Witwatersrand University Press.

Pillai, P.S. (1977). Studies on the role of antigen-binding cells in the immune response to sheep erythrocytes in the lizard, *Calotes versicolor*. Ph.D. Thesis, Madurai University, Tamilnadu, India.

Pillai, P.S. & Muthukkaruppan, V.R. (1977). The kinetics of rosette-forming cell response against sheep erythrocytes in the lizard. J. exp. Zool., *199*, 97-104.

Pillai, P.S. & Muthukkaruppan, V.R. (1982). The relationship between immune rosette forming cells and plaque forming cells in the lizard, *Calotes versicolor*. Dev. Comp. Immunol., *6*, 321-28.

Pitchappan, R.M. (1975). Studies on the development and immune functions of the thymus in the lizard, *Calotes versicolor*. Ph.D. thesis, Madurai University, Tamilnadu, India.

Pitchappan, R.M. & Muthukkaruppan. V.R. (1977a). Analysis of the development of the lizard, *Calotes versicolor*. II. Histogenesis of the thymus, Dev. Comp. Immunol, *1*, 217-30.

Pitchappan, R.M. & Muthukkaruppan, V.R. (1977b). *In vitro* properties of heterologous anti-lizard thymocyte serum. Proc. Ind. Acad. Sci. B, *85*, 1-12.

Playfair, J.H.L. (1971). Cell co-operation in the immune response. Clin. Exp. Immunol., *8*, 839-56.

Ramila, G. (1978). Studies on cellular interactions in the immune response and antigenic competition in the lizard, *Calotes versicolor*. Ph.D. Thesis, Madurai University, Tamilnadu, India.

Reite, O.B. (1973a). Redistribution of tissue histamine stores (basophil leukocytes) of turtles in response to submersion and cold exposure. Acta Physiol. Scand., *88*, 62-66.

Reite, O.B. (1973b). Effect of physical and chemical agents on the blood mast cells (basophil leukocytes) of turtles. Acta Physiol. Scand., *87*, 549-56.

Rothe, F. & Ambrosius, H. (1968). Beitrage zur Immunologie Poikilothermer Wirbeltiere. V. Die Proliferation Antikorper bildender Zellen bei Schildkrieten. Acta Biol. Med. Germ., *21*, 525-36.

Ryerson, D.L. (1943). Separation of the two acidophilic granulocytes of turtle blood, with suggested phylogenetic relationships, Anat. Rec., *85*, 25-48.

Ryerson, D.L. (1949). A preliminary survey of reptilian blood. J. Ent. Zool., *41*, 49-55.

St. Girons, M.C. (1970). Morphology of the circulating blood cells. *In* Biology of the Reptilia, eds. Gans, C. & Parsons, T.S., Vol. 3, Morphology, pp. 73-91. New York, N.Y.: Academic Press.

Schall, J.J. (1983). Lizard malaria: Parasite-host ecology. *In* Lizard Ecology. Studies on a Model Organism, eds. Huey, R.B., Pianka, E.R. & Schoener, T.W., pp. 84-101. Cambridge, MA: Harvard University Press.

Sekizawa, A., Fujii, T. & Tochinai, S. (1984). Membrane receptors on *Xenopus* macrophages for two classes of immunoglobulins (IgM and IgY) and the third complement component (C3). J. Immunol., *133*, 1431-35.

Sidky, Y.A. & Auerbach, R. (1968). Tissue culture analysis of immunological capacity of snapping turtles. J. exp. Zool., *167*, 187-96.

Sidky, Y.A. & Auerbach, R. (1975). Lymphocyte-induced angiogenesis: A quantitative and sensitive assay of the graft-vs-host reaction. J. exp. Med., *141*, 1084-1100.

Snippe, H., Williams, P.J., Graven, W.G., & Kamp, C. (1975). Delayed hypersensitivity in the mouse induced by hapten-carrier complexes. Immunology, *28*, 897-907.

Streilen, J.W. & Billingham, R.E. (1970). An analysis of the genetic requirements for delayed cutaneous hypersensitivity reactions to transplantation antigens in mice. J. exp. Med., *131*, 409-427.

Sypek, J.P., Borysenko, M. & Findlay, S.R. (1984). Anti-immunoglobulin induced histamine release from naturally abundant basophils in the snapping turtle, *Chelydra serpentina*. Dev. Comp. Immunol., *8*, 359-66.

Taylor, K. & Kaplan, H.M. (1961). Light microscopy of the blood cells of Pseudemyd turtles. Herpetologica, *17*, 186-192.

Taylor, K.W., Kaplan, H.M. & Hirano, T. (1963). Electron microscope study of turtle blood cells. Cytologia, *28*, 248-256.

Terebey, N. (1972). A light microscopic study of the mononuclear cells infiltrating skin homografts in the garden snake, *Thamnophis sirtalis* (Reptilia: Colubridae). J. Morphol., *137*, 149-60.

Wetherall, J.D. & Turner, K.J. (1972). Immune response of the lizard, *Tiliqua rugosa*. Aust. J. Exp. Biol. Med. Sci., *50*, 79-95.

Wintrobe, M.M. (1933). Variations in the size and haemoglobin content of erythrocytes in the blood of various vertebrates. Folia Haematol., *51*, 32-49.

Wolke, R.E., Brooks, D.R. & George, A. (1982). Spirorchidiasis in loggerhead sea turtles *(Caretta caretta):* Pathology. J. Wildl. Dis., *18*, 175-85.

Worley, R. & Jurd, R.D. (1979). The effect of a laboratory environment on graft rejection in *Lacerta viridis*, the European green lizard. Dev. Comp. Immunol., *3*, 653-65.

Wood, F.E. & Ebanks, G.K. (1984). Blood cytology and hematology of the green sea turtle, *Chelonia mydas*. Herpetologica, *40*, 331-36.

Wood, S.F. (1935). Variations in the cytology of the blood of geckos *(Tarentola mauritanica)* infected with *Haemogregarina platydactyli, Trypanosoma platydactyli* and *Pirahemocyton tarentolae*. Univ. Calif. Publ. Zool., *41*, 9-22.

Wright, R.K. & Cooper, E.L. (1981). Temperature effects on ectotherm immune responses. Dev. Comp. Immunol. *5*, Suppl. 1, 117-22.

Zaakarian, S. & Billingham, R.E. (1972). Studies on normal immune lymphocyte transfer reactions in guinea pigs, with special reference to the cellular contribution of the host. J. exp. Med., *136*, 1545-63.

Zapata, A., Leceta, J. & Barrutia, M.G. (1981). Ultrastructure of splenic white pulp of the turtle, *Mauremys caspica*. Cell Tissue Res., *220*, 845-55.

Zapata, A. & Solas, M.T. (1979). Gut-associated lymphoid tissue (GALT) in Reptilia: Structure of mucosal accumulations. Dev. Comp. Immunol., *3*, 477-87.

Figure 1. Paraffin section of thymus from a young snapping turtle
Chelydra serpentina. Outer region is densely packed with
lymphocytes with ligher stained medulla (m) in central region.
Bar = 67 μm.

Figure 2. Electron micrograph of the thymic cortex from a young
snapping turtle *Chelydra serpentina*. Most of the cells are
lymphocytes of various sizes. A reticular-epithelial (re) cell is
seen in the lower left corner. Bar = 2.2 μm.

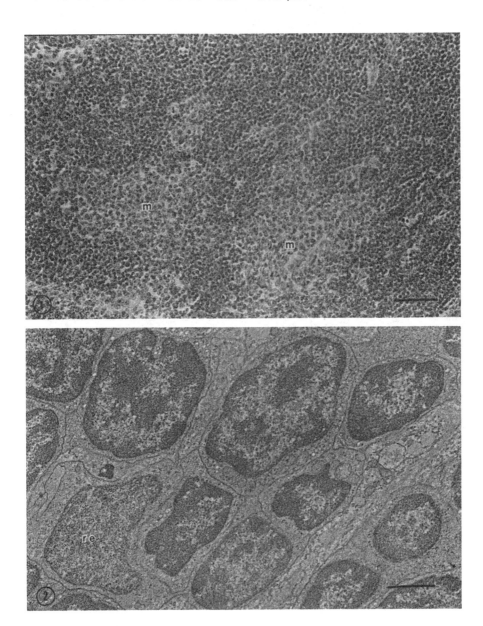

Figure 3.   Epon section of spleen from the snapping turtle,
Chelydra serpentina, showing red (r) and white (w) pulp regions.
Lymphoblasts (arrows) are randomly distributed in the white pulp.
Very dark cells in the red pulp are basophils.   Methylene blue.
Bar = 40 μm.

Figure 4.   Paraffin section of spleen from the snapping turtle,
Chelydra serpentina, showing the appearance of white pulp,
consisting of lymphocyte sheaths around arterioles (a).   The
surrounding red pulp(b) contains a number of darkly stained plasma
cells (arrows).   Methyl green-pyronin Y.   Bar = 50 μm.

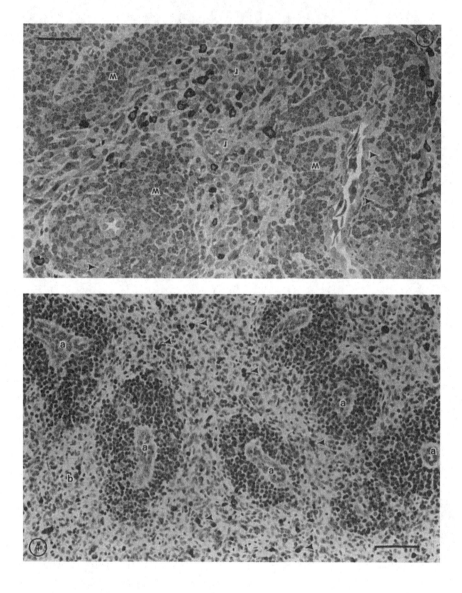

Figure 5. Epon section through a vein of an immunized spleen from the snapping turtle, *Chelydra serpentina*, containing a variety of blood cells. Most are of the lymphoid series, including lymphoblasts (lb). In addition, are erythrocytes (er), basophils (b), eosinophils (eo), heterophils (h), and monocytes (m). Bar = 13.5 μm.

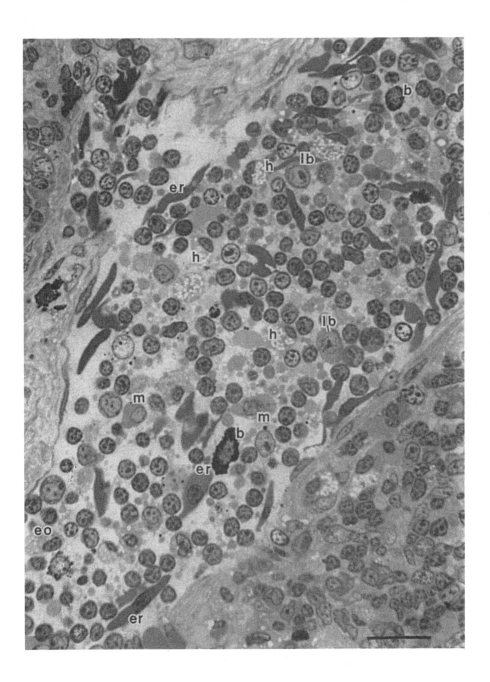

Figure 6. Giemsa stained reptilian leucocytes. Basophil from the turtle, *Chelydra serpentina* (Fig 6a). Eosinophil from the turtle, *Chelydra serpentina* (Fig. 6b). Heterophil from the lizard, *Agama agama* (Fig. 6c). Heterophil from the turtle, *Chelydra serpentina* (Fig. 6d). Thrombocyte from the caiman, *Caiman spp.* (Fig. 6e). All other cells present in these light micrographs are erythrocytes. Bars = ~ 2 μm.

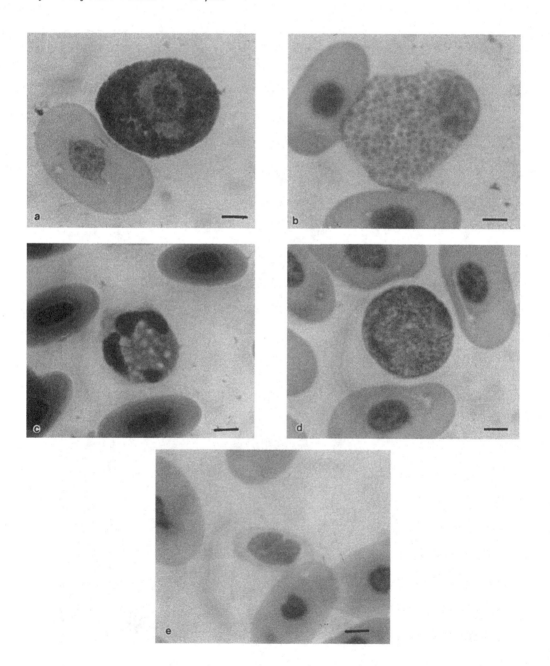

Figure 7. Electron micrograph of a red pulp sinus from the spleen
of the snapping turtle, *Chelydra serpentina*. The sinus contains
cellular elements from peripheral blood. In this section are seen
erythrocytes (er), a small lymphocyte (l), a basophil (b) and two
eosinophils (eo). Bar = 2.8 μm.

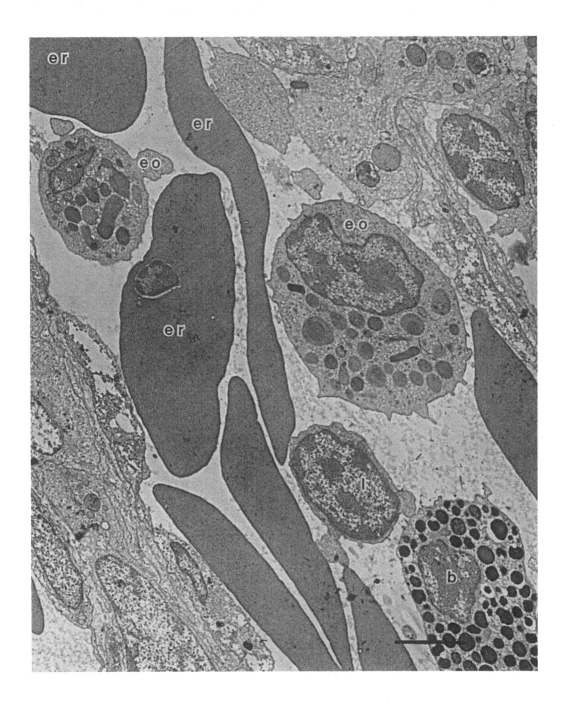

Figure 8. Electron micrograph of a monocyte from the spleen of the snapping turtle, *Chelydra serpentina*. Part of an erythrocyte is seen in the upper left. Bar = .9 μm.

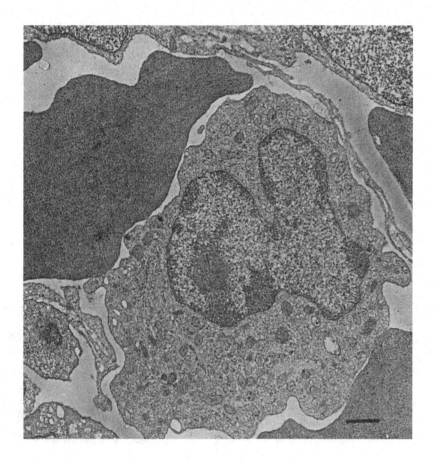

Figure 9.   Electron micrograph of a heterophil-like cell from Cook's Island Tree boa *Corallus sp.*   On the lower right and left are erythrocytes. Bar = 1 μm (courtesy of Frederic L. Frye © 1985).

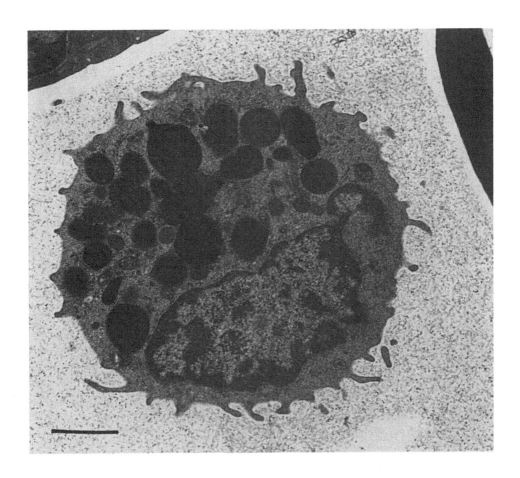

Figure 10.   Electron micrograph of a plasma cell from Cooks Island
Tree boa *Corallus sp*.   On the upper right is an erythrocyte.   Bar =
1 μm.   (courtesy of Frederic L. Frye [©] 1985).

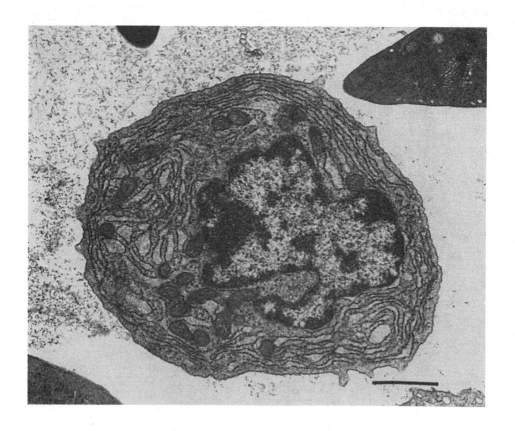

5 Birds

F. Dieterlen-Lièvre
Institut d'Embryologie du CNRS et du Collège de France
94736 Nogent sur Marne, France.

INTRODUCTION

Birds are interesting subjects in which to study various aspects of blood cell formation. Their haemopoietic system is unique, since alone among vertebrates, birds have a specific organ, the bursa of Fabricius, evolved for differentiation of B lymphocytes. Another interesting feature is the presence of a class of histocompatibility antigens on the surface of the mature red cells, which can be used as markers. Birds are particularly amenable to experimental approaches during embryogenesis, and this advantage has opened up several new avenues. For example, in birds, the phenomenon of tolerance resulting from exposure to an antigen during ontogeny was established early (Hasek and Hraba 1955) and the graft-versus-host reaction mediated by adult allogeneic lymphoid cells in immune deficient hosts, embryos for instance, was first described (Dantchakoff 1916, 1918) and explained (Simonsen 1957) using avian models. It is also in birds that the analysis of haemopoietic development has been carried out in most detail, mainly because of the availability of a simple and effective cell marking system enabling the construction of chimaeras. This involves combining cells from two closely linked species, usually the quail and the chick (Le Douarin 1973). Quail cell nuclei may be distinguished from chick by virtue of the heterochromatic DNA mass associated with the nucleolus. The two species are also similar enough to allow close to normal development of chimaeras. Using the nuclei as labels it is possible to trace the relative contribution of germ layers or rudiments to different cell lineages and to study the interactions of cells from different origins.

To date, the only comprehensive description of avian blood cells is the "Atlas of Avian Hematology" by Lucas and Jamroz (1961). The hand-drawn illustrations in this book, which depict all possible maturation forms and modifications from the standard morphology, will remain indispensable to any investigator of normal or pathological haemopoiesis in the chicken. Excellent descriptive studies have also been carried out since the beginning of the century on developmental aspects of the haemopoietic system (Dantchakoff 1909; Sabin 1920; Sandreuter 1951; Schmekel 1962).

The study of haemopoiesis, however, can no longer be restricted to the identification of cytological features, many of which were ill-defined from the start. New strategies have been developed, especially over the

last 15 years. Specific probes - usually polyclonal or monoclonal antibodies directed against surface membrane antigens - have been obtained,
thus making it possible to identify cell lineages and, in the most favour-
able instances, maturation stages. Unfortunately, dissection of avian
lineages into subpopulations, in particular where lymphocytes are
concerned, has lagged far behind the state of the art now achieved for
murine or human cells. Functional assays have been devised, with a view to
detecting and quantifying precursors cells by means of their progeny either
in vivo or in vitro. Emphasis in the present chapter will concentrate on
these more recent approaches.

For obvious reasons, the most studied species has been the chicken. Inbred
lines have been developed, and these are indispensable for studying the
immune system and obtaining genetic markers for, or within, the various
blood cell lineages. However, due to their long generation time and the
lowered fertility and viability of inbred animals, chickens are far from
being the equivalent of the laboratory mouse. Up to now, however, the
chicken is the best available avian model. Other species of birds have
been investigated only occasionally, except for the quail, which has
become so widely used in experimental embryology, that it is necessary to
understand its normal development and find genetic and cellular markers.

STRUCTURE OF THE CIRCULATORY SYSTEM

Descriptions of circulatory system of birds are to be found
in many textbooks of zoology and so the following account is by necessity
brief. Detailed descriptions of the structural and physiological aspects
of the avian circulatory system can be found in excellent reviews (e.g.
Jones & Johansen 1972; Hodges 1974; Nickel et al. 1974; Akester 1984).

The heart

The heart has a bilateral origin. Two tubes arising from
lateral splanchnic mesoderm fuse shortly after closure of the developing
foregut. The straight cylindrical tube formed by the fusion then assumes
an S-shape, in which first an interatrial septum, then an interventricular
septum, will develop.

In the adult bird, the atria are thin-walled but muscular structures acting as storage compartments for blood returning to the heart from the
systemic and pulmonary vascular beds. The right atrium is significantly
larger than the left. The ventricles are also very unevenly developed.
The wall of the left ventricle is ca. 2-3 times thicker and more muscular
than the wall of the right ventricle which it partially surrounds. The
wall of the heart comprises the outer epicardium, the myocardium and the
inner endocardium. The epicardium consists in a single layer of low
columnar to squamous mesothelial cells covering a thin layer of connective
tissues with coarse and fine collagen and some elastic fibres. The endocardium is constituted by individual striated muscle fibres which are
joined together to form a branching network by means of specialized areas
of the sarcolemma, the intercalated disks. Some transformed muscle cells,

the transitional cells, are responsible for conduction of the impulse.
Many birds have extremely large hearts in comparison with mammals.

The arterial system

The vascular system undergoes extensive reorganization during
embryonic life. The aorta arises from paired primordia which fuse. Transitory aortic arches connect the dorsal aorta to the ventral aorta which is
a short projection from the heart. The IVth right arch gives rise to the
permanent arch of the aorta while the left one disappears. This origin is
reversed by comparison with mammals in which the left IVth arch is the
permanent one. The VIth arch gives rise to the pulmonary arteries.

In the adult, blood is ejected from the right ventricle into the pulmonary
artery. Three semi-lunar valves prevent back flow from the pulmonary artery
into the right ventricle. From the left ventricle, blood is pumped into
the systemic arch (right aortic arch) and three semi-lunar valves again
prevent reflux of blood from the aorta into the left ventricle during
diastole.

All arteries of the head and neck are branches of the carotid arteries,
themselves derivatives of the embryonic dorsal and ventral cephalic
aortae. Common carotid arteries show various configurations in different
species of birds, a frequent anatomical feature being intercarotid anastomosis. This anastomosis is the avian counterpart to the mammalian
"circle of Willis" and so provides a major collateral blood supply to the
brain. The common carotid artery is one of the branches of the brachio-
cephalic trunk which also supplies blood to the wings, cranial floor of
the body cavity and ventral aspects of intercostal spaces. The descending
aorta arches over the atria of the heart, above the division of the
pulmonary trunk into left and right pulmonary arteries, and continues
along the roof of the body cavity, terminating in the tail. Along its
course it provides branches to all parts of the body below the brachio-
cephalic trunks: paired segmental arteries, a single coeliac artery, a
single cranial mesenteric artery, three pairs of renal arteries, paired
external iliac arteries, paired ischiadic arteries, a single caudal
mesenteric artery and paired internal iliac arteries.

The venous system

In general terms veins parallel the arteries. The jugular veins
are unevenly developed, the right one usually being larger than the left.
An interjugular anastomosis in the anterior region of the neck provides a
bypass system from one jugular vein into the other, as when one becomes
compressed during neck movement. The pulmonary veins enter separately or
together into the left atrium depending on the species. The entire length
of the alimentary canal caudal to the oesophagus drains into the hepatic
portal vein. The caudal mesenteric vein drains the colorectum and cloaca,
and provides a unique major anastomosis between hepatic portal and renal
portal systems. Large hepatic veins join the posterior vena cava a short
distance caudal to the heart.

The lymphatic system

Lymph from the entire body is returned to the blood in the anterior venae cavae. The cervical lymphatic vessels are paired trunks continuous caudally with the thoracic ducts. The thoracic ducts are the largest lymphatic vessels. At their junction with the venae cavae, a valve prevents entrance of blood into the lymph channels. The thoracic ducts receive cranially the cervical lymph ducts and caudally are continuous with the para-aortic lymphatic trunks.

Lymph nodes are present in only a few species of aquatic birds. When present, there are usually two pairs, or less commonly three pairs (cervical, lumbar and thoracic nodes). Excepting the spleen, birds have nodules of lymphatic tissues scattered throughout various organs rather than arranged into definite structures.

Lymph hearts are present in a few species of adult birds. They are dilations of lymphatic vessels which possess muscles in their walls. These lymph hearts pulsate and these contractions assist the flow of lymph. In the domestic fowl, lymph hearts are transitory organs during days 10 to 21 of incubation.

Structures specific to birds

The avian kidney is endowed with a portal system. This feature is characteristic of lower vertebrates and is also present in the mesonephros of embryonic mammals but not in the adult mammalian kidney. However, birds are able to switch on or off this portal system, these shunts being usually partial. The circumstances under which this shunt happens are not understood. Birds are the only class of vertebrates where such a mechanism has been uncovered.

The high metabolic rate and body temperature of birds require precise mechanisms for dissipation and conservation of heat. Heat loss is considerable through the legs, especially in aquatic and wading birds. In some of these, a "rete mirabile" i.e., vascular bundles of small arteries and veins tightly packed in a countercurrent arrangement have been described. Such structures are supposed to allow the warming of returning venous blood by warm arterial blood, reducing the heat loss markedly. The hypothesis has been put forward that the rete is bypassed, when the environmental conditions call on heat dissipation.

Another interesting structure found in birds is the pecten. It is a pigmented, highly vascular, pleated structure attached to the retina, which protrudes into the vitreous body of the eye. Its role is not known for certain but is likely to be nutritional; it would replace the retinal artery which does not exist in birds. It is also probably responsible for maintaining intraocular pressure. The pecten is supplied by a branch of the opthalmotemporal artery. Blood capillaries ramify in the pleats of the pecten among melanocytes. Endothelial cells of these capillaries display a dramatic structure. Their cytoplasm is thrown out on each face of the cell into thin, long, hairlike microprocesses.

Finally, on the ventral surface of the skin of birds is an area specially
concerned with providing warmth for the eggs. This brood patch develops
in whichever sex is responsible for incubating the eggs. The skin of this
area either undergoes a localized moult or is plucked out, depending on
the species. The blood vessels become markedly hypertrophied, under the
synergistic activity of oestrogen or testosterone, prolactin and pro-
gesterone.

FORMATION OF BLOOD CELLS

Bone marrow

        In the adult chicken, the bone marrow is the environment in
which erythrocytes and granulocytes differentiate and in which lymphocytes
mature. The spleen does not seem to have a haemopoietic function in the
adult. Evidence for this is the finding that, after sublethal irradiation
and restoration by means of fresh bone marrow cells, no colonies form in
the spleen. For this reason the chicken equivalent of the mouse spleen
Colony-Forming-Unit (CFU-S, Till & McCulloch 1961) is termed the CFU-M,
i.e., Colony-Forming-Unit in the Marrow (Samarut & Nigon 1975, 1976a)
(Fig. 1).

An important feature of chicken bone marrow is the fact that erythropoiesis, granulopoiesis and lymphopoiesis are compartmentalized within
the lumen of sinusoids and the extravascular spaces, respectively (Campbell
1967). Another specific trait is the presence of lymphoid tissue with
germinal centres (Campbell 1967; Payne & Powell 1984).

The vascular system of the marrow. In the tibia or femur, a single large
artery enters the marrow cavity through the nutritive foramen. It breaks
up into smaller branches which run along the central axis of the marrow
together with a central vein (Fig. 2). These arteries in turn bifurcate
into radiating arterioles, and into capillaries, which are in communication with sinuses. The communications between capillaries and sinuses are
not readily observed; on the other hand arterioles are never seen to
communicate directly with sinuses. The sinuses, which are the site of
erythropoiesis, are lined by very elongated endothelial cells, similar in
appearance to the reticular cells observed in the extravascular compartment. The sinuses either empty directly into the central vein or join into
larger collecting sinuses which abut on a central vein.

Erythropoiesis. A gradient of maturity exists between the sinuses, which
contain mostly erythroblasts, and the collecting sinuses and central vein.
Immature erythrocytes are still more abundant in the latter vessels than
in the peripheral blood. The most immature cells in the erythroid series
are associated with the sinus wall, while the most mature ones are located
in the centre of the vascular lumen. The haemocytoblasts are often in
close apposition to the endothelial wall of the sinus. Adherent haemocytoblasts exhibit a definite cytoplasmic polarity. The cell organelles, i.e.,
Golgi apparatus and most mitochondria, are located in the cytoplasm opposite
the site of attachment of the cell to the endothelium. The intercellular
space between endothelial and haemocytoblast plasma membranes is extremely

Fig. 1. Three erythroid colonies (arrows) developing in the bone marrow of an irradiated-reconstituted chicken. These colonies are the progeny of CFU-M, observed 6 days after injection of 5×10^7 fresh bone marrow cells into the wing vein of a 3-wk old chick which had received 760 rads + 990 rads 24h apart. The tibial marrow cylinder was stained in toto with benzidine which revealed the presence of haemoglobin in the colonies. Picture provided by Dr. J. Samarut.

Fig. 2. Semithin section of the tibial bone marrow of a 6-day chick. Haemopoietic tissue is interspersed with fat cells. A=artery, B=bone, F=fat cells, H=haemopoietic tissue, V=vein. Bar=100µm.

Fig. 3. Tibial marrow of a 6-day chick. Myeloblasts (M) in a granulopoietic area. F=fat cell. Bar = 5µm.

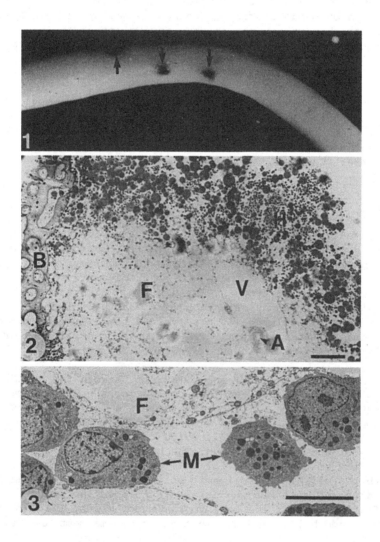

regular in width and contains electron-dense material. Basophilic
erythroblasts are interspersed with haemocytoblasts from which they can
be distinguished by their slightly smaller size and their more basophilic
cytoplasm. They also often adhere closely to the endothelial wall. Polychromatophilic erythroblasts and erythrocytes occur more centrally than
the two types of earlier precursors.

Granulopoiesis. A gradient of maturation is less readily observed within
the extravascular space. Haemocytoblasts, characterized by a relatively
large nucleus, a prominent nucleolus and a basophilic cytoplasm, locate
against the lining reticular cells, but form only restricted contacts with
them or with endothelial cells. These haemocytoblasts are very similar in
morphology to those in the lumen of the bone marrow and some may be
observed passing from one compartment into the other through the cellular
lining of the sinus wall. It is not possible to determine the direction
of passage but such micrographs indicate an exchange of cells between the
two compartments and suggest that these haemocytoblasts are pluripotential.
Numerous cells of the granuloid series, at various stages of maturation,
are intermingled among fat cells within the extravascular space (Fig. 3).
Mature and immature granuloid cells can be distinguished by scanning
electron microscopy on the basis of the abundance of surface microvilli
(Sorrell & Weiss 1980). More mature cells are smaller and have numerous
microvilli, while less mature ones are larger and display pseudopodia
rather than microvilli.

Lymphatic tissue. Masses of lymphoid tissue are segregated within the
bone marrow. The capillaries in this lymphoid tissue contain lymphocytes,
some of which appear to be in the process of traversing the capillary wall.
Scattered within the lymphoid tissue are large lymphocytes, whose
appearance is very similar to that of the haemocytoblasts within the
erythroid or granuloid compartments.

Ontogeny . The development of the bone marrow during ontogeny has been
studied in the chick embryo by Sorrell & Weiss (1980) by light, scanning
and transmission electron microscopy. They studied bone marrow from
embryos ranging between days 12 and 15 of incubation. Day 12 is the
earliest date at which marrow cells can be obtained conveniently from the
chick embryo. Marrow at this stage, presumably richer in non-committed
precursors than older marrow, has been used as a source of haemopoietic
cells (Ben Slimane et al. 1983). On day 12 in the chick embryo, the marrow
already harbours cells which can be identified as belonging to one or
another of the haemopoietic lineages and which are typically compartmentalized: the only feature which distinguishes it from older marrow is the
fact that cells are less numerous and comparatively more haemocytoblasts
are present. Blood stem cells in the bone marrow have an extrinsic origin.
Moore & Owen (1965) were the first to put forward this proposal when they
demonstrated that in parabiosed embryos there was considerable exchange of
cells, which could be traced by means of the sex chromosome marker. Using
the quail-chick marker system Jotereau & Le Douarin (1978) demonstrated
that, in early limb buds from one species grafted on the chorioallantoic
membrane of the other species, the haemopoietic and endothelial cells

were entirely of host origin. Osteoclasts were also of host origin; thus
they also probably belong to the haemopoietic lineage.

Spleen

          Studies of the avian spleen are relatively scarce (Ewart &
McMillan 1970; Nagy 1970; Hoffman-Fezer et al. 1977; Sugimura & Hashimoto
1980; Olah & Glick 1982). The most enlightening is that of Olah & Glick,
who have investigated the arrangement of vessels within the white pulp.
This complex architecture was analyzed by perfusion fixation, with and
without previous intravenous injection of colloidal carbon. Other authors
have studied the distribution of T- and B-dependent areas, relying either
on the different sizes of B and T lymphocytes (Nagy 1970), on the depletion of the different areas following neonatal thymectomy or bursectomy
(Cooper et al. 1965; Sugimura & Hashimoto 1970) or on antibodies directed
against one or the other class of lymphocytes (Hoffman-Fezer et al. 1977).
Structure (Fig. 4). The splenic arteries in the chicken enter the spleen
at the hilus, where they are surrounded by trabecular connective tissue.
Peripheral nerves and ganglion cells are present along the trabecular
arteries. The transition between the trabecular and central arteries is
marked by the disappearance of the numerous muscle layers present in the
wall of the former. By contrast, the central artery possesses a single
muscle layer along its entire length and is surrounded by the periarteriolar lymphoid sheath (PALS). The PALS consists of densely packed
lymphocytes which are thymus-dependent. Where the PALS ends, the central
artery usually makes a right-angle bend and becomes the penicilliform
capillary (PC) which is surrounded by the ellipsoid or Schweigger-Seidel
sheath. The ellipsoid projects from the white pulp into the red pulp. The
periellipsoid lymphoid tissue consists of B-cells associated with
dendritic cells. The mid portion of the PC often bifurcates three or
four times, conferring a tri- or tetralobed shape to the ellipsoid
(Fig. 5). Germinal centres, located in the vicinity of the central artery,
consist of antigen-presenting dendritic cells, which have migrated from
the ellipsoid, and of antigen-specific B cells.

Upon leaving the ellipsoid, the PC becomes the terminal capillary lined
by gradually flattening endothelial cells. The terminal capillary branches
many times before it enters the red pulp. The sinuses of the red pulp are
lined with flat, irregular, endothelial cells and communicate with one
another. While there have been some suggestions that circulation in the
avian spleen is open, the study by Olah & Glick (1982), using either
normal or perfusion fixation, clearly demonstrated the continuity of PCs
with sinuses on entering the red pulp. However, in perfused specimens,
some neighbouring endothelial cells lining the red pulp sinuses are seen
to be separated by gaps which allow extensive passage of cells between the
lumen and the red pulp cords.

The phagocytic capacities of the so-called "ellipsoid associated cells"
(EAC) have been investigated by Olah & Glick (1982) after intracardiac
injection of carbon. The EACs, the majority of which are located at the
surface of the ellipsoid, may be identified by their dense nucleus in
semithin or ultrathin sections. They are elongated cells which send

processes among the others. Within 30 min of the injection, carbon appears extracellularly in the ellipsoid, here it becomes bound to the cell membrane of EACs, which then leave the ellipsoid and intermingle with the cells of the periellipsoid white pulp. The EACs begin phagocytosing carbon 30 min after injection; the process is maximal by 2 h. Between day 3 and

Fig. 4. Diagram of the structure of the spleen. B,T=B and T dependent areas, CA=central artery, E=ellipsoid, GC=germinal centre, PALS=periarteriolar lymphoid sheath, PELT=periellipsoidal lymphoid tissue, PC=penicilliform capillary. TA=trabecular artery. Modified from Payne and Powell (1984).

Fig. 5. Spleen of a 21-day chicken. B-cell lymphoid cuffs around penicilliform capillaries (P) are stained black with an anti-B cell monoclonal antibody revealed by the per peroxidase-antiperoxidase technique. Micrograph provided by Dr. Kumbaniduwa and Coudert. Bar = 50μm

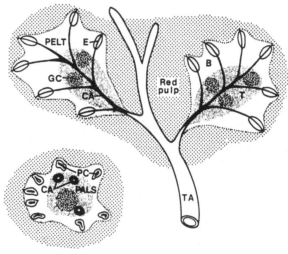

4

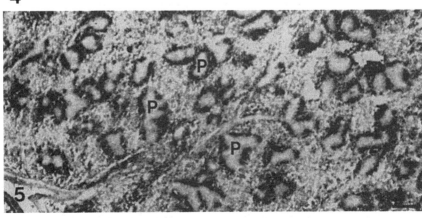

5

5 after injection, the carbon containing EACs appear in the red pulp and around the germinal centres. During this period, when depleted of EACs, the ellipsoids first appear collapsed, then progressively reacquire their complement of EACs.

Several authors have investigated the origin of lymphocytes homing to the different structures in the spleen. These studies have yielded clear-cut results, whether they relied on labelling through specific polyclonal or monoclonal antibodies, or whether the posthatching evolution of the spleen was followed in animals which had been neonatally thymectomized or bursectomized. While B-lymphocytes are present in the periellipsoidal lymphoid tissue (Fig. 6) and the germinal centres, T-lymphocytes are located in the periarteriolar zones, i.e., in the PALS (Fig. 7). Germinal centres are located within the PALS towards the root of the central arteries.

Ontogeny. Only a couple of studies are available on the ontogeny of the avian spleen and neither are recent (Delanney & Ebert 1962; Metcalf & Moore 1971). During development, the spleen carries out successively different functions. During embryonic life it is haemopoietic, with erythropoiesis and granulopoiesis both very active; the latter may become the predominant process as incubation proceeds. The characteristic splenic structure with the prominent round ellipsoids becomes apparent towards the end of embryonic life. However, granulopoiesis is still an ongoing process. During incubation, only a minimal number of B or T cells migrate into the spleen, beginning on day 12 (Grossi et al. 1977; Bruner, personal communication). In the quail, T cells carrying the AT65 antigen are found on day 9 of ontogeny (Péault et al. 1982). At hatching, a dramatic loss of the granulocytes occurs so that within 2 or 3 days these cells disappear to be replaced by lymphocytes which settle into the characteristic T- and B-dependent localizations (Figs. 8, 9).

Thymus

Structure. In birds, the thymus comprises two elongated cords located on either side of the neck. Each of these cords is made up of a succession of lobes, 7 of them in the chick, 10 in the quail. In the chick, the lobes are deeply indented into lobules, while in the quail they are not. Each cord of lobes runs closely alongside the jugular vein, its caudal extremity abutting on the thyroid; the cord then takes an oblique course towards the side of the neck, so that the cephalic extremity of the cord is located dorsal to the neck. Thymectomy, when performed immediately after hatching, is carried out by a dorsal approach, the skin of the neck being split along a median dorsal line. In each lobe, two concentric regions are clearly delineated: the cortex, where lymphoid cells are closely packed, and the medulla where they are scarcer (Fig. 10). A network of arterioles runs along the corticomedullary junction, from which capillaries are sent outwards into the cortex as far as the capsule. They drain into veins running with the arterioles between the cortex and the medulla (Payne & Powell 1984).

Figs. 6 and 7. Spleen of a 1-month chicken. The same section was double-stained with antibodies coupled respectively to fluorescein and rhodamine. Micrographs provided by Dr. B. Péault. Bars = ·200μm

Fig. 6. An anti-Ig antibody reveals fluorescein-labelled B cells in an ellipsoid.

Fig. 7. A monoclonal antibody directed against T cells reveals the rhodamine-labelled T dependent zone, i.e. the Periarterial Lymphoid Sheath. Note that the T and B localizations appear complementary.

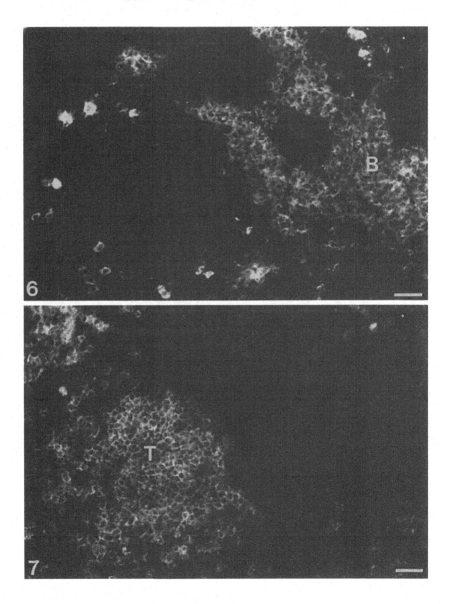

Four main cell types can be distinguished in the thymus by electron
microscopy, namely, lymphoid cells (Fig. 11), epithelial cells, macrophages
and myoid cells (Frazier 1973). Lymphoid cells and macrophages will be
considered at length in subsequent sections. Epithelial cells, which
derive from the endodermal component of the embryonic rudiment, form a
supporting network for the free cells of the thymus. They have recently

Fig. 8. Spleen of a 19-day chick embryo. Note extensive
granulopoietic tissue. Bar = 10μm.

Fig. 9. Spleen of a chicken 6 days after hatching. Lymphoid
tissue has entirely replaced granulopoietic tissue. Nuclei
display clumps of heterochromatin typical of lymphoid cells.
Bar = 10μm.

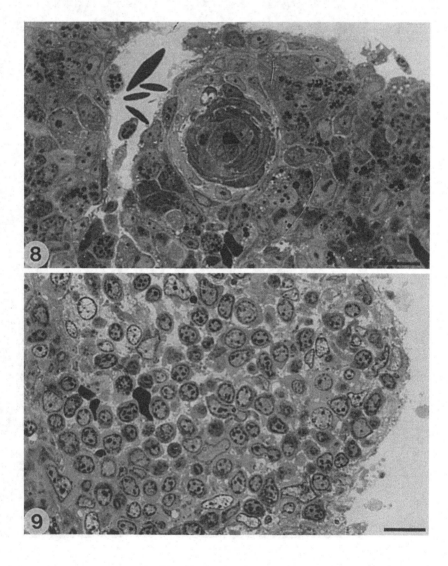

been shown to express major histocompatibility class II antigens on their surface (Guillemot et al. 1984). This was demonstrated in quail/chick chimaeric thymuses. In the normally developed organs, the intermixture of cells is such that, even with specific markers, it is difficult to identify the various types. In chimaeric thymuses obtained by transplanting the very early endodermal rudiment, the epithelial cells derive from the endoderm, while all others derive from the host. Epithelial cells, also identified by the presence of cytokeratin filaments, stained positively with a mono-clonal antibody directed against class II antigens specific for the species donating the thymic rudiment. The epithelial cells are distributed in the cortex and at the corticomedullary junction. In the cortex, these cells have long intricate processes which associate into a typical network. At the corticomedullary junction, they are arranged in more compact fashion. The fact that epithelial cells possess class II antigens indicates that they must play a role in educating T lymphocytes for MHC restriction.

Ontogeny. In all vertebrates, the thymus originates from the primitive pharynx as an epithelio-mesenchymal rudiment. There were early suggestions that the epithelium might be partly or entirely derived from ectoderm. However, explantation and ectopic grafting of the endodermal floor of the pharynx yields a variety of tissues, including thymus, demonstrating that no ectodermal contribution is necessary (see Le Douarin et al. 1984 for a review). The mesenchyme, on the other hand, derives from the neural crest. In the head, there is a specific contingent of neural crest cells which gives rise to the so-called mesectoderm. Through isotopic and iso-chronic grafting of the quail rhombencephalic primordium into the chick embryo at the stage of 6-9 pairs of somites, chimaeric thymuses were obtained in which connective tissue was quail, while epithelial, endo-thelial and lymphoid cells were chick. Mesectodermal cells contributed pericytes lining the outer aspects of vessels, interlobular connective tissue and myoid cells (Le Lièvre & Le Douarin 1975: Nakamura & Ayer-Le Lièvre 1986).

The origin of lymphocytes was debated for nearly a hundred years before being investigated by experimental rather than descriptive means. At the beginning of the century these cells were thought to arise either from the epithelium (transformation theory) or from the mesoderm (see Metcalf & Moore 1971 for a review). That lymphoid precursors originating from some extrinsic source invade the thymus rudiment was first proposed by Moore & Owen (1967a). Their argument was based on the heavy traffic of cells that they observed between chick embryos, joined by vascular connections (Moore & Owen 1965; 1967b). In their experiments, sex chromosomes were used as markers. When parabionts were of opposite sexes, chromosomal analysis following early parabiosis (day 4 or 5 of incubation) revealed high levels of chimaerism in the thymus. In contrast, little chimaerism was found in parabionts joined at later stages. An inflow of blood-borne lymphoid stem cells, invading the thymus between day 4 or 5 and day 9 of incubation, was proposed by Moore & Owen (1967a). The quail-chick chimaera method (Figs. 12-15), however, has made it possible to document in fine detail the mode of colonization of the thymus rudiment by lymphoid precursors (see Le Douarin et al. 1984 for a review). The extrinsic origin

of the thymic lymphocytes and the time in ontogeny when the thymus rudiment
is first invaded by haemopoietic cells (HC), was determined by grafting
the thymus anlage interspecifically between chick and quail embryos at
increasingly later developmental stages. The total age of the thymus (i.e.,
age at grafting time + duration of the graft) at the time of observation
was 14 days in all experimental series. It was demonstrated that the
invasion of the thymus by HC is a rapid process which starts at 5 and 6.5
days of incubation respectively in quail and chick embryos, terminating
some 24h later in the quail, and some 36h later in the chick (Le Douarin
& Jotereau 1973, 1975). This first influx of cells into the thymus is
followed by a refractory period characterized by a shut-off in thymic

Fig. 10. Chick thymus, 6 days post-hatching. C=cortex,
M=medulla. Bar = 50μm.

Fig. 11. Chick thymus, 6 days post-hatching showing a
typical thymocyte. Note large nucleus with abundant heterochromatin. Bar = 1μm.

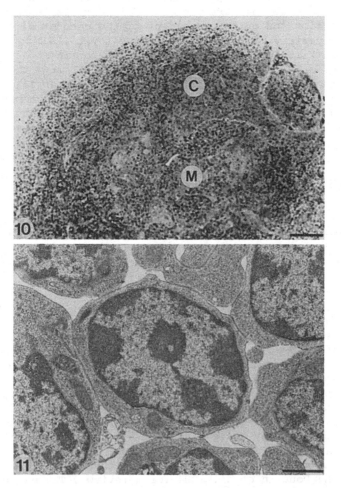

receptivity. Whether receptivity is resumed and, if so, at what stage, has been further investigated by constructing various kinds of chimaeras. In fact, evidence has been provided that thymus ontogeny involves a cyclic succession of receptive and non-receptive periods for stem cells into the embryonic and early postnatal thymus (Fig. 16).

A detailed study of these phenomena has been carried out for the quail thymus (Jotereau & Le Douarin 1982). The principle of the experiments was to graft the thymus, taken from a quail embryo after the first colonization had been completed, into two successive hosts, the first being a 3-day chick embryo and the second a 3-day quail embryo. The thymus was left in the first host for different lengths of time. It turned out that the critical event determining the species make-up of the lymphoid population at the end of the experiment occurred when the thymus was aged 11-12 days. Thus, a second influx of precursors invades the thymus between 11 and 12

Figs. 12-15. Aspects of the quail-chick marker technique.

Fig. 12. Chimaeric prelymphoid population in a 10-day embryonic chick thymus. The aorta of a 4-day quail embryo was grafted in the body wall at the level of the branchial arches in a 5-day chick embryo. Progenitors from the wall of the quail aorta (see section on haemocytoblasts) invaded the chick thymic rudiment simultaneously with host progenitors during the first wave of colonization. The quail cell nuclei display a heavy heterochromatin mass that distinguishes them from chick cell nuclei in this Feulgen-Rossenbeck stained section. Two thymic lobes are visible (dotted lines). Arrows point to the main groups of quail cells. From C. Martin and F. Dieterlen-Lièvre, unpublished. Bar = 20µm.

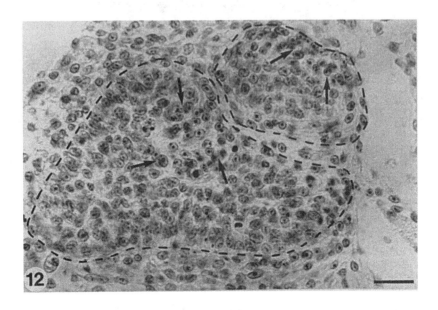

Fig. 13. 18-day chimaeric thymus. A 9-day embryonic chick thymus was grafted into a quail host. Host cells invaded the graft at the time of the second colonization wave. Cortex and medulla were colonized independently, as revealed by the peripheral ring of fluorescent cells (colonization of the cortex, arrows) and the fluorescent cells dispersed in the centre of the lobes (arrow heads). Immunofluorescence detection of quail cells was performed by means of the MB1 monoclonal antibody (Péault et al. 1983) which is specific for the quail haemangioblastic lineage. Micrograph provided by Drs. B. Péault, M. Coltey and N. Le Douarin (unpublished). Bar = 20µm.

Figs. 14 & 15. After the same experimental pattern as described in Fig. 13, the chimaeric thymus was dissociated into a cell suspension. Quail prelymphoid cells, still in the quiescent phase, immunopositive with MB1 (Fig. 14), are large cells (arrows in Fig. 15) when compared with the bulk of lymphoid cells seen in phase contrast (Fig. 15). Micrograph provided by Drs. P. Péault, M. Coltey and N. Le Douarin (unpublished). Bars = 10µm.

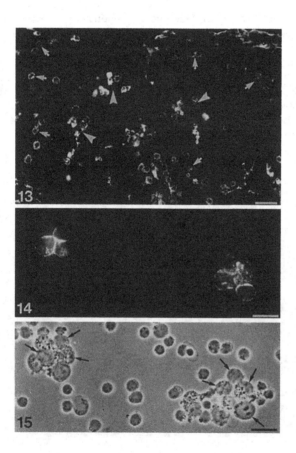

days of incubation in the quail (Fig. 16). A third wave of precursor entry
was demonstrated by similar grafting experiments performed on quail
thymuses retrieved after the second wave. This third wave was estimated
to occur between days 17 and 18 of thymus age, i.e., during the second
postnatal day of the quail. This date represents about the time limit for
this type of experiment. Indeed, these interspecific combinations are
viable only because the partners are embryonic and the immune system is
immature.

The conclusion from such experiments is that colonization of the quail
thymus is a cyclic process with a regular rhythm. The colonization waves
last 24 h and are separated by non-receptive periods of 5 days (Fig. 16).
The timing of the second wave was confirmed in parabiosis experiments
where the two partners were quail embryos, thus excluding the possibility
that the process was distorted by the species differences (Le Douarin et
al. 1984).

In the chick embryo, the pattern of seeding of the thymic rudiment by
lymphoid precursors is also cyclic. The first wave occurs from day 6.5 to
day 8 and the second wave takes place on day 12 and lasts ca. 48 h (Le
Douarin et al. 1985)(Fig. 16).

Fig. 16. A comparison of the patterns of colonization of
the avian primary lymphoid organs by stem cells (shaded areas).
The thymus receives stem cells during short recurrent periods
which obey a precise rhythm. The rhythm is specific for each
species. The bursa of Fabricius is colonized during a
distinct period of embryonic life and before hatching.

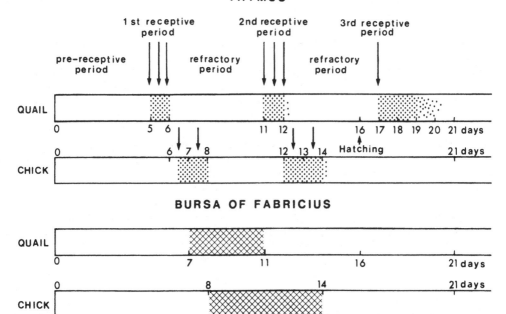

The proliferation of lymphoid precursors in the embryo also obeys a
precise spatial and temporal pattern, that has been investigated by
grafting quail thymuses in chick hosts after the thymuses had received
their first supply of precursors. Incoming cells, chick in the above
combination, were bulky with a large nucleus and a prominent (Feulgen-
negative) nucleolus; they invaded the periphery of the lobes, forming
there a single row. Scattered cells were also observed to appear
simultaneously in the medulla. These immigrant cells remained quiescent
until day 15, i.e., for 3 days after the 11-12 day colonization period.
From day 15 onward, the chick cells became more numerous and progressed
towards the centre of the lobule, finally invading the whole cortex.
Tritiated thymidine incorporation, performed from days 12 to 18, followed
the same pattern. The chick cells were devoid of isotopic labelling until
day 15. A significant proportion of these cells became labelled during
days 16 to 19. Grains were preferentially located in the external third
of the cortical area. Labelling was distributed over similar areas in
control thymuses indicating that the developmental pattern of grafted
thymuses follows the normal one closely. It is worth emphasizing the
clear evidence for simultaneous colonization of the cortex and medulla
in the chimaeric thymuses (Figs. 13-15). This is an important point of
discussion among workers investigating how the mammalian thymus functions.
Some authors are of the opinion that cortical and medullary thymocytes
represent sequential steps in the formation of thymocytes, while others
believe that the cells in the two locations are independent subsets (see
Scollay 1983).

The ontogeny of markers on the surface of chicken or quail thymocytes has
been studied in recent years by several groups. However, it has not been
possible to ascribe these markers to functional subsets so that, compared
to mammals, very little is known concerning the acquisition of differential
functions by T lymphocytes of birds. What has been studied is the appearance of T cell-specific antigens (see section on T lymphocytes). Despite
the fact that the reagents used by different investigators probably
recognize several epitopes, the results are in fairly good agreement.
There seems to be a critical time around day 12 of embryonic life when
the first thymocytes acquire surface markers. With an antiserum raised in
turkeys, Albini & Wick (1975) detected very low levels of T cells in the
9-13 day thymus; thereafter, the number of positive cells increased
rapidly, reaching 100% on day 19. Sugimoto et al. (1977a), with an antibody raised in rabbits against embryonic and neonatal thymocytes, found
that the proportion of positive cells in the thymus changed from 4-7% at
12 days to 60-75% at 13 days, attaining a plateau of 95% on day 14.
Simultaneously, the cells in the thymus acquired the typical ultrastructural phenotype, i.e., the cell diameter decreased, the cytoplasmic
basophilia disappeared, and the nucleus acquired its characteristic heterochromatin. The striking point in all observations is the very sudden
appearance of the T antigen within one day of development. Brand et al.
(1983) found that 32% of thymocytes bore the "Th-1" marker on day 14 of
incubation and 80% after day 16. Chen et al. (1984) detected 6% of "CT-1+"
thymocytes on day 12 and 54% on day 13, adult levels (<90%) being reached
on day 15. Finally, Houssaint et al. (1985) observed that the "T1OA6"
antigen appeared on day 11 and reached the adult level on day 13. These

various findings have in common that no T antigen is observed on thymocytes before day 11 and that, once the antigen is expressed, it extends
rapidly to the majority of cells in the thymus. The temporal relationship
between the colonization waves of the thymus and the appearance of the
antigen is interesting to consider. The antigen is expressed several days
after the first wave, in fact at around the time when the second one
occurs. It will be interesting to determine by experimental means whether
the first wave lymphoid precursors differ from the following in never
acquiring the antigen or whether the antigen is expressed only after a
protracted delay of maturation. In the only ontogenic study in the quail
embryo (Péault et al. 1982), "AT65$^+$" cells were detected on day 7, i.e.,
soon after the first wave and made up 90% of thymocytes at 10 days.

Bursa of Fabricius (for detailed reviews see Ivanyi 1981 &
Ratcliffe 1985).

As stressed in the introduction of this chapter, the bursa is
a unique organ which is found only in birds, and whose existence has been
crucial to our understanding of the immune system of vertebrates. First
described in the 17th century by Fabricius of Acquapendente, the bursa
was only assigned a role in 1956 (Glick et al. 1956). These workers
surgically removed the bursa of Fabricius and found that the bursectomized
young chickens became incapable of mounting an immune response against
bacteria. This discovery led to the understanding of the fundamental
dichotomy of the immune system into cellular and humoral responses carried
out by T cells differentiated in the thymus and B cells differentiated in
the bursa.

Structure. The bursa of Fabricius is a blind-ended, round-to-oval, sac-
like, dorsal diverticulum of the cloaca. Its inner surface is thrown into
elongated folds. There are about 12 such plicae, which tend to obscure
the central lumen of the organ. The walls of these folds are packed with
follicles, the whole bursa containing about 10,000 of them (Fig. 17). A
minimal amount of connective tissue separates the follicles, each of which
is constituted by a central medulla and a cortex (Fig. 18). A single layer
of epithelially-derived reticular cells and a well defined basement
membrane underlined by a network of large capillaries separates the cortex
from the medulla. A supporting network of epithelial cells, filled with
lymphoid cells, also pervades both regions of the follicles. The cortex
is somewhat more deeply staining than the medulla, being composed of
closely packed small lymphocytes (Fig. 18). Lymphoblasts and many mitotic
figures can also be seen, and signs of necrosis are frequent. The medulla
contains large lymphoblasts particularly towards its periphery, and
medium and small lymphocytes. The ultrastructural features of lymphoid
cells in both regions of the follicles are very similar (Clawson et al.
1967). The size of these cells varies from 4 to 8 μm. The smallest (lymphocytes) have only a thin rim of cytoplasm and the nucleus contains clumped
chromatin and a small nucleolus in the chicken. The largest cells
(lymphoblasts) are characterized by a more substantial amount of cytoplasm,
a lightly-staining nucleus and a prominent nucleolus. Most cells, however,
have features intermediate between these two cell types. On the whole,
"bursocytes" appear larger than thymocytes (Figs. 19 and 20).

The surface epithelium, lining the luminal faces of the plicae, is a tall columnar, pseudo-stratified type (Fig. 17), very much like that of the cloaca. Three types of secretory cells, including goblet cells, can be distinguished in it according to the nature of their

Fig. 17. One plica from the bursa of a chicken, 6 days after hatching. E=epithelium, F=follicle, CT=connective tissue. Bar = 50μm.

Fig. 18. Detail of the follicles with medulla (M) and cortex (C). Bar = 20μm.

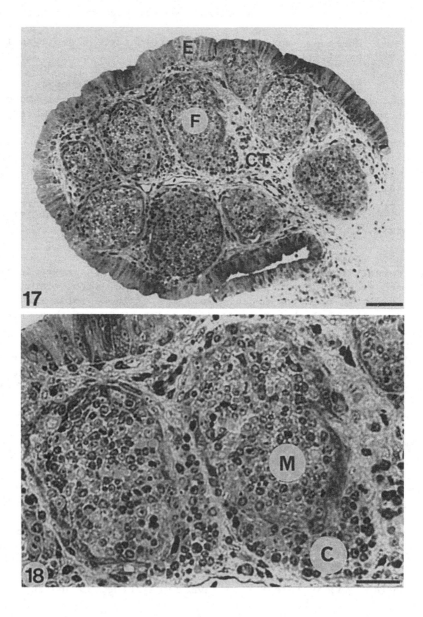

granules. At points where the apex of the follicles reaches the surface epithelium, the latter is different in appearance. The cells immediately over the follicle look undifferentiated with a pale staining cytoplasm. They are separated from surrounding epithelium by a slight indentation delimiting the "epithelial tuft" of the follicle. During the first week after hatching, the cells in the epithelial tufts exhibit a strong

Figs. 19 & 20. Comparison of typical fields in the thymus (Fig. 19) and bursa (Fig. 20). Note that the thymocytes are smaller and have more nuclear heterochromatin than the bursocytes. Bars = 2μm.

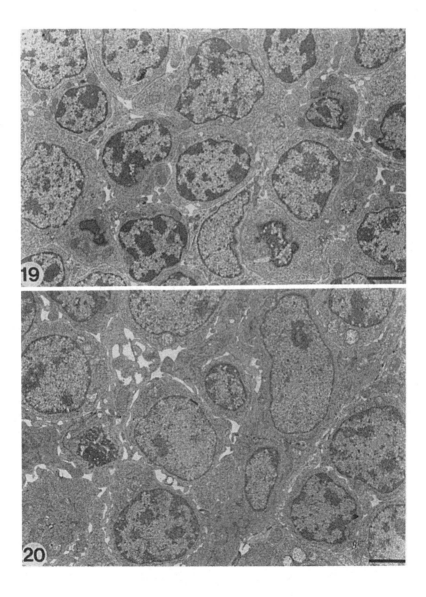

alkaline phosphatase activity, lost later in development (Ackerman &
Knouff 1959).

Experimental abrogation of bursal function. As indicated earlier, the
contribution of the bursa to a functional immune system in birds was
established by surgical bursectomy in birds (Glick et al. 1956; Cooper
et al. 1965; Ivanyi 1973). However, various other ways have been found
of suppressing the output of B lymphocytes by this organ. These comprise
treatment with the drug cyclophosphamide (Cy), that selectively destroys
B lymphocyte precursors (Lerman & Weidanz 1970), X-irradiation which also
depletes them (Weber & Weidanz 1969), and treatment with testosterone
which alters the bursal rudiment in such a way that it becomes incapable
of receiving colonizing stem cells (Meyer et al. 1959).

Bursectomy, performed just after hatching, was usually combined with
additional methods such as irradiation or an anti-μ immunoglobulin treatment: it resulted in profound life-long agammaglobulinaemia (Kincade et
al. 1970; Toivanen et al. 1972). However, an exclusive role of the bursa
in the generation of B lymphocytes has been challenged by the results of
in ovo bursectomy (Fitzsimmons et al. 1973; Jankovic et al. 1975). A
modified technique of bursectomy on day 5 of incubation has permitted
hatching and survival of the embryos in a fair proportion of cases. It was
found that suppression of the bursa, before it had a chance to process any
precursor, led to a loss of the birds' ability to produce diversified antibodies, although in some birds the serum still contained quite high levels
of Ig (Jalkanen et al. 1983; 1984).

Recent experiments, in which the extirpated bursal rudiment of a chick was
replaced by that of a quail, yielded chimaeric bursae (Belo et al. 1985).
Though similar in its principle to the experiment in which the quail bursal
rudiment is grafted in the body wall (Le Douarin et al. 1975; Houssaint
et al. 1976), this technique gave better developed organs in which the
expression of class II antigens and of a unique antigenic determinant
could be studied. This determinant termed MB1, is detected by means of a
monoclonal antibody which was obtained after immunization against the μ
chain of quail IgM (Péault et al. 1983). The MB1 determinant is present
on the surface of quail endothelial and haemopoietic cells, and is
considered restricted to that "haemangioblastic" lineage. Absent from all
chick cells, it is an excellent marker in quail/chick experiments (Fig.
13). In the chimaeric bursae, it was later shown that the bursal epithelium
also expresses MB1, precisely from the time when, and at the site where,
lymphoid precursors invaded it (Belo et al. 1985). Another finding was
that class II antigens are expressed constitutively by the endothelial
cells of perifollicular blood capillaries. The significance of these
findings is not entirely clear yet, but it is surmised that these antigens
might act as recognition signals between cells of the lymphoid lineage
and the bursal epithelium. It is interesting in this context that MB1 biosynthesis, first restricted to cells in contact with incoming haemopoietic
cells, later spreads throughout the bursal epithelium but never appears
in the mesenchyme. As to class II antigens, they remain confined to endo-
thelia which line the basement membrane of the follicles.

Using the cyclophosphamide (Cy) model, Toivanen and his coworkers
described within pre B cells a hierarchy of precursors, that emerged as
development proceeded (Toivanen et al. 1972; Toivanen & Toivanen 1973).
If the chickens that were injected neonatally with Cy, received no further
treatment, their bursa was atrophic. Full functional and morphological
reconstitution was achieved by injecting, after Cy treatment, and around
the hatching period a cell suspension from bursae of normal chickens.
These "bursal stem cells" are still immunologically incompetent, i.e.,
unable to produce antibodies, but they give rise to mature B cells if the
bursal microenvironment is there to receive them and permit their proliferation and maturation. A functional immune system is efficiently
restored, provided the injected bursal stem cells share with the host, one
major histocompatibility complex haplotype since T and B cells can co-
operate if the two lymphoid population are semi-allogeneic (Toivanen &
Toivanen 1977).

With age, bursal cells lose their capacity to restore bursal structure
but they acquire the capacity to induce functional reconstitution of the
immune system in Cy-treated recipients, even in the absence of the bursal
epithelium. Bursal development can also be impaired permanently by an
androgen treatment of young birds, a tool developed from the observation
that the bursa involutes at sexual maturity (Warner & Burnet 1961; Wolfe
et al. 1962). We shall consider here only one experiment aimed at
identifying the androgen responsive tissue (Le Douarin et al. 1980). In
this, the epithelium and the mesenchyme of the young rudiment were
separated by trypsinisation and one or the other of the tissue components
was treated in vitro with testosterone prior to reassociation. The
reconstituted rudiment was grafted into a host embryo. Treatment of the
epithelium was the critical factor. Neither mesenchyme nor lymphoid precursors (testosterone treatment of the host) were sensitive to the androgen. Furthermore, hormone receptors were only detected in the epithelium.

Ontogeny. The rudiment of the bursa appears in the chick embryo during
the fifth day of incubation as a thickening of the anal membrane or
cloacal eminence. At 10 days, the first epithelial folds begin to project
within the bursal lumen. This process initiates at the apex (more, cephalic)
of the rudiment and progresses caudalwards in the direction of the stalk.
At 12 days, epithelial folds are numerous. Vessels appear on day 11 and
invade the whole rudiment on day 12 but remain localized in the mesenchyme.
Local proliferations of the surface epithelium into the mesenchyme give
rise to epithelial buds beginning on day 12. They form the follicles where
lymphoid differentiation begins on day 15. Towards the end of the incubation period, mesenchymal cells organize concentrically around the
follicles enclosed within their basement membrane. These concentric
mesenchymal cells are the forerunners of the cortex, which develops only
after hatching.

Granulopoiesis is observed within the mesenchymal shaft of the plicae.
In the chick embryo, the first granuloblasts are observed on day 11 in
the most external region. This activity is maximal at 12-13 days, decreases
thereafter and disappears around hatching. The extent of this process is
variable among bursae of the same developmental stage. In the quail embryo,

similar processes occur slightly earlier. Epithelial buds appear at 10 days and the first lymphoid cells become recognizable at 13 days. The cortex of follicles develops after hatching. Granulopoiesis also occurs though less actively than in the chick. It is maximal on day 12 and disappears by day 15.

At the time of hatching, the bursa is a well-developed organ which rapidly grows until the 4th week. From then on, the bursal weight eventually begins to decline. The actual timing of involution of the organ usually occurs around 2-3 months.

Intensely basophilic cells are prominent in the developing bursa. They are first observed in the peripheral mesenchyme of the rudiment at 7-9 days of incubation. A few of them make contact with the epithelium at 10 days. They become very numerous in both mesenchyme and epithelium at 12 days. A prominent feature at this time is the presence of such cells in the lumen of capillaries. At 14 days, basophilic cells are very numerous inside the forming follicles. Their number decreases thereafter as lymphoid differentiation progresses (Houssaint et al. 1976). The baso-philic cells are lymphoid precursors which enter the bursal rudiment and, as in the thymus, have an external origin. It has also been proposed that these cells might arise either from the mesenchyme or the epithelium of the bursa. Their extrinsic origin was surmised by Moore & Owen (1966) in the light of their parabiosis experiments, since the bursa of parabiotic embryos were chimaeric. The pattern of colonization has been precisely documented in quail-chick combinations (Le Douarin et al. 1975; Houssaint et al. 1976). In these experiments, the endomesodermal rudiment of the bursa was taken at different embryonic days from one species, quail or chick, and grafted into the somatopleura of the other species. It was found that the bursa became seeded during only one phase, lasting from day 7 to 11 in the quail and from day 8 to 14 in the chicken (Fig. 16). The seeding process seems to decline rather than stop abruptly, as it does in the thymus. When the bursal rudiments were transplanted during the colonization period, a chimaeric lymphoid population developed but individual follicles displayed different patterns of colonization. Some were chimaeric, others were not and could harbour lymphoid cells either of donor or host species. Thus, all follicles are not colonized simultaneously and each follicle may be seeded by more than one stem cell. More than 50% of the follicles were mixed, indicating that most must be seeded by several stem cells.

In an attempt to discover more about the colonization of the avian bursa Pink et al. (1985a,b) constructed bursa chimaeras from chicken sublines carrying different alleles of a B cell surface alloantigen (Pink & Rijnbeek 1983). Using these models Pink et al. (1985a,b) arrive at the following picture of B cell differentiation in the chicken; the bursa is populated by less than 10^5 precursor cells, which do not express IgM, during days 8-15 of embryonic development. Probably these cells are not committed to expression of particular V-region combinations at the time of entry into the bursa (Lydyard et al. 1976). The extra-bursal precursor cells can no longer be detected in hatched chickens, either in situ (Ratcliffe & Ivanyi 1981; Houssaint et al. 1983) or in cell transfer experiments

(Toivanen & Toivanen 1973; Weber & Foglia 1980). Instead, Ig-bearing cells in the bursa act as precursors for B lymphocytes. These immunoglobulin-bearing bursal stem cells retain the capacity to proliferate in the bursae of CY-treated recipients. However, this capacity disappears with age (Toivanen & Toivanen 1973) and the sole source of B cells in the adult bird is a pool of self-renewing, IgM-bearing B cells with the capacity to proliferate in peripheral organs but not in the bursa.

Blood forming organs during development

During embryonic and post embryonic development in vertebrates, erythropoiesis is carried out successively in different organs. In the chicken, the organs involved are the yolk sac, which remains the main haemopoietic organ during most of the prehatching period, the spleen, which becomes erythropoietic around day 9 until day 16 to 18, with a peak around day 15, and finally the bone marrow whose function begins on day 12 and which is the permanent site of red cell formation. The role of the spleen and bone marrow has been considered in previous sections. Data on the ontogeneic development of the yolk sac and on diffuse haemopoiesis, which occurs in the embryonic mesenchyme, will be reviewed here.

Yolk sac (see also section on erythropoiesis during ontogeny). The yolk sac is the extraembryonic splanchnopleur; it is composed of two germ layers, the endoderm on its inner aspect, the mesoderm on its outer aspect. The yolk sac becomes established during the third day of incubation, when the outgrowing extraembryonic coelom splits the mesodermal layer into somatopleural and splanchnopleural components. During the first two days of incubation, two zones can be distinguished in the extra-embryonic area; the peripheral "area vitellina" which is made up of ecto-derm and endoderm only and the central "area vasculosa" consisting of three germ layers. The vascular area contains the blood islands and its boundary is delineated by a circular blood vessel, the sinus marginalis, which moves out as the extraembryonic area expands over the yolk. Between the definitive primitive streak and the headfold stages (18-24 h incubation), tight clusters of cells appear in the mesodermal layer in close contact with the endoderm in the extraembryonic region surrounding the caudal half of the embryo. These clusters of basophilic cells will form endothelium and haemoglobin-containing erythroblasts in another 10-14 h. Haemoglobin becomes detectable around the stage of 6-8 pairs of somites (Wilt 1962, 1974). By dissociating mesoderm and endoderm from the blood island presumptive region and reassociating clumps of cells from the two layers across a Millipore filter, it has been shown that the endoderm exerts a positive influence on the formation of blood islands by the mesoderm (Wilt 1965; Miura & Wilt 1970). Whether this interaction is of a trophic or a truly inductive nature has never been elucidated. Different kinds of data indicate that determination of the cells within the mesoderm which will make up the blood islands occurs much earlier than overt differentiation. When cells of the prospective area vasculosa from a definitive primitive streak embryo are cultured organotypically, haemoglobin synthesis can be detected some 16 h after starting the culture. Exposure to 5-bromodeoxyuridine (BudR), a thymidine analogue, within 6 h of explantation, prevents red cell formation (Miura & Wilt 1971). Subsequent transfer to a thymidine medium reverses the inhibition. Between 6 and 12 h after starting the culture, the tissue becomes refractory to BudR. Thus some regulatory event, crucial for terminal differentiation, occurs between 6 and 12 h. More recently, Keane et al. (1979) have been able to isolate mesenchymal cells from primitive streak stage chick embryos. In cell culture, they gave rise within 24 h to a variety of cell types including erythroid ones. It was found that these cells synthesized alpha-like globin chains nearly 24 h earlier than beta-like chains (see section on erythropoiesis during ontogeny). Some of the primitive streak mesenchymal cells were susceptible to transformation by avian erythroblastosis virus (AEV). Transformed clones of cells underwent terminal differentiation if subjected to specific inducers. They synthesized other globin chains such as alpha-D and pi. The fundamental question behind these experiments is that of the molecular mechanism underlying the developmental switches in gene activity.

Diffuse haemopoiesis, i.e., differentiation of blood cells outside of discrete organs, is prominent within the mesenchyme of the embryo during the first week of incubation (Miller 1913). In the chicken, this process occurs from day 5 to 8-9 with a peak on day 7. The blood-forming foci are scattered all over the embryo in the loose mesenchyme around the visceral organ rudiments or actually within some of them. For instance, granulo-poietic foci are usually abundant within the pancreas (Benazzi-Lentati 1932). Erythropoietic foci are especially well developed in the dorsal mesentery in close relationship with the newly-forming thoracic duct (para-aortic foci) (Figs. 21-23). In the quail embryo, diffuse haemo-poiesis is restricted to two dense symmetrical foci associated with the common cardinal veins or ducts of Cuvier (Dieterlen-Lièvre & Martin 1981) (Fig.23). These foci develop approximately during the same period as in the chick with a peak on day 6 and their disappearance is nearly completed at day 8. It is not clear whether the liver is a haemopoietic organ in avian embryos of nidifugous species. Immature erythroid forms are common in the hepatic sinusoids, as they are in all vessels of the embryo and modest granulopoietic foci develop around central veins.

BLOOD CELL STRUCTURE AND FUNCTION

Haemocytoblasts

Large round basophilic cells with a large nucleus and a prominent nucleolus occur in the bone marrow (see section on granulo-poiesis in the bone marrow). Their ultrastructure appears identical in various organs or in areas of the same organs involved in different haemopoietic processes (Figs. 24 and 25). The potentialities of these progenitors cannot be surmised from their morphology, so that the term "haemocytoblast", coined by Dantchakoff (1909), designates them conveniently. In embryos, haemocytoblasts are very frequent in blood vessels, spleen, thymus and bursa during colonization periods (Metcalf & Moore, 1971), and they make up the paraaortic foci in 6-7 day embryos.

In an experimental model constituted by quail embryos grafted on a chick yolk sac, the so-called "yolk sac chimaeras", haemocytoblasts in the paraaortic foci were quail, i.e. had an intra-embryonic origin (Dieterlen-Lièvre & Martin 1981). The wall of the quail early embryonic aorta (3-4 days of incubation) gave rise to foci of haemocytoblasts too when grafted into the dorsal mesentery of the chick embryo (Dieterlen-Lièvre 1984a,b). Cells from the wall of the aorta were also capable of colonizing an "attractive" thymus and of acquiring thymocyte surface antigens (Dieterlen-

Figs. 21-22. Para-aortic foci (arrows) of haemopoiesis in the avian embryo (Fig. 21 = 7-day chick)(Fig. 22 = 6-day quail). Ao=aorta, B=bronchi, CC=common cardinal vein. Feulgen-Rossenbeck staining. Bars = 100μm.

Fig. 23. One of the two para-aortic foci in the 6-day quail stained with monoclonal antibody QH1, specific of the quail haemangioblastic lineage (see Dieterlen-Lièvre 1984). Note fluorescent haemopoietic and endothelial cells. Bar = 20μm.

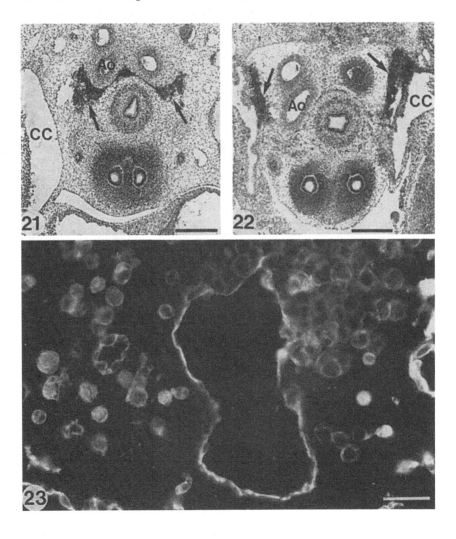

Lièvre & Martin, unpublished results) and giving rise in vitro to
colonies of macrophages, when provided with culture conditions permitting
the differentiation of this cell type (Cormier et al. 1986) (see section
on in vitro clonal assays for haemopoietic precursors). Thus, there is in
the young embryo a specific haemopoietic anlage which can be mapped to the
paraaortic zone of the mesoderm, and whose potentialities are now being

Figs. 24 & 25. Haemocytoblasts in an erythropoietic (Fig. 24)
and a granulopoietic area of bone marrow (Fig. 25). Both
have similar features; smooth contours, high nucleo-cytoplasmic
ratios, abundant ribosomes and prominent nucleoli. Bars = 2µm.

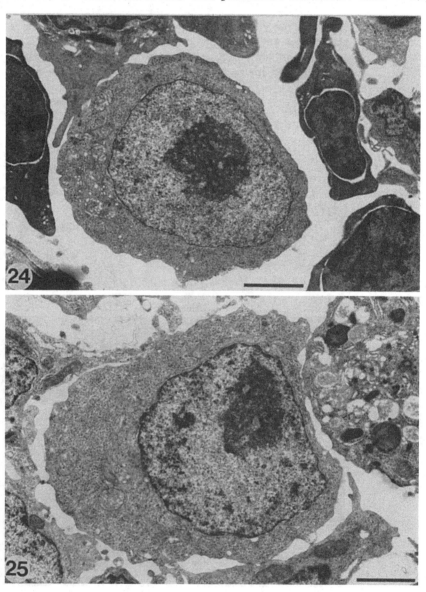

studied in the presence of various conditioned media. This haemopoietic rudiment may have a central role in the ontogeny of the haemopoietic system, since it produces precursors at the time when definitive erythropoiesis and colonization of the thymus are initiated. Alternatively, another possibility is that precursors emerging from this rudiment may in turn be relayed by others produced elsewhere at later dates. No experimental clue concerning this issue presently exists.

Erythrocytes (Figs. 26-28, 31)

The mature erythrocyte of the adult chicken has two distinctive features which it shares with most non-mammalian red cells; it retains its nucleus throughout its lifespan and its shape is a flattened ellipsoid.

The chromatin in the nucleus is in a condensed, transcriptionally inactive state, so that the actual difference between the chicken and mammalian red blood corpuscle is that, in the former, the nucleus is not extruded from the mature cell. It may be reactivated experimentally, for instance by fusion with other cells (Ringertz et al. 1971). Red cell precursors are spheroidal, becoming flattened after the final mitotic division.

Acquisition of shape. Following flattening, chicken erythrocytes become circular disks and the elliptical form is acquired with further maturation (Barrett & Scheinberg 1972). Young erythroid cells in the early stages of postmitotic maturation display a marginal band of microtubules that seem to play a critical role in the acquisition of the flat discoidal shape; these microtubules decrease in numbers as the cell matures (Small & Davies 1972; Barrett & Ben Dawson 1974). Recently, it was shown that these microtubules contain a variant of beta tubulin with unique biochemical and assembly properties (Murphy & Wallis 1983 a,b). This tubulin variant appears to be the product of a unique tubulin gene which is exclusively expressed in erythrocytes and thrombocytes (Murphy et al. 1986). An interesting pattern of expression of "general" versus "specific" tubulin was observed by these investigators using two antibodies, one which recognizes the alpha and beta subunits of tubulin common to most chicken cells, and one specific for the erythrocyte tubulin subunit. The pattern of immunofluorescence revealed that the centrosomal microtubules of early blast stages contained the general tubulin, while the erythrocyte variant appeared suddenly in midstage erythroblasts at the onset of haemoglobin synthesis, accounting for 90-95% of the tubulin in the marginal bands. This finding suggests that a developmental programme is responsible for turning off the synthesis of general tubulin and turning on the expression of a new beta tubulin gene. Purified erythrocyte tubulin is stable and produces long microtubules by means of a low rate of nucleation leading to the formation of stable tubulin oligomers (Murphy et al. 1986).

Haemoglobins. The essential feature of the erythrocyte is the presence of haemoglobin, or rather of haemoglobins. These oxygen-carrying proteins confer to the red blood cell cytoplasm a pink to orange colour after staining with the May-Grünwald-Giemsa technique. In the electron microscope, haemoglobin gives a dense appearance to the cytoplasm. Haemoglobins

of the chicken have been studied by a number of investigators (reviewed
in Bruns & Ingram 1973c), all of whom agree upon the existence of two main
electrophoretically separable components. The major one accounts for ca.
70% of the total haemoglobin, while the minor one makes up the remainder.
Several groups have detected a third haemoglobin, present as a trace
component, as well as additional bands, depending on the separation
technique and on the strain of chickens used. Various nomenclatures have
been used to designate these various bands. The most commonly adopted is
that of Ingram's group, who call the major band A (for adult) and the
minor one D (for definitive). The amino acid composition of these two
components is quite different. Each haemoglobin is constituted of 4 globin
chains, two alpha-like and two beta-like. The A and D haemoglobins share
a common beta-like chain (Table I).

Table I

Globin chain make up of the various chicken haemoglobins.
During development different haemoglobins are synthesized
at different times. Each is constituted of two alpha-like
and two beta-like globin chains. Each has a unique assortment
of globin chains. Some globin chains are restricted to the
primitive lineage (pi and pi' in the alpha-like family,
epsilon, epsilon' and ro in the beta-like family). The adult
alpha chains are found in some of the embryonic haemoglobins.
When they first appear, at 5-7d of incubation, the adult A
and D haemoglobins are synthesized in a particular ratio,
which is different from the adult ratio

Haemoglobin Designation	Globin chain assortment		Erythroid lineage
	α-like	β-like	
E	α^A	ϵ	
P	π	ρ	Primitive: Day 2
P'	$*\pi'$	ρ	to 5 of incubation
M	α^D	$++\epsilon'$	
H	α^A	β^H	Early definitive $\cong$ Day 7 to 3 weeks posthatching
A	α^A	β^A	Definitive
D	α^D	β^A	

* π' is an allele of π

++ ϵ' is very similar if not identical to ϵ

<u>Maturation stages in the erythroid lineage</u>. Immature erythroid precursors are not uncommon in the peripheral blood of birds (Figs. 26a–c) and are also recognisable in haemopoietic tissues such as the embryonic spleen (Fig. 27) and the bone marrow. The early polychromatic erythroblast is the first stage which can be definitely attributed to the erythroid lineage in blood smears of normal, healthy animals. This cell is smaller than haemocytoblasts; its circular contour is smooth, the nucleus is large and the nucleolus, which was prominent in the haemocytoblast, is sometimes still barely visible; the cytoplasm appears as a ring, medium in width, stained light blue to grey (Figs. 26a,b). This is the stage

> <u>Figs. 26a–c</u>. Appearance of erythroid precursors in smears. Bars = 20μm.
>
> <u>Fig. 26a</u>. Imprint from a 11-day chick embryo spleen. Proerythroblasts (arrows) and polychromatic erythroblasts (arrowheads) are present.
>
> <u>Fig. 26b</u>. Imprint from a 12-day chick embryo spleen. Polychromatic erythroblasts (arrowheads) have become the predominant type.
>
> <u>Fig. 26c</u>. Blood smear from a 3-week old chicken infected with the myelocytomatosis retrovirus. Basophilic erythroblasts (arrows) are very numerous in the blood of this animal. Whereas the excessive number of these cells probably reflects a pathological aspect, these erythroblasts display their typical appearance in smears: smooth contour, large nucleo-cytoplasmic ratio and strongly basophilic cytoplasm.

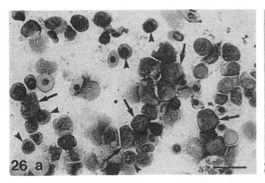
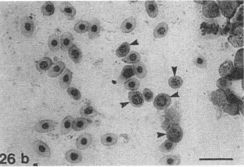
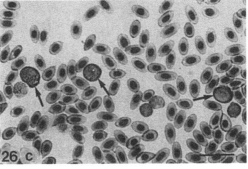

when haemoglobin begins accumulating and the ribosomes disappear (Fig. 27). Depending on the ratio of these two components, the staining affinities of erythroblasts change gradually from blue to grey and then

Fig. 27. Sequential maturation stages of erythrocytes seen in an erythropoietic area in the spleen of a 14-day chick embryo 1: proerythroblast. 2: basophilic erythroblast. 3,4,5: progressive maturation stages of polychromatic erythroblasts. 6: erythrocyte (the nucleus is not visible in this particular section). The terms used here to designate the various stages of erythroid differentiation are defined by the staining affinities of the cells on smears. The nuclear and cytoplasmic characteristics detected in these electron micrographs make it easy to identify corresponding stages. Abundant ribosomes are present in the cytoplasm of cell 2, increasing amounts of haemoglobin in cells 3 to 6 are responsible for the increasing electron density of the cytoplasmic matrix. Bar = 2μm.

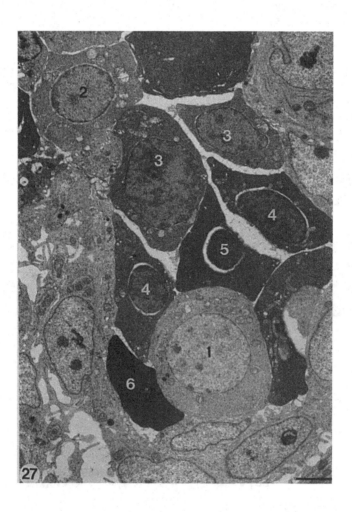

to pink in smears stained with Giemsa. Such early stages are found in the blood of young embryos, where they are released synchronously in immature form and where they mature simultaneously (Fig. 28) (see section on erythropoiesis during ontogeny). At the electron microscope level, the cytoplasm in the polychromatic erythroblast appears electron-dense and homogeneous.

Erythropoietin. As in mammals, a glycoprotein factor is necessary for the multiplication and differentiation of precursors committed to the erythroid lineage. Mammalian erythropoietin appears ineffective in promoting avian erythropoiesis (Rosse & Waldman 1966; Samarut & Nigon 1976b). Chick erythropoietin can be obtained from the blood of anaemic animals and its activity can be assayed by counting the number of erythro-cytic clones (Colony Forming Units-Erythroid = CFU-E) developing in a semi-solid medium from a suspension of haemopoietic precursors (Samarut & Nigon 1976b; Coll & Ingram 1978; Samarut 1978). Chick erythropoietin has a molecular weight of about 60,000 daltons. Its site of production in the adult is unknown. It is produced in early blastodisc cells in culture (Samarut 1979). Earlier reports about a hepatic factor (Salvatorelli 1967; Salvatorelli & Smith 1971) seem to pertain to a trophic effect rather than to a true erythropoietin-like activity.

Erythropoiesis during ontogeny. Different red cells are produced sequentially, and these differ in shape and haemoglobin content. The primitive erythroid lineage, produced by the early yolk sac blood islands, is constituted by a unique "cohort" of cells (Ingram 1972). Contrary to

Fig. 28. Blood cells of a 2-day quail embryo. They are all very large erythroblasts of the primitive series.

Fig. 29. Blood cells of a 6-day quail embryo. Large round cells (primitive erythrocytes=P), which are the progeny of the cells in Fig. 28, coexist with smaller oval definitive (D) erythrocytes. May-Grünwald-Giemsa staining. Bars = 10μm.

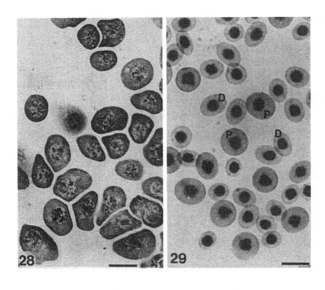

later erythrocytes, these cells are released in a very immature state
(Fig. 28): they continue dividing, and mature synchronously within the
blood vessels, between days 2 and 5 of incubation. These cells are large
and, for that reason, are often called megalocytes. They are spherical
with a round nucleus and synthesize haemoglobins which are specific to
that period of development (Table I). However, cells of the first "definitive" erythroid lineage begin entering the blood from 5 days onwards and
soon supercede the primitive cells (Fig. 29). At 7 days, primitive cells
account for only 50% of erythrocytes (Bruns & Ingram 1973a and c). From
12 days onward they are rarely encountered.

Numerous investigators have analyzed the haemoglobin switches (Fig. 30)
in the chicken embryo (e.g. d'Amelio & Salvo 1961; Fraser 1964, 1966;
Raunich et al. 1966; Manwell et al. 1963, 1966; Hashimoto & Wilt 1966;
Simons 1966; d'Amelio 1966; Godet 1967, 1974; Fraser et al. 1972; Shimizu
1972 a,b; Schalekamp et al. 1972; Bruns & Ingram 1973a,b). In the 5-day
embryo, two major haemoglobins, E and P (Table I), make up 80% of the
haemoglobins present (Fig. 30). The adult haemoglobins (A & D) appear at
that time in a particular ratio and they slowly evolve, reaching the adult
pattern some 3 weeks after hatching (Table I). During that intermediate
period there is a specific band, termed "hatching" (H) in Ingram's

Fig. 30. Development of haemoglobin patterns in the quail
embryo. Haemoglobins were separated by polyacrylamide gel
electrophoresis at pH 10.3 (the P and M components were not
resolved by this technique and the embryonic E and intermediate H band migrated to the same position). From Beaupain
et al. (1979). With the permission of Springer-Verlag.

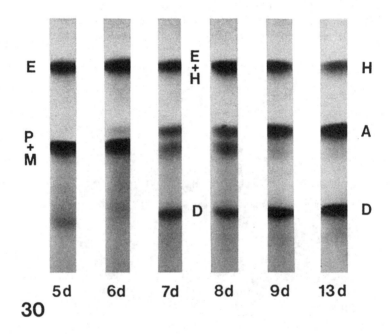

E   E   H
                     +
                     H
P   A
+
M
                     D   D

   5d   6d   7d   8d   9d   13d
30

terminology to denote its presence at that particular period of development, but sometimes designated as "foetal", by reference to the mammalian intermediate haemoglobins (Godet 1967, 1974). These various haemoglobins are constituted by the association of two alpha-like and two beta-like chains (Table I). Each embryonic haemoglobin contains at least one unique polypeptide chain.

Alpha-globin and beta-globin genes are arranged in two separate groups on the chromosomes. Globin gene clusters are one of the best studied examples of developmentally regulated multigene families. In vertebrates where they have been studied (mammals and amphibians), the order of the genes in the cluster coincides with their order of expression. The alpha-like globin genes of chickens are organized in that manner, with the embryonic pi-globin gene 5' to the adult alpha A- and alpha D-globin genes (Engel & Didgson 1980; Dodgson et al. 1981).

The gene arrangement in the chicken beta-globin gene cluster is an exception to this rule. The genes coding for the primitive ro and epsilon-globins are located at the 5' and the 3' ends of the cluster respectively. The adult beta-globin gene is between them (Dolan et al. 1981). Whatever the origin of this difference with respect to other vertebrates these findings in the chicken suggest that either the physical arrangement of the genes within a chromosomal cluster has no causal relationship to the temporal expression of these genes or that the mechanism of globin gene activation in chickens is different.

In yolk sac chimaeras the respective roles of extra- and intra-embryonic stem cells in the production of red cell progeny can be determined by assessing the proportions of chick and quail erythrocytes in the blood by means of differential immune haemolysis, using rabbit antibodies directed either against chick or quail erythrocytes (Dieterlen-Lièvre et al. 1976; Beaupain et al. 1979). Until 5 days of incubation, the blood of the chimaeras contained chick erythrocytes. From 5 days onward, quail erythrocytes appeared in the blood in increasing proportions, so that in some chimaeras at 13 days only 20% of the red cells were chick. In a homospecific model where yolk sac and embryo from two chick lines differed by histocompatibility antigens, replacement of erythrocytes derived from yolk sac stem cells by intraembryonic stem cell-derived erythrocytes was extremely rapid and regular (Lassila et al. 1982). Thus, it is clear that in the avian embryo, haemopoietic stem cells which initiate the definitive blood-forming system emerge within the embryo rather than in the primordial haemopoietic organ, i.e., the yolk sac. Similar conclusions have been reached in the amphibian embryo. This issue is still open in mammals.

Changes in chromatin structure related to globin gene activation.
Regulations responsible for sequential gene activation during haemoglobin switches have been extensively investigated in chicken erythroid cells. When the globin genes are actively expressed, the globin chromatin is preferentially sensitive to DNAase I (Weintraub & Groudine 1976; Larsen & Weintraub 1982). However, it is notable that in early chick embryo erythroid chromatin, the adult beta-globin gene is already DNAase I sensitive, although it is not transcribed until later in development (Stalder et al. 1980a). Preferential sensitivity is a consequence of the

association with two nuclear proteins, HMG 14 and HMG 17 (high mobility group nuclear proteins) (Weisbrod & Weintraub 1979; Sandeen et al. 1980; Weisbrod et al. 1980). Another feature of DNA in active regions of the chicken globin chromatin is its undermethylation, while flanking non-transcribed regions are methylated (McGhee & Ginder 1979; Weintraub et al. 1981). DNA undermethylation may constitute a signal for HMG binding.

Differentiation cell surface antigens on erythroid cells. An array of developmental antigens has been detected on the surface of erythrocytes or erythroblasts at different stages of development (Blanchet 1976; Blanchet et al., 1976; Samarut et al. 1979a; Dietert & Sanders 1977; Miller et al. 1982; Nelson et al. 1984; Trembicki & Dietert 1985). These antigens, detected by polyclonal or monoclonal antibodies, can be grossly divided into embryonic and adult, in fact it seems that the antibodies recognize many distinct antigenic determinants, the minimum number of which seems to be 13 (Nelson et al. 1984). The distribution of these antigens is by no means simple since they coexist on erythrocytes of young chickens for several months after hatching, and also because an embryonic-like antigen is still expressed on bone marrow erythroid cells in the adult. While different investigators agree that the embryonic antigenicity is associated with a 47 Kd membrane protein, different molecular weights have been obtained for the adult antigen. It is possible that the discrepancies stem from heterogeneity between the strains of birds used. Competitive radiobinding assays with B-locus specific alloantisera suggest that the adult antigens may be antigens of the polymorphic B-G locus of the chicken MHC (Miller et al. 1982). The potential interest of these antigens lies in their possible involvement in the interactions between haemopoietic cells and the microenvironment during erythropoiesis. Furthermore, if a clear-cut pattern of expression can be detected, these antigenic determinants would be useful markers to identify different lineages among successive erythroid generations.

Thrombocytes (Figs. 31-34)

The nucleated thrombocytes of 'lower' vertebrates fulfil the same function as blood platelets of mammals. Their primary role is in haemostasis, but they may also have a phagocytic function, which may be very effective in birds (Chang & Hamilton 1979). Nothing is known yet about the synthesis of serum molecules which might play roles of paramount importance in cell culture as well as in normal or cancerous growth in vivo, like mammalian Platelet-Derived-Growth-Factor. Some authors consider that avian thrombocytes are easily confused with certain other blood cells. Under the light microscope, these cells could be mistaken for erythrocytes, having a similar shape and size. They are, however, slightly smaller than the latter, elongated but with blunt ends; the cell contour is not as regular as that of the red cell and the nucleus is oval. The clear cytoplasm is pervaded by a faint bluish network (Fig. 31) and contains one or more granular inclusions, usually stained red in the May-Grünwald-Giemsa technique. The ratio of thrombocytes to erythrocytes in the blood is about 15:1000 in normal birds. Usually two to five or more thrombocytes appear aggregated together in blood smears (Fig. 31). A "trigger-like fragility" (Lucas & Jamroz 1961) is one of their functional features, so that frequently their cytoplasm is disrupted in smears. In this case several nuclei with heavily reticulated chromatin appear clumped together.

Janzarik (1980) maintains that due to damage to thrombocytes during smearing, phase contrast microscopical examination of living blood cells is the preferred method for the identification of this cell type. In *vitro*, the

Fig. 31. Blood leucocytes in the chicken. B=basophil (purple granules), E=eosinophils (pink-orange granules), Er= erythrocytes, H=heterophils (pink-orange granules), i= immature heterophils, L=lymphocyte, M=monocyte, T=thrombocytes. The blood smears are from 2-month chickens infected with avian myelocytomatosis virus. White blood cells are more numerous than in control animals but they have a normal appearance. The eosinophils (E) vary in size. Immature heterophils (i) may be distinguished from eosinophils by the fact that their round, pink-staining granules are sparse rather than tightly packed. Note the conspicuous granule, located in the indentation of the nucleus, in the cytoplasm of the monocyte (M). May-Grünwald-Giemsa. Bar = 10μm.

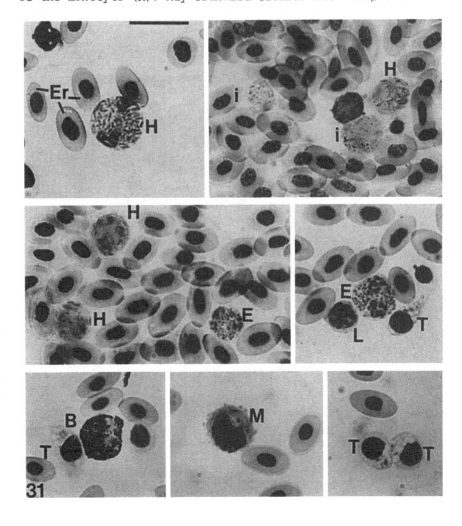

normally round-spindle shaped thrombocytes flatten and spread out over the slide to reveal a vacuolated cytoplasm (Janzarik 1980).

Under the electron microscope, thrombocytes can be confused either with erythrocytes or, according to some authors, with lymphocytes. These cells are indeed similar in size to small lymphocytes (Maxwell 1974; Janzarik & Morgenstern, 1979) but their irregular contours display pseudopodia or lobopodia and the nucleus is denser than that of lymphocytes, with marginated heterochomatin (Fig. 32). The Golgi apparatus is also well-developed. Thrombocytes share with erythrocytes the presence of a marginal

Fig. 32. Three thrombocytes (arrows) in an erythropoietic area of the chicken embryonic spleen. Note the clumped chromatin of nuclei and the presence of large dense cytoplasmic granules (G). Bar = 5µm.

Fig. 33. Perinuclear and intratubular (arrows) peroxidase activity in a chicken thrombocyte. Original micrograph provided by Professor J. Breton-Gorius. Bar = 1µm.

Fig. 34. Catalase-containing granules (arrows) in a chicken thrombocyte. From Breton-Gorius et al. (1978). With the permission of Williams & Wilkins Pubs. Bar = 1µm.

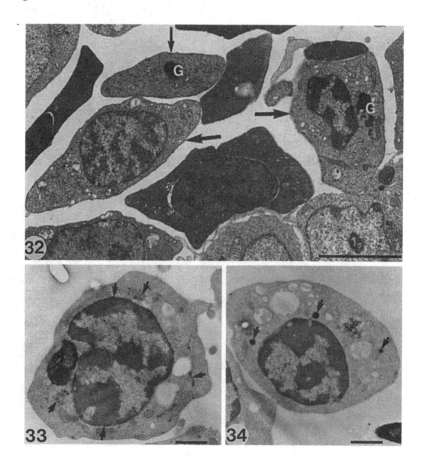

band of microtubules. The B tubulin that constitutes these microtubules is a variant restricted to these two cell types (Murphy et al. 1986). A specific type of peroxidase activity ("platelet peroxidase") and catalase activity have also been demonstrated in chicken thrombocytes (Figs. 33, 34; Breton-Gorius & Guichard 1976; Breton-Gorius et al. 1978).

A careful assessment of thrombocytes was made by Kuruma et al. (1970). These authors separated a thrombocyte-rich fraction from EDTA-treated blood. Red cells were eliminated by gentle centrifugation (100g for 1 min) followed by separation on a Ficoll gradient. The resulting suspension contained 70% so-called "spindle cells", a term often employed to designate the avian oval thrombocytes. The 30% of contaminating cells were erythrocytes and lymphocytes. The cytoplasm of these thrombocytes compared well with that of mammalian platelets. The nucleus was rich in heterochromatin associated with the nuclear membrane. A typical feature was the presence of one or more electron-dense osmiophilic bodies (Fig. 32) in the cytoplasm very similar though larger than those observed in rabbit platelets. The dense bodies seen in thrombocytes circulating in the blood appeared usually not as a single structure but as an aggregate of numerous small granules. By treating the chicken 16 h prior to sacrifice with reserpine, an agent known to induce 5-hydroxytryptamine (5-HT) release, Kumura et al. (1970) induced degranulation of some of these dense bodies. Concurrent with the morphological evidence of degranulation, the 5-HT content was found to have decreased. Isolation and purification of chicken thrombocytes have also been achieved by Traill et al. (1983) using Ficoll-Paque density gradient centrifugation.

Recently, a specific anti-thrombocyte serum has been obtained (Ries et al. 1984). Heparinized blood was incubated for 2 h at 39°C, unattached blood cells removed and adhering cells detached from the dishes by treatment with a buffered Ca^{2+} Mg^{2+}-free EDTA solution for 0.5 to 1 h at 0°C. The separated cell suspension was highly enriched in thrombocytes (87% of the cells compared with 9% monocytes, 2% granulocytes and erythrocytes). This thrombocyte-rich suspension, when injected into rabbits, induced the production of antisera which were thrombocyte-specific after absorption with bursa and thymus powder and finally with living lymphocytes. Thus, there exists a determinant unique to the surface of thrombocytes. On the other hand, these cells probably share some common antigenic determinants, at least with B lymphocytes (Janzarik et al. 1980; Traill et al. 1984). However, they should definitely be regarded as an independent cell lineage.

Chang & Hamilton (1979) are of the opinion that avian thrombocytes are important in clearing the blood of foreign material. For example, thrombocytes actively ingest colloidal carbon (Grecchi et al. 1980; Van Alten 1982), formalised Enterobacter cloaca (Chang & Hamilton 1979) but fail to internalise zymosan and sheep erythrocytes (Grecchi et al. 1980). As in fish and mammals (see Chapters 2 & 6) their phagocytic ability is questionable and should be re-examined to check for the true internalisation of such particles. The "phagocytic" ability of chicken thrombocytes has been found to vary according to age, with cells from 3-8 week old chicken displaying far greater activity than 1 week old individuals (Glick

et al. 1964). Grecchi et al. (1980) found that although chicken thrombocytes may bear complement receptors they lack receptors for the Fc portion
of homologous Ig (G?). Finally, Stinson et al. (1979) have reported that
chicken T and B cells produce a "thrombocyte inhibitory factor" which
affects the migration of these latter cells.

Granulocytes (Figs. 31, 35-38 and Table II).

There are three types of avian granular leucocytes. The most
numerous are the heterophils. The term heterophil indicates the fact that,
among vertebrates, homologous cells of this lineage contain granules with
various staining affinities. In birds, heterophils stain with eosin and
for that reason are sometimes designated as pseudoeosinophils. They are
probably equivalent to the human neutrophil which displays no granules at
the light microscope level. At the electron microscopic level, most
authors have relied for identification of the various types of avian
granulocytes on the characteristic features of mammalian granulocytes.

Heterophils. In stained blood smears, heterophils appear as circular
cells with a lobed nucleus (Fig. 31). The granules, which are the main
diagnostic feature of this type of cell, are very characteristic rods or
spindles. These granules stain bright pink, due to their affinity for
eosin. There is usually no hesitation in identifying a heterophil; however, the granules are sometimes atypical, either because they have been
dissolved during fixation or staining, or because they are immature (Fig.
31, "i"). In such cases, they may be confused with the granules of eosinophils which have the same staining affinity but which are spherical.
Immature heterophil granules are also spherical and stain reddish to
purple. In the electron microscope, the nucleus of the heterophil appears
polymorphic, consisting of one to three or more lobes joined together by
narrow bridges. The chromatin is organized into dark staining masses
associated with the nuclear membrane (Figs. 35, 36). Lobulation of the
nucleus increases with the age of the cell.

Various investigators have distinguished different types of granules in
heterophils (Campbell 1967; Maxwell & Trejo 1970; Osculati 1970; Daimon
& Caxton-Martins 1977; MacRae & Powell 1979; Maxwell 1985). The most
obvious type has an elongated shape. It is approximately 1.5μm long and
0.5μm wide and the content is homogeneous and electron-dense. As described
by Daimon & Caxton-Martins, the second type of granule is smaller (ca.
0.5μm) in diameter and its content is electron-translucent. The third
type is much smaller (0.1μm in diameter) and has a dense core separated
from a membranous envelope by an electron-lucent area. The presence of
lysosomal enzymes in these granules has been investigated. Acid phosphatase activity is only present in the large, spindle-shaped granules of the
fowl (Daimon & Caxton-Martins 1977; Osculati 1970) as is aryl sulphatase
(Osculati 1970). Thus, only these granules should be regarded as lysosomes. On the other hand, Maxwell (1984a) detected the activity of an
acid phosphatase with a specific substrate (trimetaphosphate) not only in
the large, spindle-shaped granules but also in the medium-sized, round
granules. According to this author, the two types of granules represent
different maturation stages of the same organelle. Chicken heterophils

Table II. Structure and functional characteristics of avian granulocytes

Feature	Heterophils	Eosinophils	Basophils
Differential count in blood (% of leucocytes)*	26	2.25	2.8
Mean diameter*	8.7μm	7.4μm (range 4-11μm)	8.2μm
Granules:			
1. staining properties	eosinophilic	eosinophilic	basophilic
2. shape in smears	rods and spheres	spheres	spheres
3. ultrastructure	electron-dense rods, spindles, or spheres	-homogeneous: "primary" -crystalline: "specific"	-homogeneous - stippled -honeycomb
4. enzymes present	acid-phosphatase, aryl sulphatase	peroxidase	acid-phosphatase in some stippled granules
Functions	bactericidal activity	modulate inflammatory responses in Type IV hypersensitivity	histamine release

*Data for chicken, from Hodges (1974).

have an exceptional feature which sets them apart from those of other
studied species: they are devoid of peroxidase activity (Fig. 36) (Brune
& Spitznagel 1973; Maxwell 1985) and do not produce hydrogen peroxide
during phagocytosis (Penniall & Spitznagel 1975), although they are
capable of destructive activity against a variety of microorganisms.

Eosinophils. The average diameter of these cells is 7.3 μm; however, it
can vary widely between 4-11 μm. The pale blue staining cytoplasm is
hardly visible because the cell is packed with large, round eosinophilic
granules (Fig. 31).

Description of the granules in the electron microscope follows the nomenclature adopted for mammals. Two types of granules have been distinguished
at the ultrastructural level in some avian species (see, however, Maxwell,
1985). Large spherical homogeneous granules, known as primary granules,
correspond to early stages of maturation, while mature granules contain
a crystal. These crystal-containing granules are called specific granules.
The crystal consists of arrays of a recurrent pattern which can be either
linear or a square lattice of repeating lines. The primary granules are
supposed to transform into the specific ones as maturation proceeds. In
species such as the duck and the goose (Maxwell & Siller 1972), where
eosinophil granules have a crystalline core, the identification of this
cell type is obvious. Unfortunately, the chicken (Maxwell & Trejo 1970:
Maxwell 1978a,b, 1984a, 1985), the quail, the turkey, the pigeon and
guinea-fowl (Maxwell & Siller 1972) have no such crystalloid cores in
their eosinophilic granules (Fig. 36). None of the descriptive studies
indicate the criteria by which eosinophilic granulocytes were identified
at the electron microscopical level, when enzyme cytochemistry was not
used to detect specific activities acting as markers.

As in mammals, the eosinophilia of the granules is due to a high concentration of the amino-acid arginine, which bears a strongly cationic terminal
guanidium group. A number of enzymes are present in these granules, a
prominent one being peroxidase (Maxwell 1985). Cytochemistry at the
ultrastructural level detects peroxidase, thought to have bactericidal
properties, in fowl (Fig. 36) and duck eosinophils (Maxwell 1978a; 1985).
The enzymatic activity is first present in all regions of the Golgi
apparatus, the rough endoplasmic reticulum and the perinuclear cisternae
of the myeloblasts but disappears from these organelles as maturation
proceeds. Conversely, the enzyme activity of granules increases during
cell maturation, so that in mature eosinophils the majority of granules
of both types, i.e., primary and specific granules, exhibit a positive
reaction. Acid phosphatase activity is never found in the Golgi apparatus
or RER cisternae, but is restricted to the granules, where it becomes
denser as the granule matures. An acid phosphatase with a pH optimum of
3.9, specific for the substrate trimetaphosphate, was also detected in
the mature granules of chicken and in the crystalline internum of duck
eosinophil granules (Maxwell 1984a). Aryl sulphatase, another enzyme
activity present in the granules of human eosinophils, is always very
weak, being positive only at the poles of the eosinophilic granules in
mature cells of the chicken. In the duck, the crystalloid granules are
positive for the two latter enzyme activities. Finally, Maxwell (1978b)

Fig. 35. Heterophilic granulocytes in the spleen of a 13-day
quail embryo. Note the lobulated nucleus (N) with marginated
heterochromatin. The granules vary widely in sizes and shapes.
Bar = 1μm.

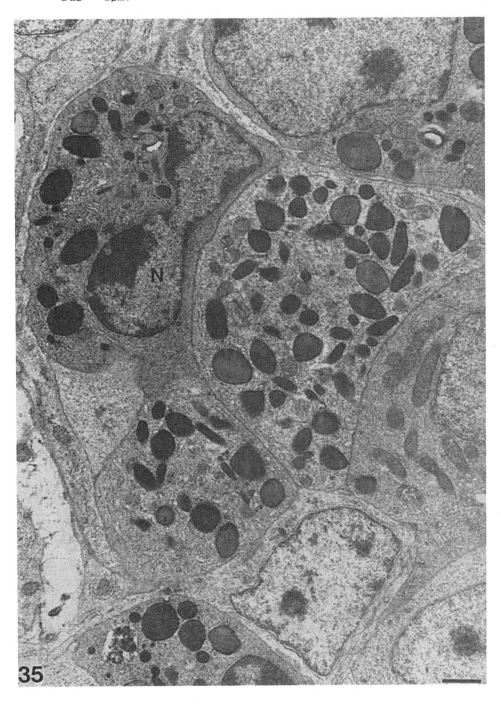

applying the ammoniacal silver reaction for cationic proteins, found it positive in the heterochromatin. As to the granules, primary granules are generally negative, while in specific granules the reaction product appears as coarse grains, which may be loosely-spread or tightly set. The presence of acid phosphatase and aryl sulphatase, also known to be present in the granules of eosinophils in mammals (see Chapter 6) makes it possible to identify these granules as primary lysosomes.

Basophils. These cells are the least well studied among the avian granulocytes. With a mean diameter of 8.2μm, they are, on average, slightly smaller than heterophils and appear circular in blood smears (Fig. 31). Their colourless cytoplasm is densely packed with medium-sized, spherical, deep purple granules. The nucleus is usually unilobular. In most avian species studied, three types of granules can be distinguished by electron microscopy (Fig. 37a). The most common type is round and electron-dense. The second type has a stippled internal structure (Fig. 37a). Finally, the third type was described as possessing a reticulated "honeycomb" arrangement (Maxwell 1973)(Fig. 37b). Human mast cell granules have this structure which is supposed to depend on the association of heparin and

Fig. 36. Chicken heterophil (H) and eosinophil (E). The preparation was treated to demonstrate peroxidase activity. Large and small granules in the heterophil are negative, whereas granules in the eosinophil all display a strong enzyme activity. Note the marginated heterochromatin in the nuclei of both cells, the heterogeneous sizes and shapes of heterophilic granules, and the circular shape of eosinophilic granules. Bar = 1μm. From Breton-Gorius et al. (1978). With the permission of Williams & Wilkins Co.

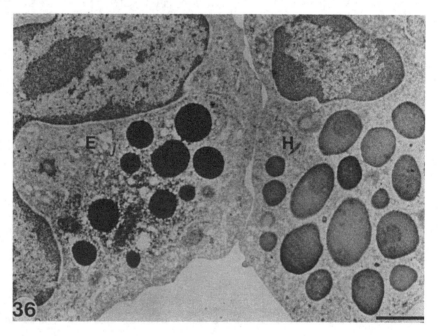

histamine. By analogy, the presence of honeycomb granules has been interpreted as a marker of avian basophils. To my knowledge, there is only one
report dealing with the cytochemistry of basophils. Daimon & Caxton-Martins
(1977) investigated acid phosphatase, peroxidase and alkaline phosphatase
activities in buffy coat specimens from the chicken. Among these, only
acid phosphatase was found in a small proportion of the large stippled
granules.

Maturation stages in the granulopoietic lineage. A detailed account of the
morphological aspect of the development of the three lineages of granulocytes has been given by Lucas & Jamroz (1961) and Maxwell (1985). These
authors distinguished a number of maturation steps which, according to
their judgement, can be distinguished more precisely than in mammals. In
their nomenclature, cells are called granuloblasts prior to the acquisition
of granules. When they acquire the very first granules which are supposed
to allow a diagnosis of lineage, they become myelocytes. A number of stages
were also described within these two categories according to the size of
the cells, the staining affinity of the cytoplasm and the appearance of the
maturing granules. The earliest stage, which Lucas & Jamroz (1961) call a
large early granuloblast, has a high nuclear:cytoplasmic ratio and a narrow
rim of cytoplasm staining deep blue with Giemsa; it is impossible to
ascertain whether it is committed to a particular lineage.

Even when the first granules appear, diagnosis remains difficult, because
immature granules of heterophils are round (Fig. 31i) and vary in colour
from pink to orange or violet and are thus very similar to early granules
of either the eosinophil or basophil series.

In the electron microscope, the progress of maturation may be evaluated
by the appearance of the nuclei, where heterochromatin clumped onto the
nuclear membrane becomes more prominent as the cells evolve towards
terminal differentiation (Fig. 38). Granules of different sizes coexist
throughout differentiation in heterophils, with rod- or spindle-shaped
granules becoming more numerous and larger as the myelocytes differentiate
(Fig. 38).

Nowadays, evaluation of haemopoiesis in mammals is only partly based on
these cytological criteria. In vitro clonal assays have been developed to
assess the potentialities and numbers of precursors in haemopoietic cell
suspensions (Metcalf 1977). Furthermore, differentiation antigens can be
identified by means of antibodies, allowing objective identification of
cell lineages or maturation steps and even sorting out of cells according
to these criteria. Concerning the avian granulopoietic lineage, this is
still a relatively unexplored area.

Granulocyte Functions. Avian heterophils probably function in a similar
way to their mammalian counterparts. There are many reports of their
participation in inflammatory reactions (e.g. Carlson & Allen 1969; Fox &
Solomon 1981; Awadhiya et al. 1982; Maxwell 1984b), and such cells are
avidly phagocytic (Carlson & Allen 1969). The granules of this cell type
have been isolated and found to be bactericidal, and lysozyme and basic

Fig. 37a. A basophilic myelocyte in the spleen of a 11-day
chick embryo. Three characteristic types of granules are
present; granules with a dense homogeneous content within a
loose-fitting bounding membrane (D), stippled granules (S)
and granules with a honeycomb structure (H). Bar = 1μm.

Fig. 37b. Chicken basophil. The granules in this cell have
the "honeycomb" appearance, which identifies the cell as a
basophil. Note the lobed nucleus with marginated heterochromatin.
A portion of a heterophil (H) is also visible. Bar = 1μm.
Original micrograph provided by Dr. J. Breton-Gorius.

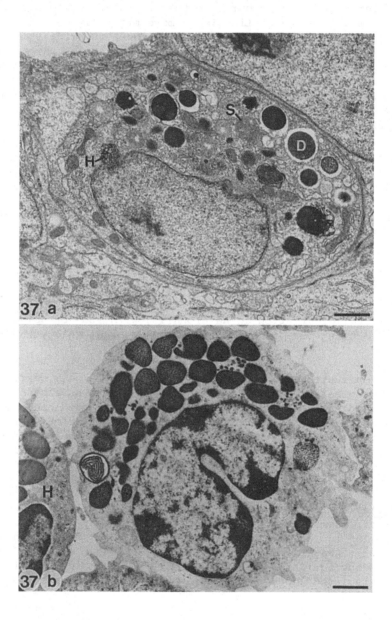

Fig. 38. Granulopoiesis in the spleen of a 15-day chick
embryo. Myeloblasts (Mb) in this electron micrograph
display round granules of various sizes and electron
densities. A myelocyte (Mc) (partly seen) contains one
large, spindle-shaped granule among round granules. Promyelocytes (P) may be identified by the beginning of
margination of the heterochromatin in their nuclei.
H=haemocytoblast, S=stromal cells. Bar = 2μm.

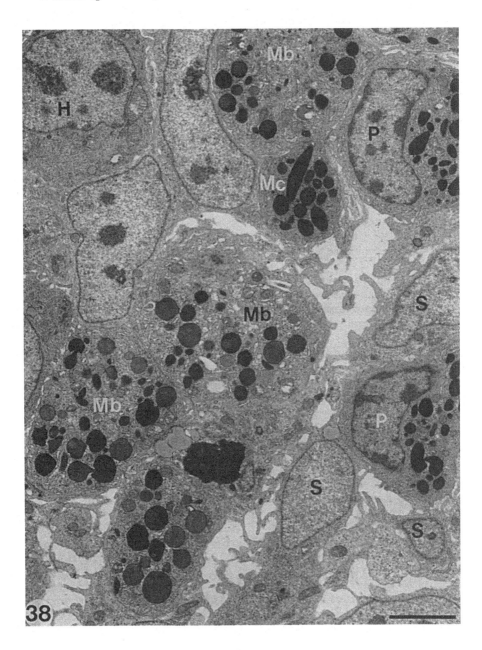

proteins seem to be responsible for this activity (Brune & Spitznagel 1983). Further details of the involvement of heterophilic granulocytes in inflammation are apparently lacking.

One characteristic,of mammalian eosinophils is their marked increase in numbers (eosinophilia) during allergic reactions such as anaphylaxis and parasitic infections. Eosinophilia has been reported as occurring in some birds during inflammation (see Fox & Solomon 1981 for a review) but apparently no studies have commented on such an event in response to helminth or protozoan infections. Indeed, cutaneous application of citraconic anhydride in birds, an agent known to induce blood eosinophilia in some mammals, failed to produce such a response in domestic fowl (Maxwell 1984b) As a result of these and other experiments, Maxwell (1984b) concluded that avian eosinophils are, unlike their mammalian counterparts, active participants in Type IV (delayed) hypersensitivity reactions. Clearly as is the situation in fish, amphibians and reptiles (see Chapters 2, 3 and 4), it should not be taken for granted that eosinophils participate in immunological responses in a similar manner to their supposed mammalian counterparts.

Avian basophils, like their mammalian equivalents, produce histamine (Chand & Eyre 1978). Furthermore, they may play a role in acute inflammation and Type I (immediate) hypersensitivity reactions (Carlson & Allen 1969; Fox & Solomon 1981).

Monocytes, macrophages and accessory cells (Figs. 31, 39-41)

These cells all probably belong to the same lineage, being respectively the blood and tissue phagocytes. They play an important role in the immune system by processing antigens and presenting them to T lymphocytes. Dendritic cells in lymphoid organs, though non-phagocytic, also bind antigen to their surface and are classified within this lineage. All these cells express not only MHC class I antigens like all cells in the body, but also high levels of class II antigens, that are responsible for their involvement in the phenomenon of MHC restriction, i.e., the process by which a foreign antigen is recognized only if associated with self class II antigens.

In blood smears stained by the Giemsa technique, chicken monocytes, which display a variety of shapes and sizes, may easily be confused with lymphocytes. The most typical are round, with a membrane thrown into blunt or filamentous blebs and a kidney-shaped nucleus (Fig. 31). The cytoplasm is usually deep blue with a reticular appearance. Often a purple staining granular area is present in the vicinity of the nucleus (Fig. 31).

Ultrastructural descriptions of avian monocytes and macrophages are relatively common (e.g. Schumacher 1965; Enbergs & Kriesten 1969; Maxwell & Trejo 1970; Nirmalan et al. 1972 Burkhardt 1980; Grecchi et al. 1980) and a few studies utilising the functional capacities of these cells for their isolation and identification have been reported (e.g. Burkhardt 1980). According to the ultrastructural descriptions, the monocyte appears as a cell with an electron-dense cytoplasm, a plasma membrane thrown into blebs or filaments, a conspicuous Golgi apparatus, numerous ribosomes and a few cisternae of rough endoplasmic reticulum. Pinocytic vesicles and small to medium granules (lysosomes) are also characteristic (Fig. 39). Macrophages can be recognized by the presence of numerous inclusions in the process of being digested (Fig. 40).

Recent studies have been based on the presence of class II antigens to identify cells of this lineage. Ewert & Cooper (1978) identified macrophages from the blood buffy coat by their property of ingesting latex particles. A subpopulation of these cells were positive with an alloantiserum directed against B-L antigens (Ia like in the chicken, see section on Major Histocompatibility Complex Antigens). Lymphoid organs also contained a population of B-L$^+$ Ig$^-$ cells which were morphologically typical monocytes and macrophages. Using a rabbit antiserum raised against blood-derived adherent cells, Peck et al. (1982) have analyzed the macrophage population from blood, bone marrow, spleen, thymus and bursa. As judged from the presence of non-specific esterase and Fc/C3 receptors, the macrophage appeared to be the predominant cell type in the adherent cell population from these sources. The majority of these cells (60-80%) expressed B-L antigens, except in the thymus where only 10% were positive. All, except the bone marrow-derived adherent cells, displayed low phagocytic capacity, perhaps reflecting the state of activation of these cells.

The phagocytic activity of avian monocytes/macrophages is well established (e.g. Glick et al. 1964; Burkhardt 1980; Grecchi et al. 1980; Van Alten 1982) though the role of serum factors in this process is less clear. For example, Van Alten (1982) concluded that neither Ig nor complement promoted the in vitro phagocytosis of erythrocytes by chicken peritoneal macrophages but that serum factors such as fibronectin might be involved. Alternatively, Grecchi et al. (1980) also using chicken peritoneal macrophages were of the opinion that Ig receptors played a role in the phagocytosis of coated heterologous erythrocytes.

Chicken splenic macrophages are also responsible for a cytotoxic effect on allogeneic cells with which such animals have been previously immunised (Palladino et al. 1980). This cytotoxicity is independent of the presence of specific antibody and may involve the activation of macrophages by factors produced by T cells.

Finally, the origin of accessory cells in the quail and chick thymus has been analyzed during ontogeny (Oliver & Le Douarin 1984; Guillemot et al. 1984). These cells were isolated by means of their glass-adherence properties and used to raise monoclonal antibodies directed against the B-L antigens of each species. Macrophages and dendritic cells (DCs) were easily distinguished from each other by their morphology, macrophages being round cells, while DCs are stellate shaped (Fig. 41). DCs often remain characteristically associated with lymphocytes after dissociation. Both types of cells were positive with anti B-L antibodies. In chimaeric thymuses obtained by heterospecific-grafting of the rudiment, it was established that the accessory cells had an extrinsic origin and that their turnover was roughly similar to that of the lymphoid population.

Fig. 39. Cell tentatively identified as a monocyte in the
spleen of a 15-day chick embryo. Note a moderate amount of
heterochromatin mostly localized along the nuclear membrane,
a well-developed Golgi apparatus (G) in the vicinity of the
nucleus, a few profiles of rough endoplasmic reticulum (ER),
medium-sized granules (Gr) and several vacuoles (V).
Bar = 1μm.
Fig. 40. Macrophage in the spleen rudiment of a 4-day chick
embryo. Numerous inclusions display contents in various
stages of digestion. Bar = 1μm.

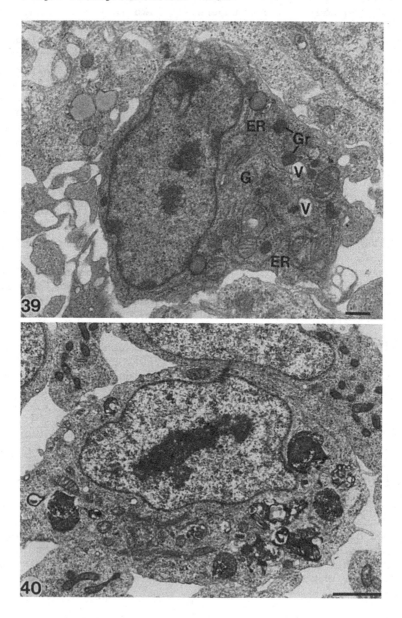

In vitro clonal assays for haemopoietic precursors

        When compared with mammalian studies, this area appears
sparsely investigated. The formation of erythroid colonies has been
investigated by Samarut and his co-workers (Samarut & Nigon 1976a,b;
Jurdic et al. 1978; Samarut et al. 1979b). In the presence of erythropoietin, usually provided by anaemic chicken serum, bone marrow cells
give rise to small colonies (5 to 150 benzidine-positive cells) which
develop within 3 days of culture (Samarut & Bouabdelli 1980). These are
the progeny of erythroid Colony-Forming-Units (CFU-E). If the medium is
supplemented with pokeweed-mitogen-spleen-cell-conditioned medium, that
provides Burst-Promoting-Activity (BPA), bursts also develop, that are
the progeny of earlier precursors, designated as Burst-Forming-Units
Erythroid (BFU-E). Chicken serum (rather than mammalian) is necessary for
the growth of avian colonies. Bursts are large, multicentric colonies,
made of 3 to 20 clusters, each cluster containing 8 to 60 erythrocytes.
The bursts appear between the third and the sixth day of cultures and
persist until the seventh day; then they progressively degenerate.
BFU-E and CFU-E are found in embryonic bone marrow and yolk sac. In the
young blastoderm, BFU-E become detectable as early as the primitive
streak stage. An antigen specific to immature red cells can be detected
on CFU-E, but not on BFU-E, showing that both progenitors are distinct
entities (Samarut & Bouabdelli 1980).

Granulocyte-macrophage colonies derived from progenitors designated as
Colony-Forming-Cells (CFC) have been studied mainly as comparison
standards for cells transformed by retroviruses (Dodge & Moscovici 1973;

        Fig. 41. Accessory cells from the quail thymus revealed
        with TAC1, an anti-class II monoclonal antibody (see
        Guillemot et al. 1984). (a) dendritic cell. (b) macrophage.
        Micrographs provided by Dr. F. Guillemot. Bars = 10µm.

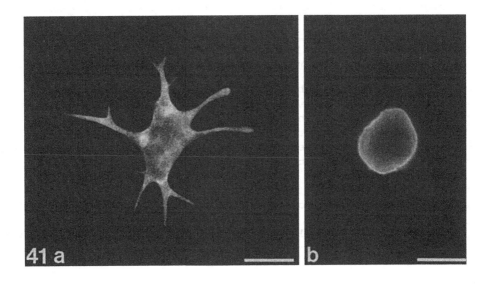

Gazzolo et al. 1979; Duprey & Boettiger 1985). Macrophage colonies readily
develop in a semi-solid medium containing chicken serum (Dodge & Hansell
1978) and "fibroblast-conditioned-medium". This is the supernatant from
9- to 11-day chicken embryonic cells, that are prepared by trypsin
dissociation of whole embryos, after eyes and viscera have been taken out.
Granulocyte colonies develop, provided serum is omitted from the medium
and fibroblast-conditioned-medium is replaced by spleen-conditioned medium
(Dodge & Sharma 1985). Macrophage-colonies are composed of scattered
cells, granulocyte colonies are dense (Figs. 42, 43) (Dodge et al. 1975).
The cells in both types of colonies contain granules but the granules in
macrophages are coarse and irregular in size, they probably represent
phagocytosed agar since they are not present when the semi-solid medium
contains methyl cellulose. Various tissues yield macrophage colonies,
such as bone marrow from the post-hatching animal or the embryo, 12-day
embryo yolk sac, embryonic spleen and wall of the 3-4d embryo aorta (see
section on haemocytoblasts). Recently a myelomonocytic growth factor was
isolated from medium conditioned by a transformed macrophage cell line
(Leutz et al. 1984). This factor, cMGF, promotes the growth of colonies
composed of macrophage-like cells and a minor proportion of granulocytes,
cMGF exists as two active species, respectively 23 and 27 kd which are
probably two glycosylated forms of a single protein moiety.

T lymphocytes

       A broad distinction should be made between thymocytes, which
are the lymphoid cells residing in the thymus, and T lymphocytes, i.e.
lymphoid cells seeded to the periphery (blood, spleen, lymph nodes), after
differentiation within the thymus. Thymocytes are small cells (2.5-5µm
in diameter, Frazier 1973) with a large nucleus. They are closely packed
in the thymus cortex, and their nucleus is heavily studded with heterochromatin masses (Figs. 11, 19). In the quail and chick, i.e. the two
species which have been combined for experimental purposes in the study
of ontogeny, the nuclei of thymocytes are strikingly different. In the
light microscope, chicken thymocytes are studded with numerous small dots
when the Feulgen-Rossenbeck technique for DNA staining is applied; quail
thymocytes have one large Feulgen-positive mass, plus several smaller
ones which adhere to the nuclear membrane (Fig. 12).

Antigenic markers (see Table III). Immunological surface markers which
have been developed for avian lymphocytes are less useful for distinguishing cell subpopulations than mammalian ones. The main reason for this
situation seems to be that most reagents developed for mouse T cells are
alloantisera raised between congeneic strains and that few truly congeneic
strains are available in birds. Antibodies presently available distinguish
essentially T lymphocytes or B lymphocytes. The distinction of subpopulations within each of these two main families has hardly begun.
Xenoantibodies, alloantibodies and monoclonal antibodies have been raised
by various investigators. By means of "ATS" (Anti-T cell Serum) raised
in turkeys, Wick et al. (1973) found that, in the spleen of 2- and 3-week
old chickens, 40 to 50% of lymphoid cells were labelled. This serum is a

Fig. 42. Macrophage colony developed in vitro from bone
marrow cells. Chicken macrophages in culture are characterized
by large vacuoles.

Fig. 43. Granulocyte colony developed in vitro in the absence
of serum. Micrographs provided by Dr. F. Cormier. Bars = 200µm.

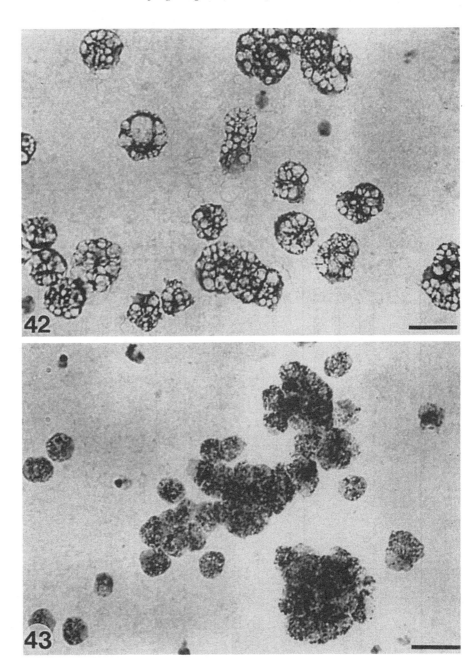

Table III. Antibodies against avian T cell surface antigens

Type of Antibody	Name	Prepared Against	Reactivity	MW of Antigen	Authors
Xeno-antisera [*]	ATS_T	Chick thymocytes	Thymocytes and T cells in spleen	N.D.[**]	Albini & Wick (1973) Wick et al. (1975).
	ATS_R.	Chick thymocytes	Thymocytes and peripheral T cells	N.D.	Boyd & Ward (1978)
	AT 65	Quail thymocytes	Thymocytes and	65kd	Pink et al. (1981)
	AT 45	Quail thymocytes	peripheral T cells	45kd	Péault et al. (1982)
Allo-antisera	CAl-T	Thymocytes between B14/14 substrains	Thymocytes and peripheral T cells		Galton & Ivanyi (1977a)
	TA	Thymocytes	Thymocytes only B14/14 B10/10		Galton & Ivanyi (1977b)
	Thl	PBL from chicken	75% thymocytes 25% PBL	N.D.	Gilmour et al. (1976)
	Ly4	lines 6 and 7	6% thymocytes 70% PBL[***]		Fredericksen et al. (1977) Fredericksen & Gilmour (1981)
Monoclonal antibodies	TD1	Quail thymocytes	80% quail cortical thymocytes <1% spleen	?	Péault et al. (1982)
	X14	Chick thymocytes	Cortical thymocytes	65kd&>	Pink & Rijnbeek (1983)
	CT1	Chick thymocytes	90% chick thymocytes; quail cortical thymocytes; 2-5% PBL	103kd + 63kd	Chen et al. (1984)
	CTla	Chick thymocytes	Chick thymocytes; quail thymocytes negative		
	T10A6	Chick thymocytes	80% of thymocytes 8-20% peripheral T cells	65kd	Houssaint et al. (1985)

*Xenoantisera were all absorbed against bursal cells. ATS_T was prepared in turkeys, the others in rabbits.
**ND = not determined. +++PBL = peripheral blood lymphocytes.

marker of thymocytes and thymus-derived lymphoid cells. In an ontogenetic
study involving chickens from the day of hatching until 24 weeks of age,
Albini & Wick (1975) showed that defined ratios of T lymphocytes
characterize the spleen, bursa of Fabricius, coecal tonsils and Harder's
gland and that the ratio in each organ was constant throughout the period
under study. In particular, T cells dominated in the spleen and among
peripheral blood lymphocytes (PBL), while B cells dominated in Harder's
gland. Recently, Traill et al. (1984) from the same group of workers were
able to distinguish subsets of peripheral T cells on the basis of the
different fluorescent intensity conferred to them in indirect immunofluorescence by a turkey antiserum and several rabbit antisera. On the
basis of these differences, PBL could be fractionated in the fluorescence
activated cell sorter. Two main subsets were tentatively identified as the
helper and suppressor/cytotoxic mammalian subsets on the basis of their
responses to concanavalin A (Con A), phytohemagglutinin (PHA), and
pokeweed mitogen (PWM) and their capacity to elicit a graft-versus-host
reaction.

Alloantisera were raised by Gilmour's group (Table III) using two highly
inbred lines of chicken that exhibit marked differences in resistance
to Marek's disease, a virus-induced T cell lymphoma. The antisera were
obtained by reciprocal immunization of the two lines with lymphoid cells
and these recognize two independent polymorphic antigens expressed on the
surface of T lymphocytes. Th-1 (= Thy 1) (Gilmour et al. 1976) and Ly-4
(Fredericksen et al. 1977). Each locus is autosomal and codominant and
has two alleles, a and b. Th-1 antisera stain the majority of thymus cells
(70 to 75%) and a minor population of PBL. Many of the Th-1$^+$ PBL are
large and medium-sized lymphocytes. In contrast Ly-4 antisera label few
thymus cells (6%) and many PBL (65-70%). Most Ly-4 PBL are small lymphocytes, with a few medium and no large cells (Fredericksen & Gilmour 1981).
Some peripheral T cells bear both antigens.

Pink and coworkers (1981) studied the antigens detected by two rabbit
antisera. These sera recognize respectively antigens of apparent molecular
weight 65 and 45 kd, which bear no resemblance to known mammalian antigens.
AT65 reacts with thymic and peripheral T cells (50% of splenic white cells
at 3 weeks) (Péault et al. 1982). In frozen sections of the thymus, AT65
labels cells both in the cortex and medulla. During development, the first
immunoreactive cells are found in the thymus on day 7 of embryonic life,
precisely 24 h after incoming precursors have colonized the thymic rudiment.

The most recently developed antibodies are monoclonal (Table III). TD1
(Péault et al. 1982) is restricted to a subpopulation of thymic lymphocytes; no cells (<1%) in the spleen are positive. The TD1$^+$ cells are
located in the cortex of the thymus. AT65 (see above and Table III) and
TD1 recognize independent determinants. In ontogeny, TD1$^+$ cells are
present only from 8 days of embryonic development onward, i.e., 24 h
later than AT65$^+$ cells.

Other monoclonal antibodies designated CT-1 and CT-1a (Chen et al. 1984)
react with two different antigenic determinants present on the same

molecule, as revealed by immunoprecipitation and blocking experiments. More than 90% of chicken thymocytes are positive (Table III); the CT-1 defined antigen is distributed over both cortex and medulla. CT-1 and CT-1a react with only 2 to 5% of PBL and splenic white cells and no cells from the bursa or the bone marrow. In the quail, CT-1 reacted with cortical but not medullary lymphocytes and CT-1a did not react with any lymphocytes. The chicken CT-1 antigen appears to be similar to the AT-65 antigen of Pink et al. (1981). It differs from it, however, in that AT-65 reacts in the quail with both cortical and medullary thymocytes. The CT-1 antigen, on the other hand, differs from AT-45 (Pink et al. 1981) and Th-1 (Gilmour et al. 1976) in tissue distribution and molecular weight. The CT-1 antigen does not appear to be an avian homologue of known mammalian T antigens.

The expression of CT-1 was studied during ontogeny (Chen et al. 1984). In 11-12 day old embryos, the youngest stage studied, the frequency of positive cells in the thymus was around 5%; it increased rapidly, reaching adult levels by day 15 of incubation. CT-1$^+$ cells were first detected in the spleen on day 13 of incubation but never reached high proportions.

Finally, the monoclonal antibody raised by Houssaint et al. (1985), T1OA6 (Table III), against chicken thymocytes recognizes thymocytes and a minor subset of peripheral T cells. In the spleen, these positive cells are located in the red pulp and in the marginal zone of the white pulp. This antibody was applied to an ontogenic study which showed that T1OA6 positive cells appear in the thymus on day 10.5 and only two days later in the spleen. The relationship of this antibody to the other monoclonal antibodies has not been established.

Response to mitogenic factors. A mitogenic response to the lectins Con A and PHA is specific for T lymphocytes in mammals, and also in birds (Weber 1970, 1973; Fredericksen & Gilmour 1983). Con A stimulation seems to be under the control of at least two major genes, one of which may be linked to the major histocompatibility complex (Miggiano et al. 1976; Pink & Miggiano 1977; Morrow & Abplanalp 1981; Pink & Vainio 1983; Fredericksen & Gilmour 1983). The PHA response is probably also under genetic control (Fredericksen & Gilmour 1983). In this latter report, reactivities against these two lectins were investigated in two inbred lines of chickens. One line is resistant to Marek's disease, while the other is susceptible. Developmental differences in the responses to the mitogens were found in the two lines, which indicated the existence of T cell subsets supporting these differences. These differences may be partly involved in the reaction to Marek's disease.

Lymphokines and interleukins. Interleukin-1 (IL-1), produced in vitro by mitogen-stimulated macrophages, promotes proliferation and activation of T and B lymphocytes. T lymphocytes, in turn, secrete IL-2 which amplifies T cell proliferation. In birds, Hayari et al. (1982), using assays similar to the ones used for mammalian cells, were able to demonstrate IL-1 in the culture medium of lipopolysaccharide-treated adherent cells from chicken spleen. Chicken T cell growth factor, IL-2, has been detected in contioned media of PBL or splenocytes stimulated with PHA or Con a in

serum-free medium (Schauenstein et al. 1982). IL-2 was detected by its
ability to stimulate the growth of T lymphoblasts in vitro. These blast
cells express IL-2 receptors (Schnetzler et al. 1983). IL-2 permits
continuous growth of chicken lymphocytes for several weeks and increases
the Con A response of chicken thymocytes. The biological properties of
this factor are similar to those of mammalian IL-2. Two peaks of activity
could be separated (9-11kDa and 19.5-21.5kDa). Under reducing conditions,
a single polypeptide of 13kDa was isolated (Schnetzler et al. 1983).

Mixed Lymphocyte Reaction (MLR). The MLR is the proliferative stimulus
that results from in vitro interactions between two lymphoid cell populations from allogeneic animals. In response to some of the foreign major
histocompatibility antigens, some cells in each population transform into
"blast cells" characterized by their large size, pronounced cytoplasmic
basophilia and active replication. The proliferative response can be
quantified by the incorporation of tritiated thymidine. By blocking mitosis
in one of the populations by pretreatment with mitomycin or irradiation,
it is possible to measure the proliferation of the other population (one-
way MLR). The genetic control and mechanism of the MLR, extensively studied
in the mouse, is interpreted as follows. T helper cells in the responding
population react to class II antigens presented by accessory cells in the
stimulating population. Thereupon, these T cells proliferate and secrete
IL-2 which, in turn, stimulates the proliferation of cytotoxic T cells. In
the chicken, dissection of the MLR into its phases and identification of
the interacting cells involved is still limited due to the difficulty
encountered by most investigators when trying to obtain reproducible
values for the stimulation. One of the basic problems is that highly inbred
or, better, congeneic strains are rarely available.

One of the earliest trials on chickens, carried out on two outbred strains
(Alm 1971), showed that spleen cells from bursectomized-irradiated
animals reacted normally in the MLR, leading to the conclusion that bursa-
dependent cells play a negligible role in this response. Miggiano et al.
(1974), using the highly inbred Prague strains CA, CB, WA and WB,
demonstrated the primordial role of the B locus, i.e., the chicken MHC, in
the genetic control of the MLR. They found that the magnitude of the
reaction depended on the strain combination used, but did observe important
variations between individual animals. Schou (1980) confirmed the role of
the MHC but showed also that a few MHC-identical individuals exhibited
strong reactions and postulated the existence of some "lymphocyte-activating
determinants", not closely linked to the B complex. However, Schou &
Simonsen (1982) later involved the implication of anti-parental reactions
when Fl cells were stimulated by homozygous parental cells. Finally, Bacon
& Lee (1981), also trying to find out why reproducibility of the MLR was
low in chickens, studied various combinations between the highly inbred
lines 6_3 and 7_2 and several less inbred strains of White Leghorns. They
found that the age of stimulating, i.e. antigen-presenting, cells was
critical, but varied with the responding strain. For instance, line 7_2
responded to 8-week cells whatever their strain, whereas cells from the
other strain responded to 14-week cells.

In our experience, an expedient procedure is to select some birds as donors of stimulating PBL (Cormier & Dieterlen-Lièvre 1985). Once the magnitude of the response induced by the cells of each donor is determined, results become readily comparable among successive experimental series involving various preparations of responding cells. It should be mentioned that the technique varies in details according to investigators but that one important point is the temperature at which it is carried out, the optimum being 40°C (Lee 1978). Lymphokine production (IL-1 and IL-2) seems to be one of the limiting factors at 37°C (Chi et al. 1984). A final point is to define conditions that allow survival of cells without exerting a mitogenic effect, that would result in a background stimulation. The same batch of serum should be selected for a whole experimental series.

B lymphocytes

B cell surface markers. Surface Ig of the IgM type is expressed on most B cells, thus allowing their convenient identification. Agammaglobulinaemia can be obtained by embryonic injection of anti-μ antibodies followed by neonatal bursectomy (Kincade et al. 1970), which indicates that all peripheral B cells in the 13-day embryo bear IgM. Immunofluorescence examination of allotype expressions on PBL, spleen or bursal cells from M-1a/M-1b heterozygous chickens demonstrated that all B cells expressed only one allotype; therefore allelic exclusion is the rule, individual chickens showing a 1:1 ratio in the proportions of cells carrying each allele. Bursal cells from hatched chickens and 17-day embryos are already allelically excluded, indicating that the phenotypic restriction occurs at a very early stage of B cell differentiation (Ivanyi & Hudson 1979; Ratcliffe & Ivanyi 1979).

An alloantigen specific to the surface of B cells has been uncovered by reciprocal immunizations with bursal cells between two inbred lines EL6 and EL7 (Gilmour et al. 1976). The autosomal locus Bu-1 controls two alleles, Bu-1a and Bu-1b; this locus is independent from MHC, blood groups or allotype loci.

Immunoglobulins. Chickens have two main immunoglobulin (Ig) isotypes, which are respectively of high (IgM) and low (IgG) molecular weights, and which resemble functionally their mammalian counterparts. The structure of IgG diverges in several aspects, though not enough to justify the IgY designation once attributed to them. An IgA isotype is found in various secretions, in particular the bile. It contains a J-chain, binds with a secretory component, and is produced by cells localized in the intestinal mucosa, all features in common with mammalian IgA. IgA synthesis is bursa-dependent and participates in the intestinal response to bacterial infection.

Allotypes have been identified for IgM and IgG. Two closely linked genetic loci, IgM-1 and IgG-1, control antigens expressed on cytoplasmic mu and gamma heavy chains, respectively. They are co-dominant alleles of single autosomal loci. The antigens coded by these loci are expressed on all IgM or IgG molecules, that is there are no alternative C allotypic markers, indicating a lack of Ig subclasses in the chicken. The G1 locus is more polymorphic (13 alleles described) than the M-1 (5 alleles) (Foppoli et

al. 1979). Various haplotypes, i.e. combinations of alleles from the two
loci have been described, and several sublines with homozygous Ig haplotypes have been established. These sublines are used in cell transfer
studies or in chimaera construction since they provide markers for B cell
populations.

Rearrangement of immunoglobulin genes. Recently, the molecular basis of
the rearrangement of the light chain gene of immunoglobulins was
investigated in the chicken (Reynaud et al. 1983, 1985; Weill et al. 1986).
Surprisingly, it was found that the same V-lambda gene (V-lambda 1) was
rearranged in most cells of the bursa of Fabricius; this phenomenon was
already apparent on day 20 of embryonic development, i.e., 8 days after
the appearance of the first IgM-positive cells in the bursa. From the
proportion of rearranged and germ line DNA hybridizing to a J-lambda probe,
it appeared that in each pre-B cell the first joining event is productive,
leading to an almost complete allelic exclusion at the rearrangement level.
Compared to the mouse or rat, this seems to indicate an accuracy and
efficiency in the rearrangement mechanism operating on the chicken light
chain gene; such properties are so exceptional that they are not easily
explained at the moment. The sequence upstream of J-lambda is deleted in
strict quantitative correlation to the extent of V-lambda gene rearrangement and is not found in another major configuration in the genome.
A simple excision deletion model would be sufficient to account for these
results, as indicated from mouse data. Both of these features, the
efficiency of the rearrangement process and the use of a deletion mechanism,
seem to proceed directly from the unusual organization of the V-lambda
locus. The V-lambda 1 gene is located in the vicinity of the J-lambda
element (1.7 kb upstream) and in the same transcriptional polarity (Fig.
44). Thus, in the chicken, the V-C gene duplication was not accompanied
by a dispersal of the locus. However, there is evidence for V-gene
amplification, the total number of genes in the V-lambda-1 hybridizing
set being estimated to be about 25. But the first three V-lambda-
hybridizing elements which were sequenced by Reynaud et al. (1985), turned
out to be pseudogenes. It appears now that the other 25 V-lambda genes
are also pseudogenes (Reynaud & Weill, personal communication). The three
pseudogenes immediately adjacent to V-lambda 1 present several remarkable
features; firstly, they are exceptionally near one another; secondly, two
of them are in the same transcriptional orientation as the active locus
and have regions of high (91%) homology with the V-lambda 1 gene; the third
one, which lies between the two others, has a reverse polarity; thirdly,
all three are truncated abruptly on the 5' side of the variable sequence
but their V-homologous part contains no crippling mutations. Consequently,
Reynaud and co-workers suggest that these pseudogenes may be used to
generate diversity. They would be reintegrated into the functional gene
pool through a gene-conversion mechanism. Thus, the chicken light chain
repertoire seems to become diversified almost exclusively from somatic
mutations, as it is probably completely generated from a single rearranged
V gene. A study by the same group indicates that the heavy chain repertoire
is similarly restricted (Weill et al. 1986). The intensive somatic
diversification appears, like the rearrangement, to possess surprising
efficiency. Indeed, it occurs during a very short space of time since the
embryonic period has been shown to be decisive for the generation of light

chain diversity (Huang & Dreyer 1978; Jalkanen et al. 1984). Furthermore, it implies little selection for functional products, since distinct specificities appear sequentially in the bursa, quasi-independently from antigen encounter (Lydyard et al. 1976). It can be concluded that the chicken light chain repertoire seems to counterbalance the use of a rather archaic genomic locus (single J; dominant V in close proximity) with elaborate somatic diversification mechanisms.

Recently, Weill et al. (1986) investigated how gene rearrangement occurs within individual follicles in relationship to their colonization by precursors. Since a sole rearrangement, involving a unique V functional gene, occurs in most B cells of the chicken, in order to evaluate the number of rearrangements in each follicle, a ploy was used to distinguish between independent, though identical (i.e., of the same V gene), re-arrangement events. A B21 chicken strain heterozygous for two polymorphic restriction sites (EcoRI and BamHI) located in the light chain locus was used in these experiments. Alleles in the germ line, as well as in the rearranged configuration, yield different restriction fragments, allowing the rearranged chromosome to be identified. Thus if there has been one rearrangement event, the restriction pattern corresponding to one or the other allele will be found. If there have been several events, the restriction patterns corresponding to the two alleles will be associated in a mixed image. Depending on the intensity of the bands, it is also possible to diagnose the proportion of rearrangements involving one or the other allele. A typical mixed pattern was observed for total bursal DNA, but when single follicles were examined in 6-week birds, many follicles

Fig. 44. Genomic organization of the chicken light chain locus in its germ line and rearranged configuration. The shaded boxes represent the coding regions whose sequence was determined: the constant (C), joining (J) and variable (V lambda 1) segments, and the three adjacent pseudogenes (pseudo V lambda 1, pseudo V lambda 2, pseudo V lambda 3). The whole V lambda subgroup (8-12 V-hybridizing elements) is clustered in a single 22 kb BglI fragment. Courtesy of Drs. Weill, Leibowitch and Reynaud.

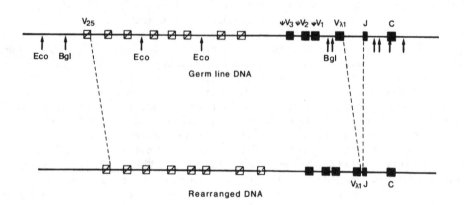

showed a monoclonal pattern. The follicles which were mixed displayed a pattern more easily explained by an unequal repartition of the two types of rearrangement. Single follicles from Cy-treated reconstituted birds, which have been shown to be of monoclonal origin at the phenotypic level (Pink & Rijnbeek 1983), showed a very similar pattern. Thus, the molecular mechanisms of Ig heavy or light chain gene rearrangements are in agreement with previous data on the commitment of colonizing precursors.

The following picture emerges from cellular and molecular data. In normal development, surface IgM-positive progenitor cells appear in the embryonic bursa during the period of colonization. Stem cells capable of reconstituting a depleted follicle 4 days after hatching are already committed to a particular Ig rearrangement. During the bursal step, progenitors accumulate and give rise to an amplified repertoire of B cell specificities. Starting from a very small number of rearrangement events, in each bursal follicle a large array of different combining sites arise through somatic mutations. Several weeks after hatching, bursal cells reach the post-bursal stage, where they can restore a depleted animal without homing to the bursa. The pool of post-bursal committed progenitor cells then assumes the renewal of the B cell population. Either they have already diversified their Ig receptors to a large extent in the bursa, or their progeny is still able to do so in extrabursal sites.

Compared to the mammalian strategy for antibody diversification, it appears that birds depend heavily on somatic mutations, occurring after V gene rearrangement. In mammals it is considered that progenitors committed to the B lineage segregate early in development in the foetal liver. These committed progenitors have not rearranged their V genes. Rearrangement occurs throughout life first in liver then in the bone marrow where the committed progenitors have moved to. The large repertoire of mammalian germ line V-genes permits the expression of a new variable region at each rearrangement event. In birds where the V-gene pool is small, the somatic mutation process ensures diversification of the already rearranged Ig genes. The rearrangement occurs only during a short period, as committed progenitors are produced in the bursa.

Major Histocompatibility Complex (MHC) antigens. The MHC was first discovered in mice as a genetic region governing the rejection of transplants of neoplastic or normal tissue. The second animal species in which an MHC was subsequently described is the chicken (Briles et al. 1950; reviewed in Longenecker & Mosmann 1981) in which the B locus was first detected as a blood group system. Most of the immunologic phenomena linked to the murine H-2 and human HLA complexes are also associated with the chicken B complex, such as resistance to neoplastic diseases (Pazderka et al. 1975; Collins et al. 1977), control of T cell-B cell interactions (Toivanen & Toivanen 1977) and susceptibility to autoimmune diseases (Bacon & Rose 1979). Such studies took advantage of the existence of numerous highly inbred chicken strains (Briles et al. 1982).

Genes of MHC complexes encode polymorphic cell surface glycoproteins involved in the control of various aspects of immune responses. In the chicken species, three genetic regions are known within the B complex

(Pink et al. 1977); i) B-F codes for class I antigens, that are classical transplantation antigens (HLA-A, B, C in man and H-2K, D, L in mouse) expressed by most of the nucleated cells and that function as restricting elements for cytotoxic T lymphocytes. ii) B-L codes for class I or Ia antigens (HLA-DR, DP, DQ in man, H-2IA, E in mouse) that restrict antigen presentation to helper T lymphocytes. These molecules are expressed at the surface of cells involved in the immune response, e.g. cells of the B lymphocyte and myeloid lineages, and also activated T lymphocytes. iii) B-G codes for class IV antigens, that are molecules only expressed by cells of the erythroid lineage. B-G antigens, that have an unknown function, are characteristic of birds. On the other hand, genes that code for complement components and are associated with the MHC in man and mouse, have not been shown to be linked with the B complex in the chicken, although there is a correlation between the serum haemolytic complement level and the haplotype.

At the biochemical level, chicken class I and class II molecules have the same characteristics as their mammalian homologues. The B-F molecule is composed of a 40,000-43,000 daltons glycoprotein non-covalently associated with beta 2-microglobulin, that is not encoded by a gene of the B complex. B-L is a non-covalent heterodimer of two glycoproteins, alpha and beta, with molecular weights of approximately 30,000 daltons, B-G appears as a complex of several proteins with molecular weights between 42,000 and 48,000 daltons (Hàla et al. 1981; Longenecker & Mosmann 1981). A high level of polymorphism of these cell surface molecules has been revealed by serological studies. Two-dimensional electrophoresis showed that B-G molecules (Miller et al. 1984) and B-L molecules (Guillemot et al. 1985) differ in both isoelectric point and molecular weight from one haplotype to another.

By comparison to murine and human MHC, the B complex is still very poorly understood. There is some indirect evidence for the existence of several genes coding for each class of B antigens. Sequential immunoprecipitation studies have shown the presence at the cell surface of two B-F (Crone et al. 1985) and two B-L antigens (Crone et al. 1981), whereas an analysis of B-L molecules by two-dimensional electrophoresis suggests that two polymorphic chains can associate with a single non-polymorphic chain (Guillemot et al. 1985).

According to a recent molecular genetic study, an even greater number of genes corresponding to the B-L beta molecule(s) could exist (Bourlet et al. 1986). A B-L beta gene has been cloned using a human class II DQ beta cDNA as a molecular probe. This B-L beta gene hybridizes in Southern blot experiments with a large number of bands that could correspond to at least 3 to 5 distinct genes. The nucleotide sequence of the 2nd exon of the cloned B-L beta gene shows a greater homology, among the human class II genes, to DQ beta and DP than to DR beta or DO beta. These results suggest that the duplication events that generated the different isotypes of mammalian class II molecules took place before the divergence of mammalian and avian ancestors.

SUMMARY AND CONCLUDING REMARKS

The amount of information accumulated on avian blood cells in recent years is so considerable that a selection has had to be made in order to fit within the scope of this review. Although such a selection is necessarily unsatisfactory, it does reflect the research areas that are currently more at the forefront. Knowledge about some types of cells, for instance granulocytes or macrophages, is still patchy. A field which remains virtually untouched by comparison to mammalian studies is that of the growth factors involved in the multiplication and maturation of haemopoietic cells. No doubt this gap will be filled soon, with the development of transformed cell lines which either secrete such factors or are targets for them. On the other hand, avian studies have their strong points, usually centred on structural or functional peculiarities, and have contributed ideas of general relevance to advanced areas of research.

With the discovery of the role of the bursa of Fabricius, an organ devoted to B cell differentiation, immunology was confronted with the crucial realisation of the dual nature of the immune system. The intriguing avian strategy for B cell amplification and antibody diversification is beginning to be understood, both at the cellular and molecular levels.

Concerning molecular biology, the haemoglobin switches, which occur during ontogeny or in transformed cells induced to differentiate, have been studied from the angles of gene organization, DNA modifications and chromatin structure. Here also, an original feature has been discovered, i.e., that the globin genes are not arranged in their order of expression during development. The reasons for this difference from the other vertebrate globin gene clusters that have been examined, are still speculative.

The avian haemopoietic system has also richly contributed to developmental biology, some of the more important points being the following: the haemopoietic cell lineage segregates early from the mesodermal germ layer, and thus has an origin independent from that of the stromal cells in the haemopoietic organs. Consequently extrinsic stem cells colonize the rudiments. The thymus rudiment receives stem cells in a precise, cyclic manner, whereas the bursa is colonized during a single period only and becomes non-receptive prior to hatching; B cell precursors disappear altogether around the same time, indicating that T and B cell progenitors may be two distinct populations; stem cells arise first in the yolk sac and later in the embryo proper, only the latter seeding the intraembryonic organ rudiments.

Some rules established with the avian model systems have subsequently been confirmed in mammals, an example being the periodic colonization of the thymus. Others provide guidelines for future experimentation in 'higher' vertebrates, the outcome of which remains open. Finally, in some cases, avian-specific mechanisms appear to have been uncovered.

ACKNOWLEDGEMENTS

I wish to thank all those who have helped me in the preparation of this chapter: Dr. Julian Smith for supervising the English style; Dr. Nicole Le Douarin for fruitful discussion and encouragement, Dr. Catherine Corbel, Dr. Bruno Péault and Dr. Jean-Claude Weill for critical reading of sections of the manuscript; Dr. Francoise Cormier and Dr. Francoise Guillemot for a contribution; Dr. Jeanine Breton-Gorius, Dr. F. Cormier, Dr. Francoise Coudert, Monique Coltey, Dr. Nicole Le Douarin, Dr. F. Guillemot, Dr. B. Péault, Dr. Jacques Samarut for providing micrographs and Dr. Jean-Claude Weill for providing a map of Ig genes. My gratitude goes to Pierre Coltey, Dominique Duchatel, Michele Klaine and Jacqueline Leroy for their excellent technical assistance, to Lydie Obert and Suzanne Roy for the preparation of the camera ready manuscript, and Bernard Henri, Bernard Louis and Sophie Tissot for preparation of the illustrations.

REFERENCES

Ackerman, G.A. & Knouff, R.A. (1959). Lymphocytopoiesis in the bursa of
    Fabricius. Amer. J. Anat. 104, 163-77.
Akester, A.R. (1984). The cardiovascular system. In Physiology and Bio-
    chemistry of the Domestic Fowl, ed. B.M. Freeman, Vol. 5,
    pp. 277-321. London: Academic Press.
Albini, B. & Wick, G. (1973). Immunoglobulin determinants on the surface
    of chicken lymphoid cells. Int. Arch. Allergy 44, 804-22.
Albini, B. & Wick, G. (1975). Ontogeny of lymphoid cell surface determin-
    ants. Int. Archs Allergy appl. Immun. 48, 513-29.
Alm, G.V. (1971). In vitro studies of chicken lymphoid cells. 3. The
    mixed spleen leucocyte reaction with special reference to the
    effect of bursectomy. Acta. Pathol. Microbiol. Scand. 79,
    359-65.
Awadhiya, R.P., Vegad, J.L. & Kolte, G.N. (1982). Eosinophil leukocytic
    response in dinitrochlorobenzene skin hypersensitivity reaction
    in the chicken. Avian Pathology, 11, 187-94.
Bacon, L.D. & Lee, L. (1981). Influence of age on reactivity of 1-way
    mixed lymphocyte cultures in young cells. J. of Immunol. 127,
    2059-63.
Bacon, L.D. & Rose, M.R. (1979). Influence of major histocompatibility
    haplotype on autoimmune disease varies in different inbred
    families of chicken. Proc. Natl. Acad. Sci. 76, 1435-37.
Barrett, L.A. & Ben Dawson, R. (1974). Avian erythrocyte development:
    microtubules and the formation of the disk shape. Dev. Biol.
    36, 72-81.
Barrett, L.A. & Scheinberg, S.L. (1972). The development of the avian
    red cell shape. J. Exp. Zool. 182, 1-14.
Beaupain, D., Martin, C. & Dieterlen-Lièvre, F. (1979). Are developmental
    hemoglobin changes related to the origin of stem cells and
    site of erythropoiesis? Blood 53, 212-25.
Belo, M., Martin, C., Corbel, C. & Le Douarin, N.M. (1985). A novel
    method to bursectomize avian embryos and obtain quail-chick
    bursal chimeras. I. Immunocytochemical analysis of such
    chimeras by using species-specific monoclonal antibodies. J.
    Immonol. 135, 3785-94.
Benazzi-Lentati, G.B. (1932). Sul focolaio eosinofilopoietico del
    pancreas di uccelli. Monitore Zool. Italiano 4-5: XLIII,
    115-8.
Ben Slimane, S.F., Houllier, F., Tucker, G. & Thiery, J.P. (1983). In
    vitro migration of avian hemopoietic cells to the thymus:
    Preliminary characterization of a chemotactic mechanism. Cell
    Diff., 13, 1-24.
Blanchet, J.P. (1976). The chicken erythrocyte membrane antigens:
    Characterization and variation during embryonic and post-
    embryonic development. Dev. Biol. 48, 411-66.
Blanchet, J.P. & Stanislawski, M. (1973). Les protéines non hémo-
    globiniques érythrocytaires du Poulet. Exptl. Cell Res. 77,
    248-54.

Blanchet, J.P., Samarut, J. & Nigon, V. (1976). Chick erythrocyte mem membrane antigens. Exp. Cell Res. 102, 9-13.

Bourlet, Y., Behar, G., Guillemot, F. & Auffray, C. (1986). Isolation and characterization of a chicken B-L immune response gene. Submitted.

Brand, A., Galton, J. & Gilmour, D.G. (1983). Committed precursors of B and T lymphocytes in chick embryo bursa of Fabricius, thymus and bone marrow. Eur. J. Immunol. 13, 449-55.

Breton-Gorius, J. & Guichard, J. (1976). Améliorations techniques permettant de révéler la peroxydase plaquettaire. Nouvelle Revue Francaise d'Hématologie. 16, 381-90.

Breton-Gorius, J., Coquin, Y. & Guichard, J. (1978). Cytochemical distinction between axurophils and catalase-containing granules in leukocytes. I. Studies in developing neutrophils and monocytes from patients with myeloperoxidase deficiency: comparison with peroxidase-deficient chicken heterophils. Laboratory Investigation, 38, 21-31.

Briles, W.E., Bumstead, N., Ewert, D.L., Gilmour, D.G., Gogusev, J., Hala, K., Koch, C., Longenecker, B.M., Nordskog, A., Pink, J.R., Schierman, L.W., Simonsen, M., Toivanen, A., Toivanen, P., Vainio, O. & Wick, G. (1982). Nomenclature for chicken major histocompatibility (B) complex. Immunogenetics 15, 441-8.

Briles, W.E., MacGibbon, W.H. & Irwin, M.R. (1950). On multiple alleles affecting cellular antigens in the chicken. Genetics 35, 633-52.

Brune, K. & Spitznagel, J.K. (1973). Peroxidaseless chicken leukocytes: isolation and characterization of anti-bacterial granules. J. Infect. Dis. 127, 84-94.

Bruns, G.A.P. & Ingram, V.M. (1973a). Erythropoiesis in the developing chick embryo. Dev. Biol. 30, 455-9.

Bruns, G.A.P. & Ingram, W.M. (1973b). Progenitor cells of the definitive erythroid series in the chick embryo. Dev. Biol. 31, 192-4.

Bruns, G.A.P. & Ingram, V.M. (1973c). The erythroid cells and haemo-globins of the chick embryo. Phil. Trans. Royal Soc. of London 266, 227-305.

Burkhardt, E. (1980). Scanning and transmission electron microscopy of glass bead column-separated monocytes from mononuclear leuko-cyte suspensions of peripheral blood of the chicken. J. Reticuloendothel. Soc. 28, 103-9.

Cain, W.A., Cooper, M.D. & Good, R.A. (1968). Cellular immune competence of spleen, bursa and thymus cells. Nature 217, 87-9.

Campbell, F. (1967). Fine structure of the bone marrow of the chicken and pigeon. J. Morph. 123, 405-40.

Carlson, H.C. & Allen, J.R. (1969). The acute inflammatory reaction in chicken skin: blood cellular response. Avian Dis., 14, 817-33.

Chand, N. & Eyre, P. (1978). Immunological release of histamine and SRS in domestic fowl. Can. J. Comp. Med., 42, 519-24.

Chang, C.F. & Hamilton, P.B. (1979). The thrombocyte as the primary circulating phagocyte in chickens. J. Reticuloendothel. Soc. 25, 589-91.

Chen, C-L.H., Chanh, T.C. & Cooper, M.D. (1984). Chicken thymocyte-specific antigen identified by monoclonal antibodies: onto-geny, tissue distribution and biochemical characterization. Eur. J. Immunol. 14, 385-91.

Chi, D.S., Bhogal, B.S., Fox, G.J. & Thorbecke, G.J. (1984). Effect of temperature and lymphokines on mixed lymphocyte and mitogen responses of chicken lymphoid cells in vitro. Dev. Comp. Immunol. 8, 683-94.

Clawson, C.C., Cooper, M.D. & Good, R.A. (1967). Lymphocyte fine structure in the bursa of Fabricius, the thymus and the germinal centres. Lab. Invest. 16, 407-21.

Coll, J. & Ingram, V.M. (1978). Stimulation of heme accumulation and erythroid colony formation in cultures of chick bone marrow cells by chicken plasma. J. Cell. Biol. 76, 184-90.

Collins, W.M., Briles, W.E., Zsigray, R.M., Dunlop, W.R., Corbett, A.C., Clark, K.K., Marks, J.L. & McGrail, T.L. (1977). The B locus (MHC) in the chicken: Association with the fate of RSV-induced tumors. Immunogenetics 5, 333-43.

Cooper, M.D., Peterson, R.D.A. & Good, R.A. (1965). Delineation of the thymic and bursal lymphoid systems in the chicken. Nature 4967, 143-6.

Cormier, F. & Dieterlen-Lièvre, F. (1985). Functional capacities of chick embryo thymocytes in the mixed lymphocyte reaction. Dev. Comp. Immunol. 9, 343-9.

Cormier, F., de Paz, P. & Dieterlen-Lièvre, F. (1986). In vitro detec-tion of cells with monocytic potentiality in the wall of the chick embryo aorta. Dev. Biol. 118, 167-75.

Crone, M., Jensenius, J.C. & Koch, C. (1981). Evidence for two popula-tions of B-L (Ia like) molecules encoded by the chicken MHC. Immunogenetics 13, 381-91.

Crone, M., Simonsen, M., Skjoldt, K., Limet, K. & Olsson, L. (1985). Mouse monoclonal antibodies to class I and class II of the chicken MHC: Evidence for at least two class I products of the B complex. Immunogenetics 21, 181-7.

Daimon, T. & Caxton-Martins, A. (1977). Electron microscopic and enzyme cytochemical studies on granules of mature chicken granular leucocytes. J. Anat. 123, 553-62.

D'Amelio, V. (1966). The globins of adult and embryonic chick hemoglobin. Biochim. Biophys. Acta 127, 59-65.

D'Amelio, V. & Salvo, A.M. (1961). Further studies on the embryonic chick hemoglobin. An electrophoretic and immunoelectrophoretic analysis. Acta Embryol. Morph. exp. 4, 250-9.

Dantchakoff, V. (1909). Untersunchungen uber die Entwicklung von Blut und Bindegewebe bei Vogeln. Arch. mikr. Anat. und Entwick. Gesch., 73, 118-81.

Dantchakoff, V. (1916). Equivalence of different hematopoietic anlages (by method of stimulation of their stem cells) I. Spleen. The Amer. J. of Anat. 20, 255-327.

Dantchakoff, V. (1918). Equivalence of different hematopoietic anlages (by method of stimulation of their stem cells) II. Grafts of adult spleen on the allantois and response of the allantoic tissues. Amer. J. Anat. 24, 128-89.

Dardick, I. & Setterfield, G. (1978). Early origins of definitive erythroid cells in the chick embryo. Tissue & Cell 10, 355-64.

DeLanney, L.E. & Ebert, J.D. (1962). On the chick spleen: origin, patterns of normal development and their experimental modifications. Contrib. to Embryol. 255, 57-86. Carnegie Inst. Washington

Dieterlen-Lièvre, F. (1975). On the origin of haemopoietic stem cells in the avian embryo: an experimental approach. J. Embryol. exp. Morph. 33, 607-19.

Dieterlen-Lièvre, F. (1984a). Emergence of intraembryonic blood stem cells studied in avian chimeras by means of monoclonal antibodies. Dev. Comp. Immunol. 3, 75-80.

Dieterlen-Lièvre, F. (1984b). Blood in Chimeras. In Chimeras in Developmental Biology, eds N. Le Douarin & A. McLaren, pp. 133-63. London: Academic Press.

Dieterlen-Lièvre, F. & Martin, C. (1981). Diffuse intraembryonic hemopoiesis in normal and chimeric avian development. Dev. Biol. 88, 180-91.

Dieterlen-Lièvre, F., Beaupain, D. & Martin, C. (1976). Origin of erythropoietic stem cells in avian development: shift from the yolk sac to an intraembryonic site. Ann. Immunol. 127C, 857-63.

Dietert, R.R. & Sanders, B.G. (1977). Evidence for multiple cell surface chicken fetal-leukemic antigens (chicken fetal antigen) in the developing chicken and other avian species. J. Exp. Zool. 202, 171-7.

Dodge, W.H. & Hansell, C.C. (1978). The marrow colony forming cell and serum colony stimulating factor of the chicken. Exp. Hemat. 6, 661-72.

Dodge, W.H. & Moscovici, C. (1973). Colony formation by chicken hematopoietic cells and virus-induced myeloblasts. J. Cell. Physiol. 81, 371-86.

Dodge, W.H., Silva, R.F. & Moscovici, C. (1975). The origin of chicken hematopoietic colonies as assayed in semisolid agar. J. Cell. Physiol. 85, 25-30.

Dodge, W.H. & Sharma, S. (1985). Serum-free conditions for the growth of avian granulocyte and monocyte clones and primary leukemic cells induced by AMV, and the apparent conversion of granulocytic progenitors into monocytic cells by a factor in chicken serum. J. Cell Physiol. 123, 264-8.

Dodgson, J.B., McCune, K.C., Rusling, D.J., Krust, A. & Engel, J.D. (1981). Adult chicken alpha-globin genes, alpha^A and alpha^D: no anemic shock alpha-globin exists in domestic chickens. Proc. Natl. Acad. Sci. USA 78, 5598-6002.

Dolan, M., Sugarman, B.J., Dodgson, J.B. & Engel, J.D. (1981). Chromosomal arrangement of the chicken beta-type globin genes. Cell 24, 669-77.

Duprey, S.P. & Boettiger, D. (1985). Developmental regulation of c-myb in normal myeloid progenitor cells. Proc. Natl. Acad. Sci. USA 82, 6937-41.

Durkin, H.G., Theis, G.A. & Thorbecke, G.J. (1972). Bursa of Fabricius as site of origin of germinal centre cells. Nature 235, 118-19.

Enbergs, H. & Kriesten, K. (1969). Zur Feinstruktur der Blutmonozyten des Haushuhns. Z. Zellforsch. mikr. Anat. 97, 377-82.

Engel, J.D. & Dodgson, J.B. (1980). Analysis of the closely linked adult chicken alpha-globin genes in recombinant DNAs. Proc. Natl. Acad. Sci. USA 77, 2596-600.

Eskola, J. (1977). Cell transplantation into immunodeficient chicken
     embryos: Reconstituting capacity of cells from the yolk sac
     at different stages of development and from the liver, thymus,
     bursa of Fabricius, spleen and bone marrow of 15-day embryos.
     Immunology 32, 467-74.

Eskola, J. & Toivanen, A. (1976). Cell transplantation into immuno-
     deficient chicken embryos: Reconstituting capacity of cells
     from the bursa of Fabricius, spleen, bone marrow, thymus and
     liver of 18-day-old embryos. Cell. Immunol. 26, 68-77.

Ewart, D.F. & McMillan, D. (1970). The spleen of the cowbird. J. Morph.
     130, 187-206.

Ewert, D.L. & Cooper, M.D. (1978). Ia-Like Alloantigens in the chicken:
     Serologic characterization and ontogeny of cellular expression.
     Immunogenetics 7, 521-35.

Fitzsimmons, R.C., Garrod, E. & Garnett, I. (1973). Immunological
     responses following early embryonic surgical bursectomy. Cell.
     Immunol. 9, 377-83.

Foppoli, J.M., Ch'ng, L.K., Benedict, A.A., Ivanyi, J., Derka, J. and
     Wakeland, E.K. (1979). Genetic nomenclature for chicken
     immunoglobulin allotypes: An extensive survey of inbred lines
     and antisera. Immunogenetics 8, 385-404.

Fox, A.J. & Solomon, J.B. (1981). Chicken non-lymphoid leukocytes In
     Avian Immunology, eds. M.E. Rose, L.N. Payne & M.B. Freeman,
     pp. 135-166, Edinburgh, Poultry Science Ltd.

Fraser, R.C. (1964). Electrophoretic characteristics and cell content
     of the hemoglobins of developing chick embryos. J. exp. Zool.
     156, 185-96.

Fraser, R.C. (1966). Polypeptide chains of chick embryo hemoglobins.
     Biochim. Biophys. Res. Comm. 25, 142-6.

Fraser, R., Horton, B., Dupourque, D. & Chernoff, A. (1972). The
     multiple hemoglobins of the chick embryo. J. Cell. Physiol.
     80, 79-88.

Frazier, J.A. (1973). Ultrastructure of the chick thymus. Z. Zellforsch.
     mikrosk. Anat. 136, 191-205.

Fredericksen, T.L. & Gilmour, D.G. (1981). Two independent T-lymphocyte
     antigen loci in the chicken, Ly-4 and Th-1. Eur. J. Immunol.
     14, 535-9.

Fredericksen, T.L. & Gilmour, D.G. (1983). Ontogeny of CON A and PHA
     responses of chicken blood cells in MHC-compatible lines 6_3
     and 7_2. J. Immunol. 130, 2528-33.

Fredericksen, T.L., Longenecker, B.M., Padzerka, F., Gilmour, D.G. &
     Ruth, R.F. (1977). A T-cell antigen system of chicken: Ly-4
     and Marek's disease. Immunogenetics 5, 535-52.

Galton, J. & Ivanyi, J. (1977a). Detection of bursa and thymus-specific
     alloantigens in the chicken. Eur. J. Immunol. 7, 457-9.

Galton, J. & Ivanyi, J. (1977b). Immunofluorescent detection of
     differentiation alloantigens (Al) in the chicken. Eur. J.
     Immunol. 7, 241-6.

Gazzolo, L., Moscovici, C. & Moscovici, M.G. (1979). Response of
     hemopoietic cells to avian acute leukemia viruses: Effects
     on the differentiation of the target cells. Cell 16, 627-38.

Gilmour, D,G., Brand, A., Donnelly, N. & Stone, H.A. (1976). Bu-1 and
    Th-1, two loci determining surface antigens of B or T lympho-
    cytes in the chicken. Immunogenetics 3, 549-63.
Glick, B., Sato, K. & Cohenour, F. (1964). Comparison of the phagocytic
    ability of normal and bursectomized birds. J. Reticuloendo-
    thelial. Soc. 1, 442-51.
Glick, B., Chang, T.S. & Jaap, R.C. (1956). The bursa of Fabricius and
    antibody production. Poultry Sci. 35, 224-5.
Godet, J. (1967). L'évolution des hémoglobines au cours du développe-
    ment du Poulet. Les hémoglobins du poussin et de l'adulte.
    C.R. Acad. Sci. 264, 2570-72.
Godet, J. (1974). HbF synthesis in chicken embryonic and postnatal
    development: studies in various explanted erythropoietic
    tissues. Dev. Biol. 40, 199-207.
Grecchi, R., Saliba, A.M. & Mariano, M. (1980). Morphological changes,
    surface receptors and phagocytic potential of fowl mono-
    nuclear phagocytes and thrombocytes in vivo and in vitro. J.
    Path. 130, 23-31.
Grossi, C.E., Lydyard, P.M. & Cooper, M.D. (1977). Ontogeny of B cells
    in the chicken. II. Changing patterns of cytoplasmic IgM
    expression and of modulation requirements for surface IgM
    by anti-µ antibodies. J. Immunol. 119, 749-56.
Groudine, M. & Weintraub, H. (1981). Activation of globin genes during
    chicken development. Cell, 24, 393-401.
Groudine, M., Holtzer, H., Scherrer, K. & Therwath, A. (1974). Lineage-
    dependent transcription of globin genes. Cell, 3, 243-7.
Guillemot, F.P., Oliver, P.D., Péault, B.M. & Le Douarin, N.M. (1984)
    Cells expressing Ia antigens in the avian thymus. J. Exp. Med.
    160, 1803-19.
Guillemot, F., Turmel, P., Charron, D., Le Douarin, N.M. & Auffray, C.
    (1985). Structure, biosynthesis and polymorphism of chicken
    MHC class II (B-L) antigens. J. Immunol. 135, 1251-7.
Haigh, L.S., Hellewell, S., Roninson, I., Owens, B.B. & Ingram, V.M.
    (1982). Control of hemoglobin expression in chick embryonic
    development. In Cell function and differentiation Part A
    pp. 35-46. New York: Alan R. Liss, Inc.
Hàla., Boyd, R. & Wick, G. (1981). Chicken major histocompatibility
    complex and disease. Scand. J. Immunol. 14, 607-16.
Hanafusa, T., Hanafusa, H. & Miyamoto, T. (1970). Recovery of a new
    virus from apparently normal chick cells by infection with
    avian tumor viruses. Proc. Natl. Acad. Sci., USA. 67, 1797-
Hasek, M., & Hraba, T. (1955). Immunological effects of experimental
    embryonic parabiosis. Nature 175, 764-5.
Hashimoto, K. & Wilt, F.H. (1966). The heterogeneity of chicken hemo-
    globin. Proc. Nat. Acad. Sci. USA. 56, 1477-84.
Hayari, Y., Schauenstein, K. & Globerson, A. (1982). Avian lymphokines
    II. IL-1 activity in supernatants of stimulated adherent
    splenocytes of chickens. Dev. Comp. Immunol. 6, 785-91.
Hodges, R.D. (1974). The histology of the fowl, London: Academic Press.
Hoffmann-Fezer, G. (1973). Histologische Untersuchungen an lymphatischen
    Organen des Huhnes (Gallus domesticus) während des ersten
    Lebensjahres. Z. Zellforsch. mikrosk. Anat. 136, 45-8.

Hoffmann-Fezer, H., Rodt, H., Gotze, R.D. & Thierfelder, S. (1977). Anatomical distribution of T and B lymphocytes identified by immunohistochemistry in the chicken spleen. Int. Arch Allergy Appl. Immun. 55, 86-95.

Houssaint, E., Belo, M. & Le Douarin, N.M. (1976). Investigations on cell lineage and tissue interactions in the developing bursa of Fabricius through interspecific chimeras. Dev. Biol. 53, 250-64.

Houssaint, E., Diez, E. & Jotereau, F.V. (1985). Tissue distribution and ontogenic appearance of a chicken T lymphocyte differencia-tion marker. Eur. J. Immunol. 15, 305-14.

Houssaint, E., Torano, A. & Ivanyi, J. (1983). Ontogenic restriction of colonisation of the bursa of Fabricius. Eur. J. Immunol. 13, 590-5.

Huang, H.V. & Dreyer, W.J. (1978). Bursectomy in ovo blocks the genera-tion of immunoglobulin diversity. J. Immunol., 121, 1738-47.

Ingram, V.M. (1972). Emrbyonic red blood cell formation. Nature 235, 338-9.

Ivanyi, J. (1973). Sequential recruitment of antibody class-committed B lymphocytes during ontogeny. European J. Immunol. 3, 789-93.

Ivanyi, J. (1981). Functions of the B-lymphoid system in chicken. Avian Immunology, ed. M.E. Rose, L.N. Payne and B.M. Freeman. pp. 63-101. Edinburgh: British Poultry Science Ltd.

Ivanyi, J. & Hudson, L. (1979). Alleleic exclusion of Ml(IgM) allotype on the surface of chicken B cells. Immunology 35, 941-5.

Jalkanen, S., Granfors, K., Jalkanen, M. & Toivanen, P. (1983). Immune capacity of the chicken bursectomized at 60 hours of incuba-tion: failure to produce immune, natural and autoantibodies in spite of immunoglobulin production. Cell. Immunol., 80, 363-73.

Jalkanen, S., Jalkanen, M., Granfors, K. & Toivanen, P. (1984). Defect in the generation of light-chain diversity in bursectomized chickens. Nature, 311, 69-71.

Jankovic, B.D., Knezevic, K., Isakovic, K., Mitrovic, K., Markovic, M. & Rajcevic, M. (1975). Bursa lymphocytes and IgM-containing cells in chicken embryos bursectomized at 52-64 hours of incubation. Eur. J. Immunol., 5, 656-9.

Janzarik, H. (1981). Nucleated thrombocytoid cells II. Phase- and interference-contrast microscopic studies on blood cells of the domestic fowl. Cell Tissue Res. 219, 497-510.

Janzarik, H. & Morgenstern, E. (1979). The nucleated thrombocytoid cells. I. Electron microscopic studies on chicken blood cells. Thrombosis and Haemostasis 41, 608-21.

Janzarik, H., Schauenstein, K., Wolf, H. & Wick, G. (1980). Antigenic surface determinants of chicken thrombocytoid cells. Dev. Comp. Immunol. 4, 123-35.

Jones, D.R. & Johansen, K. (1984). The blood vascular system of birds. In Avian Biology, ed. D.S. Farner & J.R. King. Vol III. pp. 158-85. London: Academic Press.

Jotereau, F.V. & Le Douarin, N.M. (1978). The developmental relation-ship between osteocytes and osteoclasts: A study using the quail-chick nuclear marker in endochondral ossification. Dev. Biol., 63, 253-65.

Jotereau, F.V. & Le Douarin, N.M.   (1982).   Demonstration of a cyclic
    renewal of the lymphocyte precursor cells in the quail
    thymus during embryonic and perinatal life. J. Immunol. 129,
    1869-877.
Jurdic, P., Samarut, J. & Nigon, V.   (1978).   Erythrocytic transplantable
    stem cells in young chick blastoderms. Dev. Biol. 64, 339-41.
Keane, R.W., Blindblad, P.C., Pierik, L.T. & Ingram, V.M.   (1979).
    Isolation and transformation of primary mesenchymal cells of
    the chick embryo. Cell, 17, 801-11.
Kincade, P.W., Lawton, A.R., Bockman, D.E. & Cooper, M.C.   (1970).
    Suppression of immunoglobulin G synthesis as a result of
    antibody-mediated suppression of immunoglobulin M synthesis
    in chickens. Proc. Natl. Acad. Sci. USA 67, 1918-25.
Kuruma, I., Okada, T., Katoaka, K. & Sorimachi, M.   (1970).   Ultra-
    structural observation of 5-hydroxytryptamine-storing
    granules in the domestic fowl thrombocytes. Z. Zellforsch.
    mikrosk. Anat. 108, 268-81.
Larsen, A. & Weintraub, H.   (1982).   An altered DNA conformation detected
    by SI nuclease occurs at specific region in active chick
    globin chromatin. Cell 29, 609-22.
Lassila, O., Eskola, J., Toivanen, P., Martin, C. & Dieterlen-Lievre, F.
    (1978).   The origin of lymphoid stem cells studied in chick
    yolk sac-embryo chimaeras. Nature 272, 353-4.
Lassila, O., Eskola, J. & Toivanen, P.   (1979).   Prebursal stem cells in
    the intraembryonic mensenchyme of the chick embryo at 7 days
    of incubation. J. Immunol., 123, 2091-4.
Lassila, O., Eskola, J., Toivanen, P. & Dieterlen-Lièvre, F.   (1980).
    Lymphoid stem cells in the intraembryonic mesenchyme of the
    chicken. Scand. J. Immunol., 11, 445-8.
Lassila, O., Martin, C., Toivanen, P. & Dieterlen-Lièvre, F.   (1982).
    Erythropoiesis and lymphopoiesis in the chick yolk-sac-
    embryo chimeras: contribution of yolk sac and intraembryonic
    stem cells. Blood 49, 377-81.
Le Douarin, N.M.   (1973).   A Feulgen-positive nucleolus. Exp. Cell Res.
    77, 459-68.
Le Douarin, N.M. & Jotereau, F.   (1973).   Origin and renewal of lympho-
    cytes in avian embryo thymuses. Nature New Biol. 246, 25-7.
Le Douarin, N.M. & Hotereau, F.V.   (1975).   Tracing of cells of the
    avian thymus through embryonic life in interspecific chimeras.
    J. Exp. Med. 142, 17-40.
Le Douarin, N.M., Houssaint, E., Jotereau, F.V. & Belo, M.   (1975).
    Origin of hemopoietic stem cells in embryonic bursa of Fabricius
    and bone marrow studied through interspecific chimeras. Proc.
    Nat. Acad. Sci. USA 72, 2701-05.
Le Douarin, N.M., Michel, G. & Baulieu, E.E.   (1980).   Studies of
    testosterone-induced involution of the bursa of Fabricius.
    Dev. Biol. 75, 288-302.
Le Douarin, N.M., Dieterlen-Lievre, F. & Oliver, P.D.   (1984).   Ontogeny
    of primary lymphoid organs and lymphoid stem cells. Amer. J.
    Anat. 170, 261-99.
Lee, L.F.   (1978).   Chicken lymphocyte stimulation of mitogens: a micro-
    assay with whole-blood cultures. Avian Diseases 22, 297.

Le Lièvre, C.S. & Le Douarin, N.M. (1975). Mesenchymal derivatives of the neural crest: Analysis of chimaeric quail and chick embryos. J. Embryol. Exp. Morphol. 34, 125-54.

Lemez, L. (1964). The blood of chick embryos: quantitative embryology at a cellular level. In Advances in Morphogenesis, eds. M. Abercrombie and J. Brachet, Vol. 3 pp. 97-245. New York: Academic Press.

Lerman, S.P. & Weidanz, W.P. (1970). The effect of cyclophosphamide on the ontogeny of the humoral immune response in chickens. J. Immunol. 105, 614-9.

Leutz, A., Beug, H. & Graf, T. (1984). Purification and characterization of cMGF, a novel chicken myelomonocytic growth factor. EMBO J. 3, 3191-7.

Longenecker, B.M. & Mosmann, T.R. (1981). Structure and properties of the the major histocompatibility complex of the chicken. Speculations on the advantages and evolution of polymorphism. Immunogenetics 13, 1-3.

Lucas, A.M. & Jamroz, C. (1961). Atlas of Avian Hematology. Agriculture monograph 25, U.S. Department of Agriculture, Washington.

Lydyard, P.M., Grossi, C.E. & Cooper, M.D. (1976). Ontogeny of B cells in the chicken. I. Sequential development of clonal diversity in the bursa. J. Exp. Med. 144, 79-97.

MacRae, E.K. & Powell, R.E. (1979). Cytochemical reaction for cationic proteins as a marker of primary granules during development in chick heterophils. Histochemistry 60, 295-308.

Manwell, C., Baker, C.M.A. & Betz, T.W. (1966). Ontogeny of haemoglobin in the chicken. J. Embryol. exp. Morph. 16, 65-81.

Manwell, C., Baker, C.M.A., Rolansky, J.D. & Foght, M. (1963). Molecular genetic of avian proteins. II. Control genes and structural genes for embryonic and adult haemoglobins. Proc. Natl. Acad. Sci. USA 53, 1147-54.

Martin, C., Beaupain, D. & Dieterlen-Lièvre, F. (1978). Developmental relationships between vitelline and intra-embryonic haemopoiesis studied in avian "yolk sac chimaeras". Cell Diff. 7, 115-30.

Maxwell, M.H. (1973). Comparison of heterophil and basophil ultrastructure in six species of domestic bird. J. Anat. 115, 187-202.

Maxwell, M.H. (1974). An ultrastructural comparison of the mononuclear leucocytes and thrombocytes in six species of domestic bird. J. Anat. 117, 69-80.

Maxwell, M.H. (1978a). Electron cytochemistry of developing and mature eosinophils in the bone marrow of the fowl and the duck. Histochemical J. 10, 63-77.

Maxwell, M.H. (1978b). The development of eosinophils in the bone marrow of the fowl and the duck. J. Anat. 125, 387-400.

Maxwell, M.H. (1984a). The distribution and localisation of acid trimetaphosphatase in developing heterophils and eosinophils in the bone marrow of the fowl and the duck. Cell Tissue Res. 235, 171-6.

Maxwell, M.H. (1984b). Histochemical identification of tissue eosinophils in the inflammatory response of the fowl (Gallus domesticus). Res. Vet. Sci., 37, 7-11.

Maxwell, M.H. & Siller, W.G. (1972). The ultrastructural character-
        istics of the eosinophil granules in six species of domestic
        bird. J. Anat. 112, 289-303.
Maxwell, M.H. & Trejo, F. (1970). The ultrastructure of white blood
        cells and thrombocytes of the domestic fowl. British Veterinary
        J. 126, 583-92.
McGhee, J.D. & Ginder, G.D. (1979). Specific DNA methylation sites in
        the vicinity of the chicken beta-globin genes. Nature, 280,
        419-20.
Metcalf, D. (1977). Hemopoietic colonies. In vitro cloning of normal
        and leukemic cells. Springer Verlag Berlin, 227 pp.
Metcalf, D. & Moore, M.A.S. (1971). Haemopoietic Cells. North Holland
        Publ. Co., Amsterdam.
Meyer, R.K., Rao, M.A. & Aspinall, R.L. (1959). Inhibition of the
        development of the bursa of Fabricius in the embryos of the
        common fowl by 19-nortestosterone. Endocrinology 64, 890-7.
Miggiano, V., Birgen, I. & Pink, J.R.L. (1974). The mixed leukocyte
        reaction in chickens. Evidence for control by the major
        histocompatibility complex. Eur. J. Immunol., 4, 397-401.
Miggiano, V., North, M., Buder, A. & Pink, J.R.L. (1976). Genetic
        control of the response of chicken leukocytes to a T-cell
        mitogen. Nature 263, 61-3.
Miller, A.M. (1913). Histogenesis and morphogenesis of the thoracic duct
        in the chick: Development of blood cells and their passage
        to the blood stream via the thoracic duct. Amer. J. Anat. 15,
        131-98.
Miller, M.M., Goto, R. & Abplanalp, H. (1984). Analysis of the B-G anti-
        gens of the chicken MHC by two-dimensional gel electrophoresis.
        Immunogenetics 20, 373-86.
Miller, M.M., Goto, R. & Clark, D. (1982). Structural characterization
        of developmentally expressed antigenic markers on chicken
        erythrocytes using monoclonal antibodies. Dev. Biol., 94,
        400-14.
Miura, Y. & Wilt, F.H. (1970). The formations of blood islands in
        dissociated-reaggregated chick embryo yolk sac cells. Exp.
        Cell Res. 59, 217-26.
Miura, Y. & Wilt, F.H. (1971). The effects of 5-bromodeoxyuridine on
        yolk sac erythropoiesis in the chick embryo. J. Cell Biol.
        58, 523-32.
Moore, M.A.S. & Owen, J.J.T. (1965). Chromosome marker studies on the
        development of the haemopoietic system in the chick embryo.
        Nature, 208, 989-90.
Moore, M.A.S. & Owen, J.J.T. (1966). Experimental studies on the
        development of the bursa of Fabricius. Dev. Biol. 14, 40-51.
Moore, M.A.S. & Owen, J.J.T. (1967a). Stem cell migration in developing
        myeloid and lymphoid systems. Lancet, 2, 658-59.
Moore, M.A.S. & Owen, J.J.T. (1967b). Experimental studies on the
        development of the thymus. J. Exp. Med. 126, 715-23.
Morrow, P.H. & Abplanalp, H. (1981). Genetic control of T-lymphocyte
        mitogenesis in chickens. Immunogenetics 13, 189-200.

Murphy, D.B. & Wallis, K.T. (1983a). Brain and erythrocyte microtubules from chicken contain different beta-tubulin polypeptides. J. Biol. Chem. 258, 7870-75.

Murphy, D.B. & Wallis, K.T. (1983b). Isolation of microtubules protein from chicken erythrocytes and determination of the critical concentration for tubulin polymerization in vitro and in vivo. J. Biol. Chem. 260, 12293-301.

Murphy, D.B., Grasser, W.A. & Wallis, K.T. (1986). Immunofluorescence examination of beta-tubulin expression and marginal band formation in developing chicken erythroblasts. J. Cell Biol. 102, 628-35.

Nagy, Z.A. (1970). Histological study of the topographic separation of thymus-type and bursa-type lymphocytes and plasma cell series in chicken spleen. Zentralblat. Veterinarmed. 17, 422-9.

Nakamura, H. & Ayer-Le Lievre, C. (1986). Neural crest and thymic myoid cells. Current topics in developmental biology 20, 111-5.

Nelson, C.H., Kline, K., Ellison, J.R. & Sanders, B.G. (1984). Hemato-poietic differentiation cell surface antigen switching in the bone marrow of different aged chickens. Differentiation 26, 36-41.

Nickel, R., Schummer, A. & Seiferle, E. (1977). Anatomy of the domestic birds. Verlag Paul Parey, Berlin, Hamburg.

Nirmalan, G.P., Atwal, O.S. & Carlson, A.C. (1972). Ultrastructural studies on the leukocytes and thrombocytes in the circulating blood of Japanese quail. Poultry Sci. 51, 2050-4.

Olah, I. & Glick, B. (1982). Splenic white pulp and associated vascular channels in chicken spleen. Amer. J. Anat. 165, 445-80.

Oliver, P.D. & Le Douarin, N.M. (1984). Avian thymic accessory cells. J. of Immunol. 132, 1748-55.

Osculati, F. (1970). Fine structural localisation of acid phosphatase and arysulfatase in the chick heterophil leucocytes. Z. Zellforsch. mikrosk. Anat. 109, 398-406.

Palladino, M.A., Chi, D.S., Blyznak, N., Paolino, A.M. & Thorbecke, G.J. (1980). Cytotoxicity to allogeneic cells in the chicken. I. Role of macrophoges in the cytotoxic effect on ^{51}Cr-labeled red blood cells by immune spleen cells. Dev. Comp. Immunol. 4, 309-22.

Payne, L.N. & Powell, P.C. (1984). The lymphoid system. In Physiology and biochemistry of the domestic fowl. ed. B.M. Freeman, pp. 277-321. London: Academic Press.

Pazderka, F., Longenecker, B.M., Law, G.R.J. & Ruth, R.F. (1975). The major histocompatibility complex of the chicken. Immunogenetics 2, 101-30.

Péault, B.M., Coltey & Le Douarin, N.M. (1982). Tissue distribution and ontogenetic emergence of differentiation antigens on avian T cells. Eur. J. Immunol., 12, 1047-50.

Péault, B.M., Thiéry, J.P. & Le Douarin, N.M. (1983). A surface marker for the hemopoietic and endothelial cell lineages in the quail species defined by a monoclonal antibody. Proc. Natl. Acad. Sci. USA, 80, 2976-80.

Peck, R., Murthy, K.K. & Vainio, O. (1982). Expression of B-L (Ia-like) antigens on macrophages from chicken lymphoid organs. J. Immunol. 129, 4-5.

Penniall, R. & Spitznagel, J.K. (1975). Chicken neutrophils: Oxidative metabilism in phagocytic cells devoid of myeloperoxidase. Proc. Natl. Acad. Sci. USA, 72, 5012-15.

Pink, J.R. & Miggiano, V.C. (1977). Complementation between genetic variants affecting the response of chicken leukocytes to concanavalin A. J. Immunol. 119, 1796-9.

Pink, J.R. & Rijnbeek, A.M. (1983). Monoclonal antibodies against chicken lymphocyte surface antigens. Hybridoma 2, 287-96.

Pink, J.R. & Vainio, O. (1983). Genetic control of the response of chicken T lymphocytes to concanavalin A: cellular localization of the low responder defect. Eur. J. Immunol. 13, 571-5.

Pink, J.R.L., Droege, W., Hàla, K., Miggiano, V.C. & Ziegler, A. (1977). A three-locus model for the chicken major histocompatibility complex. Immunogenetics 5, 203-16.

Pink, J.R.L., Fedecka-Bruner, B., Coltey, M., Péault, B.M. & Le Douarin, N.M. (1981). Biochemical characterization of avian T lymphocyte specific antigens. Eur. J. Immunol. 11, 517-20.

Pink, J.R.L., Ratcliffe, M.J.H. & Vainio, O. (1985a). Immunoglobulin-bearing stem cells for clones of B (bursa-derived) lymphocytes. Eur. J. Immunol. 15, 617-20.

Pink, J.R., Vainio, O. & Rijnbeek, A-M. (1985b). Clones of B lymphocytes in individual follicules of the bursa of Fabricius. Eur. J. Immunol. 15, 83-7.

Ratcliffe, M.J.H. (1985). The ontogeny and cloning of B cells in the bursa of Fabricius. Immunology Today 6, 223-7.

Ratcliffe, M.J.H. & Ivanyi, J. (1979). Allelic exclusion of surface IgM allotypes on spleen and bursal B cells in the chicken. Immunogenetics 9, 149-56.

Ratcliffe, M.J.H. & Ivanyi, J. (1981). Allotype suppression in the chicken. II. Suppression in homozygous chickens with anti-allotype antibody and allotype-disparate B cells. Eur. J. Immunol. 11, 296-300.

Raunich, L., Callegarini, C. & Cucchi, C. (1966). Ricerche sopra le hemoglobine elettroforeti camente lente durante lo sviluppo embrionale del pollo. Ricerca scient. 36, 203-6.

Reynaud, C-A., Anquez, V., Dahan, A. & Weill, J-C. (1985). A single rearrangement event generates most of the chicken immuno-globulin light chain diversity. Cell 40, 289-91.

Reynaud, C-A., Dahan, A. & Weill, J-C. (1983). Complete sequence of a chicken lambda-light chain immunoglobulin derived from the nucleotide sequence of its mRNA. Proc. Natl. Acad. Sci. USA 80, 4099-103.

Ries, S., Käufer, I., Reinacher, M. & Weiss, E. (1984). Immunomorphologic characterization of chicken thrombocytes. Cell Tissue Res. 236, 1-3.

Ringertz, N.R., Carlsson, S.A., Ege, T. & Bolund, L. (1971). Detection of human and chick nuclear antigens in nuclei of chick erythrocytes during reactivation in heterokaryons with Hela cells. Proc. Natl. Acad. Sci. USA, 68, 3228-32.

Rosse, W.F. & Waldmann, T.A. (1966). Factors controlling erythropoiesis in birds. Blood 27, 654-61.

Sabin, F. (1920). Studies on the origin of the blood-vessels and of red-blood corpuscles, as seen in the living blastoderm of chicks during the second day of incubation. "Contributions to Embryology" Carnegie Inst. Publ., vol.9, 213-62.

Salvatorelli, G. (1967). L'influence favorable du foie embryonnaire sur l'hématopoïèse in vitro dans la moelle osseuse de l'embryon de poulet. J. Emb. exp. Morph. 17, 359-65.

Salvatorelli, G. & Smith, J. (1971). Effet de diverses fractions d'extraits de foie embryonnaire de poulet sur l'érythropoïèse in in vitro. C.R. Acad. Sc., 222, 869-72.

Samarut, J. (1978). Isolation of an erythropoietic stimulating factor from the serum of anemic chicks. Exp. Cell Res. 115, 123-6.

Samarut, J. (1979). Erythropoietin-like factor production by young chick blastoderms. Dev. Biol. 70, 278-82.

Samarut, J. & Bouabdelli, M. (1980). In vitro development of CFU-E and BFU-E in cultures of embryonic and post-embryonic chicken hematopoietic cells. J. Cell. Physiol. 105, 553-63.

Samarut, J. & Nigon, V. (1975). Potentialités hématopoïétiques de tissus embryonnaires et adultes analysées par greffe a des poussins irradiés. J. Embryol. exp. Morph. 33, 259-78.

Samarut, J. & Nigon, V. (1976a). Properties and development of erythropoietic stem cells in the chick embryo. J. Embryol. exp. Morph. 36, 247-60.

Samarut, J. & Nigon, V. (1976b). In vitro development of chicken erythropoietin-sensitive cells. Exp. Cell Res. 100, 245-8.

Samarut, J., Blanchet, J.P. & Nigon, V. (1979a). Antigenic characterization of chick erythrocytes and erythropoietic precursors: Identification of several definitive populations during embryogenesis. Dev. Biol. 72, 155-66.

Samarut, J., Jurdic, P. & Nigon, V. (1979b). Production of erythropoietic colony-forming units and erythrocytes during chick embryo development: an attempt at modelization of chick embryo erythropoiesis. J. Embryol. exp. Morph. 50, 1-20.

Sandeen, G., Wood, W. & Felsenfeld, G. (1980). The interaction of high mobility proteins HMG 14 and 17 with nucleosomes. Nucl. Acid Res. 8, 3757-78.

Sandreuter, A. (1951). Vergleichende Untersuchungen uber die Blutbildung in der Ontogenese von Haushuhn (Gallus gallus L.) und Star (Sturnus v. vulgaris L.). Acta Anat. 11, 1-61.

Schalekamp, M., Schalekamp, M., Van Goor, D. & Slingerland, R. (1972). Re-evaluation of the presence of multiple haemoglobins during the ontogenesis of the chicken. J. Embryol. exp. Morph. 28, 681-713.

Schauenstein, K., Globerson, A. & Wick, G. (1982). Thymic cell growth factor in supernatants of mitogen stimulated chicken spleen cells. Dev. Comp. Immunol. 6, 533-41.

Schmekel, L. (1962). Embryonale und frühe postembryonale Erythropoiese in Leber, Milz, Dottersack und Knochenmark der Vögel. Rev. Suisse Zool., 69, 559-615.

Schnetzler, M., Dammen, A., Nowak, J.S. & Franklin, R.M. (1983). Characterization of chicken T cell growth factor. Eur. J. Immunol. 13, 560-6.

Schou, M.   (1980).   A micromethod of chicken MLE: Technical aspects and
    genetic control. Tissue Antigens 15, 373-80.
Schou, M. & Simonsen, M.   (1982).   The major histocompatibility complex
    of outbred chicken. II. Analysis of the typing response in
    mixed lymphocyte culture stimulated by homozygous typing cells.
    Tissue Antigens 20, 320-6.
Schumacher, A.   (1965).   Zur submikroskopischen Struktur der Thrombozyten,
    Lymphozyten und Monozyten des Haushuhnes (Gallus domesticus).
    Z. Zellforsch. mikr. Anat., 66, 219-32.
Scollay, R.   (1983).   Intrathymic events in the differentiation of T
    lymphocytes: a continuing enigma. Immunol. Today 4, 282-6.
Shimizu, K.   (1972a).   Ontogeny of chicken hemoglobin. I. Electrophoretic
    study of the heterogeneity of hemoglobin in development. Dev.
    Growth Differ. 14, 43-5.
Shimizu, K.   (1972b).   The ontogeny of chicken hemoglobin. II. Chromato-
    graphic study of the heterogeneity of hemoglobin in develop-
    ment. Dev. Growth Differ. 14, 281-95.
Simons, J.A.   (1966).   The ontogeny of the multiple molecular forms of
    hemoglobin in the developing chick under normal and
    experimental conditions. J. Exp. Zool. 162, 219-30.
Simonsen, M.   (1957).   The impact on the developing embryo and newborn
    animal of adult homologous cells. Acta Path. Microbiol. Scand.
    40, 480-500.
Small, J.V. & Davies, H.G.   (1972).   Erythropoiesis in the yolk sac of
    the early chick embryo: an electron microscope and micro-
    spectrophotometric study. Tissue & Cell, 1972, 4, 341-78.
Sorrell, J.M. & Weiss, L.   (1980).   Cell interactions between hemato-
    poietic and stromal cells in the embryonic chick bone marrow.
    Anat. Rec. 197, 1-19.
Stalder, J., Groudine, M., Dodgson, J.B., Engel, J.D. & Weintraub, H.
    (1980a).   Hb switching in chickens. Cell 19, 973-80.
Stalder, J., Groudine, M., Dodgson, J.B., Engel, J.D. & Weintraub, H.
    (1980b).   Tissue-specific DNA cleavages in the globin
    chromatin domain introduced by DNAase I. Cell 20, 451-60.
Stehelin, D., Varmus, H.E., Bishop, J.M. & Vogt, P.K.   (1976).   DNA
    related to the transforming genes(s) of avian sarcoma viruses
    is present in normal avian cells. Nature 260, 170-3.
Stinson, R.S., Mashaly, M.M. & Glick, B.   (1979).   Thrombocyte migration
    and release of thrombocyte inhibitory factors (ThrIF) by T
    B cells in the chicken. Immunology, 36, 769-72.
Sugimoto, M., Yasuda, T. & Egashira, Y.   (1977a).   Development of the
    embryonic chicken thymus. I. Characteristic synchronous
    morphogenesis of lymphocytes accompanied by the appearance
    of an embryonic thymus-specific antigen. Dev. Biol. 56,
    281-92.
Sugimoto, M., Yasuda, T. & Egashira, Y.   (1977b).   Development of the
    embryonic chicken thymus. II. Differentiation of the epithelial
    cells studied by electron microscopy. Dev. Biol. 56, 2933-05.
Sugimura, M. & Hashimoto, Y.   (1980).   Quantitative histological studies
    on the spleen of ducks after neonatal thymectomy and
    bursectomy. J. Anat. 131, 441-52.

Till, J.E. & McCulloch, E.A. (1961). A direct measurement of the radiation sensitivity of normal mouse bone marrow cells. Rad. Res., 14, 213-221.

Toivanen, P. & Toivanen, A. (1973). Bursal and postbursal stem cells in chicken. Functional characteristics. Eur. J. Immunol. 3, 385-95.

Toivanen, A. & Toivanen, P. (1977). Histocompatibility requirements for cellular cooperation in the chicken: Generation of germ germinal centers. J. Immunol. 118, 431-6.

Toivanen, P., Toivanen, A. & Good, R.A. (1972). Ontogeny of bursal function in chicken. III. Immunocompetent cells for humoral immunity. J. Exp. Med. 136, 816-31.

Traill, K.N., Bock, G., Boyd, R. & Wick, G. (1983). Chicken thrombo-cytes. Isolation, serological and functional characterisa-tion using the fluorescence activated cell sorter. Dev. Comp. Immunol., 7, 111-25.

Traill, K.N., Bock, G., Ratheiser, K. & Wick, G. (1984). Ontogeny of surface markers on functionally distinct T cell subsets in the chicken. Eur. J. Immunol. 14, 61-7.

Trembicki, K.A. & Dietert, R.R. (1985). Chicken developmental antigens: analysis of erythroid populations with monoclonal antibodies. J. Exp. Zool. 235, 127-34.

Van Alten, P.J. (1980). Ontogeny of RES function in birds. In The Reticuloendothelial System A Comprehensive Treatise Vol. 3, Ontogeny and Phylogeny, eds. N. Cohen and M.M. Sigel, pp. 659-85. New York: Plenum Press.

Warner, N.L. & Burnet, F.M. (1961). The influence of testosterone treat-ment on the development of the bursa of Fabricius in the chick embryo. Aust. J. Biol. Sci. 14, 580-7.

Weber, W.T. (1970). Mixed lymphocyte interaction and PHA response of chicken spleen cells in a chemically defined medium. J. Reticuloendoth. Soc. 8, 37-54.

Weber, W.T. (1973). Direct evidence for the response of B and T cells to to pokeweed mitogen. Cell Immunol. 9, 482.

Weber, W.T. (1975). Avian B lymphocyte subpopulations: Origins and functional capacities. Transplant. Rev. 24, 8-158.

Weber, W.T. & Alexander, J.E. (1978). The potential of bursa-immigrated hematopoietic precursor cells to differentiate to functional B and T cells. J. Immunol. 121, 653-7.

Weber, W.T. & Ewert, D.L. (1982). The effect of monoclonal anti-Ia serum on the migration and differentiation potential of avian embryonic bone marrow cells. Immunobiology 163, 366-7.

Weber, W.T. & Foglia, L.M. (1980). Evidence for the presence of pre-cursor B cells in normal and in hormonally bursectomized chick embryos. Cell. Immunol. 52, 84-94.

Weber, W.T. & Weidanz, W.P. (1969). Prolonged bursal lymphocyte deple-tion and suppression of antibody formation following irradia-tion of the bursa of Fabricius. J. Immunol. 103, 537-43.

Weill, J.C., Reynaud, C.A., Lassila, O. & Pink, J.R.L. (1986). Re-arrangement of chicken immunoglobin genes is not an ongoing process in the embryonic bursa of Fabricius. Proc. Natl. Acad. Sci. USA 83, 3336-40.

Weintraub, H. & Groudine, M. (1976). Chromosomal subunits in active genes have an altered conformation. Science 93, 848-58.

Weintraub, H., Larsen, A. & Groudine, M. (1981). Alpha-globin gene switching during the development of chick embryos: expression and chromosome structure. Cell 24, 333-44.

Weintraub, H., Beug, H., Groudine, M. & Graf, T. (1982). Temperature-sensitive changes in the structure of globin chromatin in lines of red cell precursors transformed by ts-AEV. Cell 28, 931-40.

Weisbrod, S. & Weintraub, H. (1979). Isolation of a subclass of nuclear proteins responsible for conferring a DNase I-sensitive structure on globin chromatin. Proc. Nat. Acad. Sci. USA, 76, 630-4.

Weisbrod, S., Groudine, M. & Weintraub, H. (1980). Interaction of HMG 14 and 17 with actively transcribed genes. Cell 19, 289-301.

Wick, G., Albini, B. & Milgrom, F. (1973). Antigenic surface determinants of chicken lymphoid cells 15, 237-49.

Wilt, F.H. (1962). The ontogeny of chick embryo hemoglobin. Proc. Nat. Acad. Sci. 48, 1582-90.

Wilt, F.H. (1965). Regulation of the initiation chick embryo hemoglobin synthesis. J. Mol. Biol. 12, 331-41.

Wilt, F.H. (1974). The beginnings of erythropoiesis in the yolk sac of the chick embryo. Ann. New York Acad. Sci. 241, 99-112.

Wolfe, H.R., Sheridan, S.A., Bilstad, N.M. & Johnson, M.A. (1962). The growth of lymphoidal organs and the testes of chickens. Anat. Rec. 142, 485-93.

6 Mammals

Richard T. Parmley
Departments of Pediatrics and Pathology, University of Texas Health
Science Center at San Antonio, 7703 Floyd Curl Drive, San Antonio,
Texas 78284, USA

INTRODUCTION

This chapter provides a summary which emphasizes the
morphological and cytochemical features of mammalian blood cells. The
circulatory system is only briefly mentioned and is meant to outline the
apparatus through which blood cells travel and are distributed.
Similarly, an extensive literature on the cytokinetics and functions of
blood cells exists and is covered in less detail within the text.
Consequently, many of the references are review articles which will
provide more definitive detail and specific references. Although the
morphology of mammalian blood cells has been studied in detail in some
animal species, numerous species have not been similarly studied. Not
surprisingly, considerable variation exists between animal species,
presumably as a result of evolution or environmental demands.
Cytochemical investigations of blood cells have provided a link between
biochemistry, morphology, and function of blood cells. Thus, specific
enzyme activities or biochemical components can be localized to a
specific organelle thereby allowing better identification and assignment
of functional properties to that organelle. The cytochemical properties
of blood cell organelles also appear to vary considerably among
species. Furthermore, the development of new cytochemical methods and
the virtual explosion in the availability of specific monoclonal
antibodies for immunostaining has resulted in frequent updates in
classification and function of blood cells.

STRUCTURE OF THE CIRCULATORY SYSTEM
Blood vessels

The heart pumps blood cells through a complex system which
includes arterial vessels, capillaries, and venous vessels. The blood
volume is approximately 70 cc kg^{-1} in humans but varies with age and
animal species. All blood vessels are lined with simple epithelial
cells (the endothelium) through or between which pass nutrients,
metabolic by-products, and cells. The endothelium overlies a basal
lamina which provides an additional permeability barrier containing type
IV collagen, proteoglycans, fibronectin, entactin and other large
macromolecules (Laurie et al. 1982).

Arterial vessels. Vessels carrying oxygenated blood from the left heart can be divided into elastic and muscular arteries and arterioles. Elastic arteries are large vessels, the walls of which contain abundant elastic fibres in addition to smooth muscle cells. These arteries include the aorta and the pulmonary, subclavian, carotid, innominate and common iliac arteries. Muscular arteries receive blood from elastic arteries and distribute it to specific body regions or organs. Smooth muscle cells in the vessel wall control lumen size by contraction and relaxation. Arterioles are very small, often less than 300 μm in diameter and contain a single or double layer of muscle cells.

The walls of arterial vessels can be divided into the tunica intima, tunica media, and the tunica adventitia. The intima includes endothelial cells and basal lamina, both of which provide a continuous barrier. Myoendothelial junctions are present in terminal arterioles and can penetrate the basal lamina. The media contains elastic fibres, smooth muscle cells and an extracellular matrix which includes collagen and proteoglycans. The adventitia contains fibroblasts and an extracellular matrix, and sometimes mast cells and macrophages. The walls of large blood vessels may also contain lymphatics, small blood vessels, and nerves.

Capillaries. Arterioles end in a complex network of small vessels with a luminal diameter of 5 - 10 μm. These vessels contain only an endothelium, basal lamina and scattered pericytes. Ultrastructurally, endothelial cells may appear continuous, fenestrated or discontinuous (Bruns and Palade 1968; Clementi and Palade 1969; Simionescu et al. 1976; Simionescu and Simionescu 1983).

The most common type of capillary has a continuous endothelium 0.2 - 0.3 μm thick and a basal lamina 0.2 - 0.5 μm thick. Although endothelial junctions are present, distinct gap junctions are not seen. Continuous capillaries are found in muscular tissue, connective tissue, certain exocrine glands, the central nervous system, and a variety of other tissues.

The endothelium of fenestrated capillaries contains transcellular circular openings, 60 - 80 nm in diameter called fenestrae. The fenestrae are usually covered by a membrane diaphragm, 4 - 6 nm thick. These may be randomly distributed or in patches. Fenestrated capillaries are found in renal glomeruli, renal tubules, the gastrointestinal tract, endocrine glands, choroid plexus and the ciliary body.

Capillaries with a discontinuous endothelium generally line sinusoids or vascular sinuses such as is seen in the liver. These capillaries contain large open gaps and often lack underlying basal lamina.

Venous vessels. Blood cells return to the heart through venous vessels which include venules and larger veins. In addition to receiving blood from capillaries, arteriovenous anastamoses are present to varying degrees in the microvascular bed (Simionescu and Simionescu 1983).

The immediate postcapillary venules are pericytic venules, which vary in luminal diameter from 10 - 50 μm, and contain pericytes. The pericytic venules transport blood 50 - 200 μm to muscular venules with a thin media containing one or two layers of smooth muscle cells. Endothelial cells in venules are usually continuous but contain scattered fenestrae. The underlying basal lamina is also continuous except in areas where it is penetrated by myoendothelial junctions. The endothelial cells of muscular venules contain communicating (or gap) junctions in addition to the occluding type (or endothelial) junctions seen in pericytic venules and capillaries.

Venules empty their contents into veins, which through a progressive network of increasing vessel size, transport blood back to the right heart. Like arteries, veins also contain a tunica intima, media and adventitia, although the walls of veins are much thinner than arteries. Endothelial cells in veins also overlie a basal lamina and form both occlusive and gap junctions. Elastic fibres are present in some veins, especially propulsive veins which push blood against the force of gravity, however, even in these veins, the elastic interna is not as clearly distinct as in arteries.

Lymphatics
        Fluid and cells in interstitial spaces drain into lymphatic capillaries, vessels and lymph nodes. The accumulated lymph is eventually returned to the venous system through the thoracic duct. The primary function of lymphatic vessels is to recover fluids that escape into connective tissue spaces from blood vessels, whereas lymph nodes appear to filter this fluid and direct T and B cell responses.

Lymphatic capillaries vary from a few to several hundred micrometers in luminal diameter (Weiss 1983). The lymphatic capillary endothelium is flattened with a zonula adhaerens connecting endothelial cells, which form a generally continuous layer. Unlike blood vessels, the basal lamina of lymphatics is poorly developed and not distinctly identified. Anchoring filaments extend from the surrounding matrix to the outer membrane of endothelial cells. The lymphatic capillaries are especially prominent under body surfaces such as the skin, mucous membranes, gastrointestinal and respiratory tracts. The central nervous system and bone marrow notably lack lymphatic capillaries.

The lymphatic capillaries empty into collecting vessels which resemble veins. Although these collecting vessels have a tunica intima, media, and adventitia they are much more difficult to delineate than their venous counterpart. Lymphatic vessels often appear beaded as a result of valves or cusps which prevent back-flow of lymph.

Lymph nodes lie across the collecting lymphatic vessels. Afferent lymphatic vessels deliver lymph through specific entry sites in the capsule. An efferent lymphatic vessel leaves the lymph node at the hilus where arterial and venous vessels enter and exit. Within the cortex are a number of lymphatic nodules, which contain densely packed B lymphocytes. Primary nodules appear homogeneous, whereas secondary nodules contain a mantle or crescent of small lymphocytes with a central germinal centre containing larger activated lymphocytes involved in B cell differentiation. During inflammation, the germinal centre contains activated T and B lymphocytes as well as macrophages which may enclose a phagocytic inclusion or tingible body consisting of degraded nuclear material (Weiss, 1983). The nodules within lymph nodes are surrounded by a less well defined diffuse or tertiary cortex or paracortical region containing T lymphocytes(Hsu et al. 1983). A variety of T and B lymphocytes exist and have been characterized in blood and lymph nodes using monoclonal antibodies. The characterization and distribution of these cells is discussed in more detail in the lymphocyte section of this text. Radial sinuses project into the cortex and hilus from the peripheral subcapsular sinus. These sinuses are lined by reticular cells and varying numbers of macrophages and plasma cells depending upon the degree of immunologic stimulation. The sinuses drain into the efferent lymphatic vessel in the hilus.

Less well defined collections of lymphocytes may form in most parts of the body during inflammation. Aggregates of lymphoid tissue line parts of the intestinal tract forming Peyer's patches, which presumably process ingested antigenic material. Similar collections are associated with the tonsils and the respiratory tract.

Spleen

The spleen is a highly specialized dynamic structure with a diversity of functions. Virtually all blood is circulated through the spleen which 1) filters and removes microorganisms and abnormal cells, 2) provides both primary and secondary immunologic responses, 3) contributes to erythrocyte, platelet, lymphocyte and monocyte differentiation, 4) provides a reservoir for storage of blood cells and 5) is a site for haematopoiesis in foetal and young animals. The extent to which the spleen carries out these functions varies with the animal species and age (Hartwig and Hartwig 1985; Weiss 1985). Removal of the spleen in the human infant results in a significant risk for overwhelming bacterial infection, particularly with Streptococcus pneumoniae; whereas removal of the spleen in an adult carries less risk (Krivit 1977). Functional asplenia can be detected in humans by the observation of erythrocyte Howell-Jolly bodies, or nuclear remnants, normally removed by the spleen. The spleen may become massively enlarged in disease states in which there are large numbers of abnormal circulating cells as occurs in leukaemia or abnormalities of red cell shape. In dogs and cats unlike humans, the spleen is capable of storing up to one third of the blood volume, which can be released by contraction of a muscular capsule during times of stress (Song and Groom 1971).

Blood enters the spleen by the splenic artery located at the hilus, passes through a trabecular framework, and exits by the splenic vein at the hilus (Knisely 1936). Blood entering, passes through trabecular and then central arteries surrounded by a collection of predominantly T lymphocytes called the white pulp, which is comparable to the cortex of lymph nodes. Within the white pulp are scattered nodules containing B cells (Eikelenboom et al. 1985; van Ewijk and Nieuwenhuis 1985). During

an immune response, these cells migrate to the periphery of the sheath
and secrete antibody. The central arteries terminate in splenic cords
or capillaries in the red pulp after passing through a marginal zone
between the red and white pulp (Weiss et al. 1985; Fujita et al.
1985). The red pulp is drained by venous sinuses and eventually the
splenic vein (Knisely 1936). In the splenic cords of the marginal zone
and red pulp, blood cells are exposed to a variety of environmental
stresses related to concentration of blood cells in this area.
Defective cells or old cells are destroyed and removed by the
reticuloendothelial cells or macrophages (Crosby 1959; Weiss and
Tavassoli 1970). Also within the red pulp, haematopoiesis can be
observed on occasion in rodents but is seen normally only in foetal or
infant humans (Seifert and Marks 1985). Splenic capillaries and basal
lamina are fenestrated with unusually large openings. This allows for
egress of cells out of the system and allows access of macrophages to
the defective cells.

ORIGIN AND FORMATION OF BLOOD CELLS
Site of formation of blood cells
A variety of tissues may contribute to the formation and
differentiation of blood cells. These include the bone marrow, lymph
nodes, thymus, spleen and liver. The sites of haematopoiesis vary with
animal species and developmental age (Gilmour 1941; Seifert and Marks
1985).

In the human embryo, haematopoiesis, particularly erythropoiesis can be
detected in the yolk sac during the second week of life. During the
mesoblastic period, these blood islands become connected and eventually
surrounded by endothelial cells. By the sixth week of life,
haematopoiesis is present within the liver. During the third through
the seventh month of gestation, the liver is the major site of
haematopoiesis. The spleen (Seifert and Marks 1985) and to a lesser
extent the lymph nodes also become sites of haematopoiesis. The bone
marrow becomes a major site of haematopoiesis during the third trimester
and produces greater than 90% of blood cells at birth (Gilmour 1941).
In infants, red marrow is prominent in flat and long bones, whereas, in
adults active haematopoiesis occurs predominantly in the skull, sternum,
vertebrae, ribs, pelvis and proximal portions of the long bones.

In the adult human, marrow tissue comprises 3.4 to 5.9% of body weight
or 1600 to 3700g (Mechanik 1926). Donahue et al. (1958) have estimated
approximately 18×10^9 nucleated cells kg^{-1} of body weight. The marrow
appears somewhat gelatinous containing 20 to 40% fat in sites of active
haematopoiesis.

Nutrient arteries are centrally located within the marrow and terminate
in peripheral capillary beds. Post capillary venules may coalesce to
form sinusoids with blood flow toward the centre of the marrow cavity.
Haematopoiesis takes place outside and not within the venous sinusoids.

Adventitial cells (reticular cells) send cell processes between
haematopoietic cells. Although their exact function is not known, they

are capable of producing reticulin and do not appear to be phagocytic.
Adventitial cells may also accumulate fat. These cells have not been
demonstrated to be marrow stem cells.

After maturation and development is complete, haematopoietic cells enter
the circulation through fenestrations in endothelial cells lining venous
sinusoids (Campbell 1972), or may pass directly through endothelial
cytoplasm (De Bruyn 1981). A significant amount of deformability is
required for mature cells to pass through these openings. Presumably,
both mechanical factors as well as humoral factors contribute to
movement of the maturing cell through the parenchyma to the endothelial
cell lining the sinus.

In addition to forming myeloid and erythroid cells, the bone marrow
appears to be the bursa equivalent for B lymphocyte formation in mammals
(Kincade and Phillips 1985). It also releases stem cells which are
subsequently processed by the thymus to form T lymphocytes. The thymus
develops before birth, appears to complete much of its function early in
life, and in humans becomes involuted after the first decade of life.
The thymus contains multiple lobules with a cortex area rich in
lymphocytes and a medullary area containing prominent epithelial cells
(Bearman et al. 1978; Kendall 1981). It is within this microenvironment
that initial T cell differentiation occurs with release of T cells into
the circulation. The lymphocytes then accumulate in peripheral
lymphatics where differentiation is completed.

The stem cell
The pluripotential stem cell has the capability of giving
rise to each of the haematopoietic cell lines (Till et al. 1964;
Quesenberry and Levitt 1979). Stem cells forming colonies in the spleen
after infusion into an irradiated mouse are referred to as CFU-S whereas
those forming colonies in agar are termed CFU-C (Figure 1). The latter
cells are committed stem cells and require colony stimulating factor,
CSF, a substance produced in part by monocytes. CFU-GM forms mixed
granulocyte macrophage colonies; CFU-G forms granulocytes only and CFU-
EO forms compact eosinophil colonies. Erythoid colony forming cells
include erythropoietin dependent CFU-E (Figures 2,3) and erythropoietin
independent burst forming units (BFU-E) which are derived from an
earlier stem cell than CFU-E and forms a larger less tightly packed
colony (Ogawa et al. 1976). Megakaryocyte colonies, CFU-M (Levine 1982)
and basophil-mast cell colonies (Leary and Ogawa 1984) have also been
identified. Mast cells can also be cultured from connective tissue
(Ishizaka et al. 1977), unlike basophils, and appear to be extremely
long lived cells (Padawer 1974) in contrast to short lived basophils and
other myeloid cells (Boggs 1967). Osteoclasts appear to arise from stem
cells which also form monocyte-macrophage colonies (Gothlin and Ericsson
1976). Osteoblasts are not hematopoietic cells and appear to arise from
different stem cells (Amsel and Dell 1972).

Cell cycle and differentiation
Replicating haematopoietic cells pass through a cell cycle
consisting of G1, S, G2 and M phases (Boggs 1967; Mazia 1974). A GO

phase also exists in which cells are in a resting, non dividing, phase, from which they can be recruited as needed into the cell cycle. In G1 phase the cell prepares for DNA replication with RNA and protein synthesis lasting approximately 12 to 20h. DNA synthesis occurs over a period of 8h during S phase. Cells in S phase readily incorporate tritiated thymidine. A short, premitotic phase termed G2 phase preceeds mitosis or M phase which requires 30 min. The mitotic index or incidence of mitotic cells in human marrow is 44.8/1,000 cells capable of division for erythrocytes and 20.2/1,000 cells for granulocytes (Cronkite et al. 1959).

The marrow can be divided into 1) the stem cell compartment, 2) the mitotic compartment and 3) the maturation and storage compartment (Cronkite et al. 1959). The vast majority of the cells are in the latter two compartments. Transit time through the mitotic compartment averages 5 to 6 days and usually involves at least four divisions with simultaneous differentiation. Post-mitotic cells spend an average of 7 days in the maturation compartment (Boggs 1967). In the granulocytic series, myeloblasts, promyelocytes and myelocytes are part of the mitotic compartment, whereas metamyelocytes, band and segmented neutrophils are in the maturation compartment. Proerythroblasts, basophilic and polychromatophilic erythroblasts are in the mitotic compartment whereas orthochromatic erythroblasts, reticulocytes and erythrocytes are in the maturation compartment. In humans, myeloid cells outnumber erythroid cells by a 2.3:1 ratio. (Table I). Late cells with mature cytoplasmic organelles are the only types normally found in the circulation and vary to some extent among species (Table II).

Table I: Cell Types in Adult Human Marrow Aspirates

	Mean (%)	Range (%)
Myeloblasts	0.9	0.2 - 1.5
Neutrophilic Promyelocytes	3.3	2.1 - 4.1
Neutrophilic Myelocytes	12.7	8.2 - 15.7
Neutrophilic Metamyelocytes	15.9	9.6 - 24.6
Band Neutrophils	12.4	9.5 - 15.3
Segmented Neutrophils	7.4	6.0 - 12.0
Eosinophilic Myelocytes	0.8	0.2 - 1.3
Late Eosinophils	2.6	0.6 - 5.9
Basophils	<0.1	0 - 0.2
Mast Cells	<0.1	0 - 0.2
Proerythroblasts	0.6	0.2 - 1.3
Basophilic Erythroblasts	1.4	0.5 - 2.4
Polychromatophilic Erythroblasts	21.6	17.9 - 29.2
Orthochromatic Erythroblasts	2.0	0.4 - 4.6
Lymphocytes	16.2	11.1 - 23.2
Plasma Cells	1.3	0.4 - 3.9
Monocytes	0.3	0 - 0.8
Megakaryocytes	<0.1	0 - 0.4
Histiocytes	0.3	0 - 0.9
Myeloid to Erythroid Ratio	2.3	1.5 - 3.3

Data taken from Wintrobe (1974), based on evaluation of marrow aspirates
from 12 healthy men.

STRUCTURE OF BLOOD CELLS
Haematopoietic cells can be distinguished by specific
structural features unique to each blood cell type. Developmental
stages are described as early, mid, and late depending upon the point in
time that the cell is studied. Generally, cells in the mitotic
compartment are at early or mid stages of development, whereas those in
the post-mitotic compartment are late cells. The nucleus, cytoplasmic
matrix, and organelles such as lysosomal granules can exist as mature,
or immature forms which generally correspond with cell development.
Both mature and immature organelles, however, can occur in early or late
stages of development. Furthermore, in pathologic conditions,
maturation and development may be asynchronous.

In normal animals, only late cells are observed in the circulation. The
quantity and distribution of these cells varies to some extent among
different animal species (Table II). In addition to distinct
morphological characteristics the cells demonstrate unique cytochemical
differences (Tables III, IV).

Erythroid cells (Figures 2-11)
The earliest identifiable erythroid cell is the proerythroblast or pronormoblast. This cell has a diameter from 14 -19 μm
in marrow smears, and 10 - 12 μm in sectioned plastic blocks for
electron microscopy. The nucleus appears round with predominantly

TABLE II: Blood Counts of Selected Mammalian Species

Species	Hgb	Hct	RBC	WBC	N	E	B	L	M	Platelet	Number of Animals
Human Adults	16 (men) 14 (women)	47 (men) 42 (women)	5.4 (men) 4.8 (women)	7.0	53	3	1	36	7	260	
Cat	10.5	---	7.24	17.2	59	7	0	33	1	232	52
Chimpanzee	12.3	41.6	5.11	12.0-22.0	53	7	0	40	0	290	10
Cow	13.5	40	6.96	10.2	26	7	0	58	8	---	2;10
Dog	13	38	6.49	11.2	74	2	0	20	4	621	60
Elephant	10.2	34.8	2.47	8.8	34	6	1	53	6	---	14
Guinea Pig	14.5	47.7	5.75	10.8	42	5	0	45	8	500	4;500
Horse	---	---	6.0-8.5	6.0-12.0	57	4	1	30	8	---	318
Mouse	---	39	---	6.6	28	0	0	65	7	698	44
Rabbit	13	40	6.29	7.9	43	2	4	42	9	500	61;900
Rat	13.0	39.4	6.60	8-15	27	8	0	67	5	330	73;200
Sheep	12.9	35.7	10.50	5.0	28	8	0	62	2	350	4

Abbreviations used: Hgb, haemoglobin concentration in g dl^{-1}; Hct, haematocrit, volume packed red cells l/l; RBC, red blood cell count in number x 10^{12}; WBC, white blood cell count in number x 10^9; N, % neutrophils; E, % eosinophils; B, % basophils; L, % lymphocytes; M, % monocytes; platelet count in number x 10^9.

Data represent mean values or ranges taken from: Scarborough (1931) rats, rabbits, horse, guinea pig; Landsberg (1940) for cats; Wintrobe (1933) for cows and guinea pigs; Wintrobe et al. (1936) for chimpanzees, rabbits, rats, and sheep; Delaune (1939) for cows; Mayerson (1930) for dogs; Nirmalan et al. (1967) for elephants; Petri (1933) for mice; Wintrobe (1974) and Santa Rosa Hospital, San Antonio, TX USA for humans.

TABLE III: Enzyme Cytochemistry of Human Blood and Marrow Cells

	Acid phosphatase	Chloroacetate esterase	-Napthyl esterase	Peroxidase		Alkaline phosphatase	Gluco-saminidase	Lysozyme
				Without cyanide	With cyanide			
Neutrophil								
Myeloblast	0-3+	0-3+	0-1+	0-3+	0	0	0	0
Promyelocyte	4+	4+	0-1+	4+	0	0	0-1+	0-3+
Myelocyte	3+	4+	0-1+	4+	0	0-1+	0-1+	3+-4+
Segmented	3+	4+	0-1+	4+	0	1-4+	0-1+	4+
Eosinophil								
Myelocyte	3+	0	0-1+	4+	3+	0	ND	0
Segmented	3+	0	0-1+	4+	3+	0	ND	0
Basophil	1+	0-1+	0	0-2+	0	0	ND	0
Mast cell	1+	4+	ND	ND	ND	ND	ND	0
Monocyte	3+	0-1+	4+	0-3+	0	0	4+	2-3+
Macrophage	3-4+	0-1+	4+	0-3+	0	1+	ND	2-3+
Megakaryocyte	3+	0	4+	0	0	0	3+-4+	0
Platelet	1+	0	1-3+	0	0	0	ND	0
Lymphocyte	0-2+	0	0-2+	0	0	0	0-1+	0
Plasma cell	0-2+	0	0-2+	0	0	0	0-2+	0
Erythroblast	0	0	0	0	0	0	ND	0

Data from Yam et al. (1971); Beckstead et al. (1981); Reed and Bennett (1975); Mason et al. (1975). ND, not done or reported; Staining graded 0 (absent staining); ± (occasional staining), 1+ (weak staining) to 4+ (intense staining)

TABLE IV: Nonenzymatic Cytochemistry of Human Blood and Marrow Cells

	Alcoholic PAS (vic-glycol)		HID (sulphate)	Aldehyde fuchsin (sulphate)	Biebrich scarlet pH 9.5 basic protein	Sudan black	Oil red 0	Lactoferrin	Prussian blue
	Diastase	No diastase							
Neutrophil									
Myeloblast	0-1+	0-1+	0-2+	0-2+	0-1+	0-3+	0	0	0
Promyelocyte	2+	1+	3+	3+	3+	4+	0	0	0
Myelocyte	1+	2+	1+	1+	0-2+	4+	0	0-2+	0
Segmented	1+	4+	0	0	0	4+	2+	2-4+	0
Eosinophil									
Myelocyte	3+	3+	4+	4+	4+	3+	0	0	0
Segmented	1+	3+	2+	3+	4+	3+	2+	0	0
Basophil	3+	3+	4+	4+	3+	ND	ND	0	0
Mast cell	3+	3+	4+	4+	ND	ND	ND	0	0
Monocyte	0-1+	0-1+	0	0-1+	ND	0-2+	0-2+	0	0-1+
Macrophage	0-3+	3+	0-1+	0-1+	ND	0-3+	0-2+	0	1+-4+
Megakaryocyte	2+	3+	2+	2+	ND	0-1+	ND	0	0
Platelet	2+	3+	2+	2+	2+	0-1+	ND	0	0
Lymphocyte	0-1+	0-3+	0-1+	0-1+	ND	0	ND	0	0
Plasma cell	0-2+	0-3+	0-1+	0-1+	ND	0-1+	ND	0	0
Erythroblast	0	0	0	0	ND	0	ND	0	0

Data from Dunn and Spicer (1969); Parmley et al. (1983b); Sheehan and Storey (1947); Coimbra and Lopes-Vas (1971); Mason et al. (1975).

ND, not done or reported; Staining graded 0 (absent staining), to 4+ (intense staining).

dispersed or reticulated nuclear chromatin containing 1 - 2 nucleoli. The cytoplasm is scanty but circumferential in marrow smears. The basophilia in Wright's stained preparations corresponds to the presence of numerous polyribosomes seen at the ultrastructural level (Figure 4). The cytoplasm lacks stainable haemoglobin at this stage and contains several rod shaped mitochrondria, but is relatively free of other organelles. A small Golgi apparatus can be seen ultrastructurally in some cell profiles and corresponds to an eccentric small pale area in Wright's stained preparations. The cell contains rare granules which sometimes contain ferritin particles. From 2 to 7 catalase positive peroxisomes per cell profile may be stained using alkaline diaminobenzidine (Breton-Gorius and Guichard 1975). The cell surface is indented by caveolae which may contain ferritin particles. These are internalized forming pinocytotic vesicles 0.05 - 0.2 μm in diameter.

The second distinct stage of erythroid development is the basophilic erythroblast or normoblast (Figures 6,7). This cell is slightly smaller than the pronormoblast and contains more abundant deeply basophilic cytoplasm which at the ultrastructural level appears packed with polyribosomes (Figure 6). At the light microscope level, the nuclear chromatin is more coarse without distinct nucleoli. Ultrastructurally, the heterochromatin appears more prominent than in pronormoblasts. Ferritin can be identified in surface caveolae and pinocytotic vesicles (Bessis and Breton-Gorius 1959) (Figure 7). The cytoplasm continues to lack stainable haemoglobin, although peroxisomes can be stained with alkaline diaminobenzidine (Breton-Gorius and Guichard 1975).

Basophilic erythroblasts form polychromatophilic erythroblasts varying from 8 - 12 μm in diameter in marrow smears (Figure 5). The cells are easily identified in Wright's stained preparations by the faint pink or acidophilic cytoplasm superimposed upon a moderately basophilic background. The former material corresponds to the appearance of haemoglobin which is ultrastructurally stained with diaminobenzidine (Figure 8). Cytoplasmic ferritin and siderosomes may be present.

The polychromatophilic erythroblast divides and develops into ortho-chromatic erythroblasts, which lack the ability of further mitotic division. In Wright's stained preparations, these cells vary from 7 - 10 μm in diameter and contain pycnotic dense nuclei and eosinophilic cytoplasm with minimal basophilia (Figure 5). Ultrastructurally, the nucleus contains predominantly condensed chromatin and a small ring shaped nucleolus. The cytoplasm has an abundant amount of benzidine-reactive haemoglobin with scattered polyribosomes. A few siderosomes containing ferritin particles may be present. Ferritin particles may also occasionally be seen on surface caveolae.

After expulsion of the nucleus, the remaining reticulocyte may be distinguished from other mature erythrocytes by the presence of scattered ribosomes and rare mitochondria seen with the electron microscope. Also present are surface caveolae, as well as pinocytotic vesicles and autophagic vacuoles. Ribosomes appear polychromatophilic in Wright's stained preparations. Reticulocytes can be vitally stained

using new methylene blue. Normally less than 2% of circulating human
red cells are reticulocytes.

Fully mature erythrocytes have a homogeneous interior without identifiable organelles. The haemoglobin has pseudo-peroxidase activity and
ultrastructurally demonstrates increased density after staining with
3,3' diaminobenzidine. Human erythrocytes average 7.5 µm in diameter
and have a distinct biconcave appearance when visualized with the
scanning electron microscope (Bessis 1973) (Figure 9). The majority of
other mammalian erythrocytes also have a similar appearance, although
red cells of camels and llamas are noted for their oval or elliptical
appearance (Ponder et al. 1928).

Erythrocyte surfaces have a strong negative charge as a result of their
content of sialic acid (Eylar et al. 1962). Some of this is
attributable to glycophorin, a major sialoglycoprotein on erythrocyte
surfaces. This can be stained with cationic reagents such as dialyzed
iron or colloidal iron (Marikovsky and Danon 1969). Senescent red cells
lose this staining. This loss of negative surface charge may result in
increased clearance by the reticuloendothelial system. Erythroblast and
reticulocyte surfaces also contain transferrin receptors which
presumably mediate transferrin iron uptake needed for haemaglobin
synthesis. Both transferrin and ferritin can be demonstrated on the
surface and in pinocytotic vesicles of these cells (Sullivan et al.
1976; Parmley et al. 1983a). Fully mature erythrocytes lack surface
staining for transferrin receptors, transferrin or ferritin, as would be
expected with completion of haemoglobin synthesis. Human erythrocytes
stain weakly with concanavalin A (Figure 9), whereas immature
erythrocytes stain more intensely. In contrast, rabbit erythrocytes
stain intensely with concanavalin A (Figures 10,11) regardless of
maturation state and also aggregate with this lectin indicating an
increase in Con A reactive hexoses when compared with humans (Parmley et
al. 1973, 1979a).

Erythrocyte surfaces have also been examined extensively using freeze
fracture and shadowing methods. Both an internal and external face are
identified when the membrane is split. Intermembrane particles are
readily seen on the external face and correspond to proteins within the
membranes (Stolinski and Brethnach 1975).

Neutrophilic leucocytes (Figures 12-24)
        Neutrophilic leucocytes and other granulocytes are derived
from myeloblasts which lack morphologic features that predict their
differentiation into one of the granulocytic lines. Myeloblasts vary
from 10 - 15 µm in diameter on marrow smears and 10 -12 µm in sectioned
tissues. The nucleus contains abundant dispersed chromatin and 2 - 4
distinct nucleoli averaging 1.5 µm in diameter. The cytoplasm appears
grey-blue in Wright's stained preparations and circumferentially
surrounds the nucleus. Ultrastructurally, numerous polyribosomes are
present and some segments of endoplasmic reticulum (Scott and Horn
1970a). Myeloblasts can be further distinguished from erythroblasts by
their lack of ferritin containing surface caveolae and fewer pinocytotic

vesicles and content of more endoplasmic reticulum. There are scattered segments of rough endoplasmic reticulum and variably prominent Golgi lamellae.

The development of the first cytoplasmic granules allows definitive identification of differentiation of myeloblasts into neutrophilic leucocytes. The ultrastructural morphology of developing human (Scott and Horn 1970a; Ackerman 1971; Bainton et al. 1971), feline (Ackerman 1968, Ward et al. 1972), mink (Davis et al. 1971a), mouse (Oliver and Essner 1975), rat (Yamada and Yamauchi 1966), and ruminant (Baggiolini et al. 1985) neutrophils; and rabbit and guinea pig (Bainton and Farquhar 1966; Wetzel et al. 1967a; Brederoo and Daems 1977) heterophils has been studied in detail. Less extensive studies have been performed on neutrophils from other mammalian species. All of these studies have distinguished early, mid and late developmental stages, which correspond to promyelocytes, myelocytes and late neutrophils in the non-mitotic compartment (metamyelocytes, band and segmented forms) respectively.

Early neutrophilic leucocytes or promyelocytes include a morphologic spectrum of cells involved in primary granulogenesis (Bainton and Farquhar 1966; Wetzel et al. 1967a; Ackerman 1968). These cells may contain a few to abundant primary or azurophilic granules (Figure 12). In Wright's stained smears, the cells vary from 14 - 20 μm in diameter, whereas the diameter averages 12 - 14 μm in sectioned tissues. The round nucleus contains predominantly dispersed nuclear chromatin and 1 - 2 nucleoli. The cytoplasm is abundant and contains numerous segments of dilated rough endoplasmic reticulum and polyribosomes accounting for the basophilia seen in light microscopic preparations. Ultrastructurally, the Golgi apparatus is prominent and appears as a clear perinuclear zone in light microscopic preparations. Numerous vesicles and spherules can be observed budding from the Golgi apparatus and coalescing to form condensing vacuoles or granule precursors. Variable numbers of primary granules in different stages of maturation and/or condensation can be observed throughout the cytoplasm (Figure 14). Secondary or specific granules are not observed at the promyelocyte stage of development.

Neutrophilic myelocytes vary from 10 - 15 μm in diameter in marrow smears (Figure 13) and 10 - 12 μm in sectioned tissue. The cells contain a slightly indented nucleus with predominantly dispersed nuclear chromatin and 1 or 2 small nucleoli. The abundant cytoplasm contains numerous cytoplasmic granules, some of which lack azurophilia and represent specific or secondary granules not found in promyelocytes. The cytoplasm is less basophilic than promyelocytes and corresponds ultrastructurally to the presence of less dilated rough endoplasmic reticulum and fewer polyribosomes. The Golgi apparatus of neutrophilic myelocytes is rather prominent, appearing as a clear zone in light microscopic preparations, and an area with distinct lamellae and budding vesicles forming secondary granules in ultrastructural preparations (Bainton and Farquhar 1966; Wetzel et al. 1967a).

Metamyelocytes, band, and segmented neutrophils represent non-dividing cells in which a progressive condensation of nuclear chromatin occurs

(Figure 12). Nucleoli are indistinct. In fully developed cells, an average of three nuclear lobes may be identified (Figure 13). The Golgi apparatus is much less prominent than early and mid myeloid cells and the cytoplasm contains very little endoplasmic reticulum. There are sparse mitochondria and no sizeable increase in granules compared to myelocytes (Figure 14). Glycogen is prominent and may be stained with the PAS method at the light microscope level and periodate-thiocarbohydrazide silver proteinate (PA-TCH-SP) at the electron microscope level (Fittschen et al. 1983) (Figures 15,16).

At least three distinct types of granules have been identified in neutrophils and include primary, secondary and tertiary granules formed in early, mid and late cells respectively (Scott and Horn 1970a; Murata and Spicer 1973; Parmley et al. 1983b; Brederoo et al. 1983). Within these groups, there is significant heterogeneity and morphologic overlap in several species (Shannon and Zellmer 1982). Recently, Rice et al. have isolated 13 distinct granule fractions from human neutrophils, further demonstrating the complex heterogeneity of neutrophil granules. In addition to granules, cytoplasmic vesicles containing a variety of cytochemically reactive substances have been identified. Some of these vesicles eventually fuse with cytoplasmic granules whereas others have independent functions.

In humans, primary granules vary from 0.2 - 0.4 µm in diameter (Figure 14). Mature granules are very electron-dense with a central lucent crystalloid sometimes being present (Figure 14). A subpopulation of primary granules tend to be larger and nucleated under some staining conditions (Brederoo et al. 1983). The morphology of primary granules in rats (Calamai and Spitznagel 1982), mink (Davis et al. 1971a), cats (Ward et al. 1972), mice (Oliver and Essner 1975) guinea pigs (Brederoo and Daems 1977) and baboons (Parmley, unpublished observation) is very similar, whereas rabbit heterophil primary granules are larger varying from 0.3 - 0.8 µm in diameter (Figure 17). Bovine peroxidase-positive granules (Figure 20), presumed to be primary granules, are very sparse in segmented neutrophils and much smaller (0.1 - 0.2 µm) than the other granule types in this animal (Gennaro et al. 1983). A subpopulation of small (< 2 µm) peroxidase positive granules has also been identified in human neutrophils (Ackerman and Clark 1971b). Ultrastructural cytochemistry and biochemical studies have localized peroxidase (Figures 19,20), acid phosphatase (Figures 21,22), aryl sulphatase, elastase and proteases in endoplasmic reticulum, Golgi vesicles and primary granules of promyelocytes and cytoplasmic granules of polymorphonuclear leukocytes (Wetzel et al. 1967b; Ackerman 1968; Bainton and Farquhar 1968; Dunn et al. 1968; Bretz and Baggiolini 1974; Spitznagel et al. 1974; Clark et al. 1980; Calamai and Spitznagel 1982).

Primary granules vary from deeply azurophilic to neutral in Wright's stained preparations with the most intense staining observed in the immature granules. This affinity for cationic dyes is reflected at the ultrastructural level by intense staining of sulphate in immature primary granules using dialyzed iron and high iron diamine (HID) (Hardin and Spicer 1971; Parmley et al. 1980, 1983b) (Table IV). The HID staining can be enhanced with TCH-SP (Parmley et al. 1983b) (Figures 23,24). This staining appears to represent glycosaminoglycans including chondroitin and heparan sulphate (Olsson 1969; Parmley et al. 1983b). Autoradiographic studies have similarly demonstrated $^{35}SO_4$ incorporation into Golgi lamellae and primary granules of promyelocytes (Payne and Ackerman 1977; Parmley et al. 1980). In addition to demonstrating increased anionic glycoconjugate staining, immature primary granules stain intensely with pyroantimonate (Hardin et al. 1969) a reagent which precipitates a variety of monovalent and divalent cations including calcium. Cationic proteins may also be stained in immature but not mature primary granules (Dunn and Spicer 1969). At least some of the cationic proteins present in neutrophils appear to be esterases (Odeberg et al. 1975). Vicinal glycol containing glycoconjugates which include glycoprotein enzymes can also be demonstrated ultrastructurally in immature primary granules using the PA-TCH-SP method (Parmley et al. 1980; Fittschen et al. 1983) (Figure 16). Mature primary granules lack or have markedly decreased stainable sulphate, cations, and vicinal glycols. Acid phosphatase staining also appears masked in mature primary granules (Figure 22) (Wetzel et al. 1967b, Bainton and Farquhar 1968). This loss of staining reflects continued modification of granule contents during maturation and either results from removal and/or complexing or masking of granule components. However, the unmasking of sulphate and acid phosphatase with phagocytosis of latex beads and bacteria (Parmley et al. 1986) indicates that these granule components are retained in mature granules in a cytochemically unreactive form. The cytochemical observation of a similar intragranular distribution and masking of anionic glycoconjugates and acid phosphatase (Hardin and Spicer 1971; Parmley et al. 1983b) together with the biochemical observation that glycosaminoglycans inhibit lysosomal enzymes (Avila and Convit 1976) has resulted in speculation that primary granule glycosaminoglycans function to inactivate some enzymes and to facilitate their storage in lysosomes.

The onset of secondary granule genesis appears to be the most consistent morphologic feature which identifies neutrophilic myelocytes. In humans, the granules vary from 0.1 - 0.3 μm in diameter, are moderately dense and may be elongated in appearance (Figure 14). Although in humans the granules overlap in size with primary granules, they appear to have less osmiophilia and less affinity for metal counterstains in morphologic preparations. The secondary granules of rabbit heterophils are distinctly smaller, 0.25 - 0.5 μm, than larger primary granules, which vary from 0.3 - 0.8 μm in diameter (Bainton and Farquhar 1966; Wetzel et al. 1967a) (Figure 17). In most species, secondary granules outnumber primary granules by a ratio of 3 or 4 to 1.

Secondary granules are most notably known for their content of glyco-proteins (Noseworthy et al. 1975), lactoferrin (Spitznagel et al. 1974; Pryzwansky et al. 1979), Vitamin B_{12} binding protein, NADPH oxidase, and cytochrome B (Borregaard et al. 1983; Borregaard and Tauber 1984). In rabbits, secondary granules demonstrate alkaline phosphatase staining (Wetzel et al. 1967b; Bainton and Farquhar 1968) (Figure 18), which is confined to cytoplasmic vesicles in humans (Borgers et al. 1978).

Lysozyme and collagenase have been reported in both primary and/or
secondary granules by various investigators (Robertson et al. 1972;
Bretz and Baggiolini 1974; Spitznagel et al. 1974; West et al. 1974;
Ohlsson et al. 1977). Secondary granules lack peroxidase (Dunn et al.
1968) (Figure 19), acid phosphatase (Wetzel et al. 1967b) (Figure 22),
elastase (Clark et al. 1980), acidic glycoproteins (Hardin and Spicer
1971), sulphated glycosaminoglycans (Parmley et al. 1983b) (Figure 24),
cations (Hardin et al. 1969) and cationic proteins (Dunn and Spicer
1969) found in primary granules.

Ultrastructural studies have identified intense matrix staining for
glycoprotein in secondary granules, which is not found in mature primary
granules (Parmley et al. 1980; Fittschen et al. 1983). Immature and
mature secondary granules can be distinguished by vicinal glycol
staining of granule membranes in the former but not the latter granules.

Tertiary granules are found in late neutrophils where they appear to be
actively synthesized (Murata and Spicer 1973). These granules are
smaller than primary and secondary granules varying from 0.1 - 0.2 μm in
diameter in humans. They are moderately electron-dense and stain for
acid phosphatase, aryl sulphatase, (Murata and Spicer 1973), and
sulphated glycosaminoglycans (Parmley et al. 1980, 1983b). Using
cytochemical stains an average of 3 to 4 granules per cell profile may
be identified although some morphologic studies have suggested a higher
frequency for this granule type (Scott and Horn 1970a). The Golgi
apparatus of late cells also stains for acid phosphatase and sulphated
glycosaminoglycans, and sulphate incorporation into the Golgi region and
eventually tertiary granules has been ultrastructurally demonstrated
using cytochemical (Figure 24) and radioautographic methods (Payne and
Ackerman 1977; Parmley et al. 1980).

A unique granule type has been identified in bovine neutrophils by
Gennaro et al. (1983). This moderately dense granule is larger than
primary and secondary granules (Figure 20) and contains both highly
cationic proteins and lactoferrin but lacks peroxidase, proteinases, and
vitamin B_{12} binding protein.

In addition to cytoplasmic granules, a variety of vesicles are present
in late neutrophils. Some of these contain hydrolases being transported
from the Golgi region to forming tertiary granules (Murata and Spicer
1973; Parmley et al. 1983b). Some vesicles appear derived from the
plasmalemma and may contain endocytosed material (Abrahamson and Fearon
1983) and/or NADH oxidase (Briggs et al. 1975). Other vesicles appear
to contain elastase-like enzymes (Clark et al. 1980). Alkaline
phosphatase, which strongly stains late neutrophils (Okun and Tanaka
1978), is confined to vesicles in human neutrophils (Spitznagel et al.
1974; Borgers et al. 1978) but can be localized in secondary granules of
rabbit heterophils (Wetzel et al. 1967b; Bainton and Farquhar 1968).
Lipid droplets may also be identified within the neutrophil cytoplasm
(Coimbra and Lopes-Vas 1971) (Table IV).

Marked interspecies differences have been observed for a variety of
granule and vesicle components. Myeloperoxidase is markedly decreased
in bovine (Gennaro et al. 1983) and goat (Rausch and Moore 1975)
neutrophils. Lysozyme is practically undetectable in Rhesus monkeys,
cats, cows, goats, sheep and hamsters (Rausch and Moore 1975). Alkaline
phosphatase is markedly reduced in Rhesus monkeys, cats, and mice
(Rausch and Moore 1975).

Eosinophilic leucocytes (Figures 12,25-35)
Eosinophilic leucocytes are derived from myeloblasts and can
be identified ultrastructurally by their content of crystalloid or
specific granules and their precursors. The nuclear and cytoplasmic
maturation of early, mid and late eosinophils generally resembles that
observed for neutrophils. Late eosinophils, however, usually only
develop two segmented nuclear lobes (Figures 12, 25). Late eosinophils
contain sparse endoplasmic reticulum, cytoplasmic ribosomes,
mitochondria and a moderate amount of glycogen which is not as prominent
as that seen in late neutrophils. Early eosinophils vary from 12 - 15
µm in diameter in marrow smears whereas late eosinophils vary from 10 -
12 µm in diameter. The size is somewhat smaller in sectioned tissue.

The earliest identifiable eosinophil granule varies from 0.3 - 0.7 µm in
diameter and has a homogeneous dense matrix which lacks characteristic
crystalloids (Figures 26,27) (Wetzel et al. 1967a). In Wright's stained
marrow preparations, these granules are most prominent at the earliest
stage of development and appear somewhat basophilic. Ultrastructurally,
these immature granules demonstrate an affinity for cationic dyes,
including high iron diamine (HID), indicating the presence of sulphated
glycoconjugates (Figures 28-30) (Parmley et al. 1982). The removal of
some stained material with chondroitinase suggests the presence of at
least some chondroitin sulphate. In addition to increased sulphate
staining, immature granules also demonstrate strong, diastase resistant,
PA-TCH-SP staining of glycoprotein (Parmley et al. 1982) (Figures 31-
33). Pyroantimonate also stains cations in immature but not mature
eosinophil granules (Hardin and Spicer 1970). Like neutrophils,
immature eosinophil granules demonstrate acid phosphatase staining
(Table III, Figure 34) (Wetzel et al. 1967b). This staining is masked
in mature granules but can be unmasked during phagocytosis (Bass et al.
1981). Both mature and immature eosinophil granules stain intensely
with diaminobenzidine indicating the presence of peroxidase, which can
also be stained in the endoplasmic reticulum and Golgi apparatus of some
early cells (Dunn et al. 1968).

With maturation, granules become intensely eosinophilic as a result of
their content of basic protein (Dunn and Spicer 1969, Lewis et al. 1978)
(Table IV). Ultrastructurally, the basic protein takes on a distinct
periodic crystalline shape which varies in morphology among different
animal species (Figures 26,27,35). In man, guinea pigs, rats, mice,
dogs, mink, and monkeys, one to two distinctly angulated or trapezoidal
crystals are present in each granule (Miller et al. 1966; Hudson 1967;
Scott and Horn 1970b; Davis et al. 1971b; Hung 1972; Hudson et al.
1976). In rabbits, the crystalloids are needle-like (Figure 27) and the

granule appears somewhat longer than in other species (Wetzel et al.
1967a; Bainton and Farquhar 1970). Cats appear to have an unusual
doughnut or cylindrical shaped crystalloid (Figure 35) (Ward et al.
1972; Presentey et al. 1980). Horse eosinophils lack distinct
crystalloids (Henderson et al. 1983). Unlike their immature precursor
granules, crystalloid granules in rabbits and humans lack stainable
sulphate, glycoprotein and cations (Parmley et al. 1982). Recent
studies using immunogold techniques indicate that crystalloid granules
contain catalase in addition to cathepsins and peroxidase, indicating
both lysosomal and peroxisomal activities not found in most other cell
types (Yokata et al. 1984).

Late eosinophils also produce a second granule type which varies from
0.1 - 0.3 µm in diameter. The granule is round and moderately dense.
Cytochemically, acid phosphatase activity distinct from that observed in
crystalloid granules is present (Parmley and Spicer 1974), whereas
peroxidase and sulphate staining are absent. Acid phosphatase can also
be localized in Golgi lamellae synthesizing these granules (Figure
34). The granules are first observed at the myelocyte stage of
development and generally number 5 or less in circulating eosinophil
profiles, but the number of these granules appears to dramatically
increase in tissue eosinophils. In addition to these acid phosphatase-
positive small granules, a variety of other microgranules and vesicles
with variable cytochemistry may be identified in morphologic
preparations (Ward et al. 1972; Schaefer et al. 1973). Studies to date
have not indicated that all of these structures are lysosomes, and
presumably many small granules or vesicles are involved in a number of
different metabolic activities.

Basophilic leucocytes (Figures 12, 36-40)
    Basophilic leucocytes develop from myeloblasts. The
morphology of nuclear and cytoplasmic maturation in early, mid and late
basophils is similar to that seen in neutrophils and eosinophils. Late
basophils contain an average of three nuclear lobes with sparse
endoplasmic reticulum and mitochondria and a moderate amount of
glycogen.

Both early and late basophils are identified by their characteristic
cytoplasmic granules which appear very basophilic in Wright's stained
preparations (Figure 36). Ultrastructurally, the granules vary from 0.3
- 1 µm in diameter. In humans (Zucker-Franklin 1967; Hastie and Chir
1974) and rabbits (Wetzel et al. 1967a) the granules are filled with
particulate material which often appears in a coiled thread pattern
(Figure 37). The granule content is more homogeneous in cats (Figure
38) (Ward et al. 1972), mice (Dvorak et al. 1982b) and guinea pigs
(Murata and Spicer 1974; Dvorak et al. 1972,1977, 1982a).

The affinity of basophils for cationic dyes is readily seen at the
ultrastructural level with dialyzed iron (Hardin and Spicer 1971) and
specific sulphate staining with HID (Parmley et al. 1979b) (Figures
39,40; Table IV). The latter staining presumably represents sulphated
glycosaminoglycans present within these granules (Orenstein et al.

1978). Using specific fluorescent dyes, histamine may also be demonstrated within basophils (Catini et al. 1977).

Basophil specific granules contain a variety of enzymes including proteases and esterases (Dvorak et al. 1977). Peroxidase activity has also been demonstrated within specific granules in some studies (Ackerman and Clark 1971a; Parmley et al. 1979b), however, its staining is variable and not as consistently demonstrable as that observed in other myeloid cells (Table III). Acid phosphatase may be detected in basophil specific granules using a modification of Gomori's method (Komiyama and Spicer 1974; Parmley et al. 1979b). Thus, basophil specific granules resemble, to some extent, eosinophil specific granules and neutrophil primary granules in their content of glycosaminoglycans, proteases, peroxidase and acid hydrolase reactivity.

A small nucleoid-containing granule averaging 0.1 μm in diameter has been observed in human basophils (Hastie and Chir, 1974), however, its functional and cytochemical properties have not been investigated. Cytoplasmic vesicles are also frequently observed in basophils (Ward et al. 1972). Ultrastructural studies suggest that at least some of these may be involved in transport of material from specific granules to the extracellular space in a process called piecemeal degranulation (Dvorak et al. 1976).

Monocytes (Figures 41-43)
          Promonocytes vary from 14 - 18 μm in diameter in marrow smears and 12 - 15 μm in sectioned tissue. The nucleus is indented with dispersed nuclear chromatin and 2 or 3 prominent nucleoli. The cytoplasm appears bluish-grey with sparse granulation in Wright's stained preparations. Ultrastructurally, the endoplasmic reticulum appears less dilated and the cytoplasmic granules smaller (0.05 - 0.20 μm in diameter) and more pleomorphic than in promyelocytes.

Monocytes vary from 11 - 14 μm in diameter in Wright's stained preparations (Figure 41) and average 10 μm in sectioned tissue. The folded nucleus contains more condensed nuclear chromatin and the nucleoli are not as readily identified as in promonocytes. The cytoplasm is moderate in amount and lacks basophilia. Sparse segments of endoplasmic reticulum and pleomorphic granules can be identified ultrastructurally (Figure 42). Mitochondria are more numerous (10-20 per cell profile) and appear less elongated and more vesiculated than neutrophil mitochondria. The cytoplasm contains sparse glycogen particles in contrast to neutrophils. Monocyte surfaces have numerous villous projections and stain for both acidic and neutral glycoconjugates using a variety of methods.

Monocyte granules appear azurophilic in Wright's stained preparations. They contain cytochemically demonstrable acid phosphatase, peroxidase (Figure 43), lysozyme, cathepsins, and non-specific esterase (Nichols and Bainton 1973; Sannes et al. 1979; Spicer et al. 1979; Ackerman and Douglas 1980; Beckstead et al. 1981) (Table III). Ultrastructurally, peroxidase and acid hydrolases can also be demonstrated in the endoplasmic reticulum, Golgi apparatus and forming granules of promonocytes (Figure 43) (Nichols et al. 1971; Nichols and Bainton 1973). Monocytes also produce peroxidase negative granules.

Monocyte granules contain both vicinal glycol containing glycoconjugates which can be stained with the PA-TCH-SP method (Spicer et al. 1979) and acidic glycoconjugates, which can be stained with dialyzed iron (Parmley et al. 1977b). In humans and rabbits the latter glycoconjugates appear to be predominantly carboxylated and lack sulphate groups since they rarely demonstrate HID staining and do not incorporate radiosulphate in comparable quantities to that observed in myeloid cells (Parmley et al. 1977b, 1980).

Macrophages (Figures 44,45)
Circulating monocytes migrate into all body tissues and transform into macrophages. These cells have a variety of morphologies and cytochemical properties depending upon the location and material engulfed or cleared by the macrophages (Figure 44; Table III). The majority of macrophages contain one nucleus with coarse nuclear chromatin and one or two nucleoli. Some macrophages and/or monocytes coalesce to form multinucleated giant cells.

Acid phosphatase is detectable in the lysosomes of all macrophages. Similarly a variety of proteases have been localized in macrophage granules (Sannes et al. 1977). Peroxidase is present in some macrophages and its synthesis can be induced by surface adhesion (Bodel et al. 1978). Peroxisomes may also be demonstrated in macrophages using alkaline diaminobenzidine staining (Eguchi et al. 1979). Vicinal glycol containing glycoconjugates can be localized in Golgi lamellae cytoplasmic vesicles and granules of most macrophages (Spicer et al. 1979). Sulphated glycoconjugates can be localized in cytoplasmic granules of some macrophages.

Ferritin particles are present in the cytosol as well as within membrane bound secondary lysosomes. Degraded ferritin or haemosiderin is also detectable within secondary lysosomes of many macrophages (Figure 45). Ferritin characteristically has a quaternary structure at high magnification and is capable of binding over a thousand ferric iron cations in contrast to transferrin which binds only two iron cations (Munro and Linder 1978). Both ferritin and transferrin iron can be stained with acid ferrocyanide (Parmley et al. 1978, 1979b), however, only the former is detectable as Prussian blue at the light microscope level (Table IV). Ultrastructurally, ferrocyanide reactive ferritin often appears as a cuboidal crystal (Figure 45), whereas transferrin is much smaller without a distinct crystalline shape.

Megakaryocytes and platelets (Figures 46-51)
The megakaryoblast varies from 6 - 24 μm in diameter, has a compact lobed nucleus and basophilic cytoplasm in Wright's stained marrow preparations (Levine 1982). Ultrastructurally, these cells have some demarcation membranes and a few cytoplasmic granules. The promegakaryoblast varies from 14 - 30 μm in diameter, has a horseshoe-

shaped, lobulated nucleus with a central cytoplasmic clear or
eosinophilic area. A moderate number of granules and demarcation
membranes can be identified with the electron microscope. The granular
megakaryocyte is one of the largest marrow cells ranging from 16 - 56 µm
in diameter with a multilobed nucleus and a more eosinophilic than
basophilic cytoplasm (Figure 46). Ultrastructurally, the cell contains
extensive demarcation membranes and abundant granules (Figure 47). The
granular megakaryocyte evolves into a mature megakaryocyte of similar
size but with a compact highly lobulated nucleus and an eosinophilic
cytoplasm. Distinct areas of platelet formation can be visualized.
During the transition from megakaryoblast to mature megakaryocyte the
nuclear ploidy changes from 2N to as much 32N in some cells (Levine
1982).

Platelets vary from 2 - 4 µm in diameter, have moderately eosinophilic
fine granular cytoplasm and lack nuclear material. There is significant
heterogeneity in platelet density, volume, and ultrastructure. Less
dense, small platelets, with fewer granules may represent older
platelets (Corash et al. 1977).

At least two granule types, dense bodies and alpha granules, may be
identified within platelets (Figure 48) and granular megakaryocytes
(White and Gerrard 1976). The dense bodies containing serotonin and
adenine nucleotides vary from 0.1 - 0.2 µm in diameter and appear
extremely dense even in unstained, unosmicated preparations (Figure
48). The dense core material is surrounded by a peripheral lucent
zone. The dense material presumably represents calcium and stains
intensely with pyroantimonate (Spicer et al. 1969). The alpha granule
also varies from 0.1 - 0.2 µm in diameter and can be identified by its
moderately dense content with an eccentric nucleoid in uranyl acetate-
lead citrate stained preparations (Figure 48). This alpha granule is a
lysosomal granule with acid phosphatase activity (Murata et al. 1973).
The eccentric nucleoid also contains acidic glycoconjugates, possibly
chondroitin sulphate, demonstrable with dialyzed iron (Spicer et al.
1969) and HID (Figure 49) staining. Cathepsin D has also been localized
in lysosomes distint from alpha granules (Sixma et al. 1985). In
addition to acid hydrolases, megakaryocytic cells also contain acetyl
cholinesterase (Jackson 1973) which allows identification and
differentiation of cells in this lineage from other blood cell types.

Megakaryocytes and platelets contain two distinct membrane systems, the
open canalicular system and the dense tubular system (Figures
48,50,51). The outer membrane is continuous with an open canalicular
system which penetrates the platelet interior. The plasma membranes are
rich in acidic glycoconjugates that react with cationic reagents such as
dialyzed iron and colloidal thorium (Behnke 1968; Spicer et al. 1969),
but lack high iron diamine staining for sulphate (Figure 49) (Parmley et
al. unpublished observation). The open canalicular system and platelet
surface contain abundant hexoses which stain with PA-TCH-SP and
Concanavalin A (Parmley et al. 1973). The surface of megakaryocytes
and platelets also express HLA antigens and specific megakaryocyte-
platelet antigens which may be stained with monoclonal antibodies

(Thurlow et al. 1985). Freeze fracture studies indicate that the open
canalicular system is fenestrated (White and Clawson 1980). Both
fibrinogen and factor VIII can be stained on platelet and megakaryocyte
surfaces and antibodies to surface glycoproteins have been used for
specific immunostaining. The dense tubular system of platelets and
megakaryocytes is not continuous with the surface and can be positively
stained with diaminobenzidine when tannic acid and/or formaldehyde is
incorporated into the fixation procedure (Figure 50) (Breton-Gorius and
Guichard 1972). The reactivity appears to be associated with
prostaglandin production, however, the oxidase involved in the reaction
has not been clearly demonstrated and does not appear to be
cyclooxygenase (Gerrard et al. 1976). The dense tubular system also
contains glucose-6-phosphatase activity as does the endoplasmic
reticulum of leucocytes (Nichols et al. 1984). Similarly, presumed
protein sulphydryl groups may be stained in the dense tubular system
(Figure 51) using the osmium zinc iodide reaction as Clark and Ackerman
(1971) have used to localize reactivity in the endoplasmic reticulum of
other blood cells.

The discoid shape of platelets appears dependent upon circumferentially
distributed microtubules. Platelets may also have randomly dispersed
microtubules. Both circumferential and random microtubules may mediate
internal transformation with aggregation (White 1983). Cooling of
platelets prior to fixation results in dissolution of microtubules and
should be avoided in morphological studies. Other structures within the
platelet include PAS positive glycogen particles, mitochondria,
cytoplasmic vesicles and a small collection of Golgi membranes.

Lymphocytes and Plasma Cells (Figures 52-62)

Lymphocytes vary widely in morphology depending upon the
degree of maturation and activation (Tokuyasu et al. 1968; Parmley et
al. 1977a). Immature and activated cells appear larger, 9 - 15 μm in
diameter, have dispersed nuclear chromatin with 2 or 3 micronucleoli,
and a moderate amount of cytoplasm (Figure 56,57). Ultrastructurally,
nucleoli in resting cells are small, segregated, and appear ring-shaped
with a central less dense nucleolar organizing region and surrounding
dense fibrillar component (Figures 53,54), whereas nucleoli in activated
cells are larger with dispersion of the nucleolar organizing regions and
fibrillar components (Figure 55). Cytoplasmic polyribosomes may be
abundant in activated cells, however, endoplasmic reticulum is not
usually prominent. Occasional scattered granules may been seen. Mature
or inactive cells appear smaller, 6 - 9 μm in diameter, with sparse
cytoplasm and condensed nuclear chromatin (Figures 52-54). The majority
of circulating lymphocytes are small lymphocytes. Although in humans 70
- 80% of circulating lymphocytes are T lymphocytes, T and B cells cannot
normally be distinguished using morphological criteria alone and require
specific identification with immunostaining methods (Table V).

A variety of membrane markers are available for distinguishing T and B
cells (Grossi and Greaves 1981). Subsets of B cells have binding sites
for IgM and complement as well as surface membrane immunoglobulin which
can be stained with immunofluorescent and immunoperoxidase methods at

the light and electron microscope level respectively. Also using monoclonal antibodies, Ia antigen and other B cell specific antigens may be demonstrated on the cell surface. T cells may be distinguished by their surface binding to sheep erythrocytes and their reactivity with a variety of monoclonal antibodies against T cell antigens which vary with specific lymphocyte subsets (Table V). Specifically, using monoclonal antibodies immature thymocytes, helper, suppressor and natural killer T lymphocytes may be identified (Table V). Similarly, immature T cells expressing T9 and T10 may be distinguished from mature thymocytes which lack these markers.

Cytoplasmic granules may be more prominent in some subsets of T cells. These granules contain acid phosphatase, alpha-napthyl esterase and beta-glucuronidase activity (Grossi and Greaves 1981). The granules always lack peroxidase and sudan black staining seen in myeloid cells (Tables III, IV). Large granular lymphocytes, which include natural killer cells, also demonstrate acid hydrolase and esterase activity (Grossi et al. 1982), however, the latter enzyme appears to be of the monocytic type rather than the lymphocytic type as evidenced by its inhibition with NaF. Many of these granules contain chondroitin and heparan sulphate which can be stained with HID at the ultrastructural level (Figure 58) (Parmley et al. 1985). These cells also appear to have surface glycosaminoglycans which distinguish them from other blood cells (Parmley et al. 1985).

The cytoplasmic matrix of pre B cells contains immunoglobulin heavy chains which may be stained with immunofluorescent and immunoperoxidase methods (Table V) (Cooper 1981). These cells can be identified within the marrow but are not found among circulating lymphocytes. The nuclear matrix of pre-thymocytes, thymocytes as well as some very early B cells may contain terminal deoxynucleotidyl transferase activity which can be stained with immunofluorescent and immunoperoxidase methods. These early cells are also not normally seen in the circulating blood.

B lymphocytes may transform and mature into immunoglobulin producing plasma cells (Figures 59-62). These cells vary from 9 - 14 μm in diameter contain an eccentric round nucleus with peripherally clumped chromatin and deeply basophilic cytoplasm. Ultrastructurally, the cytoplasm contains abundant dilated rough endoplasmic reticulum with a prominent perinuclear Golgi apparatus (Figure 60) (Movat and Fernando 1962). These secretory organelles are intensely involved in synthesis of immunoglobulin which can be stained with immunofluorescent and immunoperoxidase methods. Individual plasma cells may demonstrate IgM, IgG, IgA, IgE or IgD synthesis. Occasionally the synthesized protein crystallizes in cisternae of endoplasmic reticulum forming Russell bodies (Figures 61,62). In addition to cytoplasmic staining, surface staining for immunoglobulin is also present in approximately 50% of cells (Grossi and Greaves 1981).

Mast cells (Figures 63,64)
Mast cells are long lived cells (Padawer 1974) with a unique morphology and distribution which distinguishes them from basophils

TABLE V: Monoclonal Antibody Staining of Lymphoid Cells

Cell Type	Monoclonal Antibody[2]	Location of Stained Cells
T cell	OKT6; Leu 6 (CD1)[3]	70% of thymocytes; primarily cortical thymus
	OKT11; Leu 5 (CD2)	99% of T cells in blood and thymus; E rosette receptor
	OKT1; Leu 1 (CD5)	95% of thymocytes; > 95% blood T cells paracortical region of lymph node; periarteriolar sheath of spleen
	OKT3; Leu 4 (CD3)	80-95% E-rosette positive lymphocytes; 10% of thymocytes (70% of medullary thymocytes) paracortical region of lymph node; periarteriolar sheath of spleen
	OKT4; Leu 3 (CD4)	40-60% of peripheral blood lymphocytes; 95% of thymocyte helper/inducer antigen in cortical and medullary thymocytes and germinal centers of secondary follicles
	OKT8; Leu 2 (CD8)	20-40% of peripheral blood lymphocytes; 85-95% of thymocyte T-suppressor/cytotoxic antigen in cortical thymocytes; 20-30% medullary thymocytes and thymocytes in the spleen red pulp.
	OKT9	Transferrin receptor, immature thymocytes, activated T cells
	TAC (CD25)	Interleukin 2 receptor of activated T cells
	HNK1; Leu 7	Natural killer (NK) cells; 15% of peripheral blood mononuclear cells; cells rarely observed in thymus but frequent in germinal centres of secondary follicles
	TdT	Terminal deoxynucleotidyl transferase in immature thymocytes and lymphoid precursors; cortical thymocytes
B Cell	IgA_1; IgA_2	IgA subclasses, < 1% peripheral blood lymphocytes
	IgG	IgG heavy chain, 0-2% of peripheral blood lymphocytes
	IgM	IgM heavy chain, > 95% of peripheral blood B lymphocytes
	B1	Pre-B and all mature B cells
	J5, CALLA	Less than 5% normal bone marrow and fetal liver pre-B cells
	Ia	Pre-B and mature B cells; activated T cells; haematopoietic progenitors; monocytes; Langerhans cells

1 Data from reviews by Grossi and Greaves (1981); Hsu et al. (1983); Knowles et al, (1985).

2 OK monoclonal antibodies from Ortho Diagnostic, Leu and TAC antibodies from Becton Dickinson.

3 New nomenclature based on cluster of differentiation (CD) antigens from the 1st and 2nd workshops on the Human Leukocyte Differentiation Antigens (Bernard et al. 1984)

(Murata and Spicer 1974; Parmley et al. 1975). Mast cells vary from 9 -
16 μm in diameter, have a single nucleus with dispersed nuclear
chromatin, and abundant deeply basophilic granules. The cells are
predominantly distributed in parenchymal and connective tissue and are
not found in the circulation. To date, there is no definitive data
demonstrating maturation of mast cells into basophils. In the embryonic
rat Combs et al. (1965) have identified four stages of mast cell
development beginning with a lymphocyte-like cell. More recently,
Czarnetzki et al. (1984) have ultrastructurally demonstrated the
development of mast cells from monocytes or monocyte-like cells.

Mast cell granules vary from 0.3 - 0.8 μm in diameter (Figures 63,64).
The granules are generally smaller, rounder, more abundant and variable
than those seen in basophils (Murata and Spicer 1974; Parmley et al.
1975). Granule morphology varies with maturation (Combs 1966) and
activation (Dvorak et al. 1983). Mast cell granules have a strong
affinity for cationic dyes as a result of their content of heparin and
possibly other glycosaminoglycans (Uvnas 1974). The incorporation of
radiosulphate and removal of the sulphate staining with nitrous acid
further confirms the presence of N-sulphated heparin (Combs et al.
1965). Basic protein is also present within the granules (Uvnas 1974)
and can be histochemically stained (Combs et al. 1965). Proteases may
be stained in heterophagic granules of mast cells (Sannes and Spicer
1979). Komiyama and Spicer (1975a) have demonstrated acid phosphatase
within mast cell granules involved in endocytic activities. Endogenous
peroxidase can also be demonstrated in mast cell granules under specific
preparative conditions (Christie and Stoward 1978).

Osteoclasts (Figures 65,66)
Osteoclasts are giant cells, which may be derived from
monocytes or their precursors. They average 50 m in diameter and
contain between 10 and 20 round nuclei (Gothlin and Ericsson 1976; Lucht
1980) (Figure 65). Osteoclasts tend to be smaller in rodents than in
other mammals. The cytoplasm is more eosinophilic than basophilic. A
ruffled border containing villous projections of cytoplasm can sometimes
be identified at higher magnification using the light microscope. The
ruffled border interfaces with areas of bone resorption.

Ultrastructurally, the nuclei of osteoclasts appear round, have variable
diameters, with a moderate amount of dispersed chromatin, and 0 to 2
nucleoli. The nuclei are usually located at the cell pole opposite the
ruffled border. The cytoplasmic organelles are not distinctive and
include multiple mitochondria, a moderate amount of non-dilated rough
endoplasmic reticulum, a prominent Golgi apparatus, several lysosomes
varying from 0.05 - 0.3 μm in diameter, vesicles, vacuoles and residual
bodies. The lysosomes, residual bodies, and some vesicles and vacuoles
demonstrate acid phosphatase reactivity as well as staining for vicinal
glycol containing glycoconjugates (Gothlin and Ericsson 1976; Lucht
1980; Takagi et al. 1982). Residual bodies or secondary lysosomes may
also contain sulphated HID reactive glycoconjugates which presumably
represent bone and cartilage proteoglycans in various stages of
degradation (Takagi et al. 1982). Osteoclasts may contain several

centrioles. The ultrastructure may change dramatically with parathyroid hormone stimulation which increases bone resorption and is reflected by an increase in endocytic organelles and an increase in dense granules within mitchondria (Lucht 1980).

The ruffled border distinguishes osteoclasts from other cell types (Figure 66). Ultrastructurally cytoplasmic projections or folds, 0.1 - 0.3 μm in width and several micrometers in length, extend into areas of bone and calcified cartilage resorption. The projections are generally free of cytoplasmic organelles. Endocytic caveolae and vacuoles are abundant at the base of the villi. These structures may contain degraded collagen, proteoglycans or calcium apatite. Also some degradation of proteoglycans and collagen appears to occur prior to engulfment. Both HID stained proteoglycans and PA-TCH-SP stained collagen appear significantly altered in an area extending several micrometers from the ruffled border (Takagi et al. 1982). Degradation of matrix material near the ruffled border implies secretion of lysosomal enzymes into this area. The localization of acid phosphatase in this area (Lucht 1980) is consistent with this hypothesis.

FUNCTION OF BLOOD CELLS
Erythrocytes
The primary function of the erythrocyte is to transport oxygen to all body tissues. It also transports CO_2 out of the tissues. These functions are primarily accomplished through haemoglobin, which is over 100 times more effective in binding oxygen than other plasma proteins (Riggs 1965). Haemoglobin is retained within membrane-bound red cells in great quantities which would tremendously increase osmotic pressure if released into the plasma. Thus retaining haemoglobin in red cells allows the blood viscosity to remain relatively low.

In humans, each gram of fully saturated haemoglobin binds 1.34 ml of oxygen. In a normal environment the oxygen tension ranges from 100 mm Hg in arterial blood to 35 mm Hg in venous blood. The P_{50} is the point at which 50% oxygen saturation occurs and averages 27 mm Hg. Oxygen binding to and dissociation from haemoglobin is influenced by a variety of factors within the blood and the red cell (Thomas 1974). Oxygen affinity is increased at alkaline pH and reduced by acidity, a phenomenon known as the Bohr effect. Thus in relatively acidic venous blood more oxygen is released than in arterial blood. Phosphorylated compounds especially 2,3-diphosphoglycerate (DPG) generated by anaerobic glycolysis, will decrease the oxygen affinity by binding to deoxygenated haemoglobin (Benesch et al. 1968). Consequently, stored or banked blood, in which 2,3-DPG is decreased, has increased oxygen affinity and is less functional, in that oxygen is not readily released to hypoxic tissues. Similarly, foetal haemoglobin less readily binds 2,3-DPG and more readily binds oxygen. This facilitates oxygen transfer from maternal to foetal blood (Oski 1972).

The relatively small size of mammalian red cells compared with the larger nucleated cells in other vertebrates presumably facilitates transport through blood vessels. Furthermore, the smaller, non-nucleated cells, have greater utilizable surface area for gaseous exchange. The biconcave shape exhibited by red cells of many mammals further minimizes the haemoglobin-surface distance during gas exchange. The increased plasticity of non-nucleated biconcave cells also improves movement through blood vessels thereby increasing oxygen delivery (Bessis 1973).

The red cell membrane acts as a barrier against entry of a variety of solutes. However, monosaccharides readily cross the membrane and provide an energy source through anaerobic glycolysis. Generated ATP is utilized in the sodium-potassium pump which maintains a high potassium, low sodium interior in most species. Failure of the pump results in increased sodium and water retention, intracellular swelling, loss of the biconcave appearance and ultimately haemolysis.

Metabolic pathways also exist for protecting haemoglobin and the membrane from oxidative damage from a variety of environmental insults. When haemoglobin iron is oxidized (from ferrous to ferric iron) methemoglobin is formed. Haemoglobin is maintained in a reduced form by methemoglobin reductases (also known as NADH diphorase) and reducing agents such as glutathione, ascorbate, and nicotinamide-adenine dinucleotide (NADH) generated by anaerobic glycolysis (Embden-Myerhof pathway) (Jaffe 1964; Jacob and Jandl 1962). Oxidation can also result in formation of disulphide bridges which precipitate or cross link haemoglobin and/or membrane proteins. This is prevented by glutathione reductase which maintains glutathione in the reduced state and catalyzes the reduction of oxidized glutathione by nicotinamide-adenine dinucleotide phosphate (NADPH) (Beutler and Yeh 1963). This latter component is generated by aeorobic glycolysis through the pentose phosphate pathway. This pathway is, also known as the hexose monophosphate shunt, which normally accounts for 10% of glycolysis in erythrocytes, whereas anaerobic glycolysis accounts for the other 90% of glucose metabolism (Murphy 1960).

Neutrophilic leucocytes
Neutrophils are phagocytic cells whose major function is to seek out and destroy microorganisms, particularly bacteria. This requires both an ability to migrate to the location of the invading organism, and once there, to participate in the destruction and removal of the organism.

Chemotaxins attract neutrophils to sites of inflammation or bacterial invasion (Wilkinson 1982). The chemotaxins are produced or released by bacteria, other inflammatory cells, or damaged tissues involved in the inflammatory reaction. Specific examples of chemotaxins include cleavage fragments of C5a and metabolites from lipo-oxygenation of arachidonic acid such as leukotriene B_4 (Goetzl 1983). Synthetic chemotaxins such as formylmethionine (N-fMet-Leu-Phe or FMLP) are used in vitro to study neutrophil movement. Many of the synthetic chemotaxins resemble bacterial by-products.

Neutrophil surfaces contain a variety of specific receptors which bind these chemotaxins. Oligopeptide receptors, including the FMLP receptor, exist in low and high affinity forms resulting in a dose dependent response (Synderman and Pike 1984). An increase in FMLP receptors and a decrease in affinity is observed with secretagogues such as calcium ionophore A23187 and phorbal esters, which translocate granule membranes to the cell surface (Gallin and Seligmann 1984). Receptors for complement derived factors or anaphylatoxins such as C5a have not been well defined. LTB_4 similarly binds to a distinct subset of surface receptors eliciting chemotaxis in low concentrations, whereas high concentrations of LTB_4 result in lysosomal degranulation (Valone 1984). After chemotaxins bind to the neutrophil surface, they initiate a number of reactions that initially include transmethylation and alteration of membrane phospholipid with subsequent activation of phospholipase C, protein kinase C and cAMP dependent protein kinase pathways (Snyderman and Pike 1984). Polarization or orientation is one of the first morphologic changes observed in vitro as the neutrophil prepares to move toward a higher concentration of chemotaxin (Malech et al. 1977). Granules become located in the front of the neutrophil and the nucleus may be found at the rear of the cell. Pseudopodia are prominent at the leading edge and appear to facilitate cell movement. Ultrastructurally, actin-myosin filaments are prominent just beneath the surface of the leading edge. Cations, particularly Ca^{++}, accumulate in submembranous areas at the leading edge and may be visualized with pyroantiomonate at the ultrastructural level (Cramer and Gallin 1979). Microtubules radiating from a centriole maintain a skeletal framework responsible for cell shape. The actual movement of cells can be observed with a light microscope using a Boyden chamber, in which neutrophils in a glass chamber accumulate on a filter as they migrate toward a chemotaxin released on the other side of that filter.

In vivo neutrophil migration begins with margination from the circulation or attachment to the luminal surface of an endothelial cell. The neutrophils become firmly anchored on the endothelial surface and begin movement through endothelial junctions. The basement membrane becomes the next obstacle through which the neutrophil passes. This is possibly facilitated by release of enzymes from the neutrophil which create an opening in the basement membrane. The neutrophil then migrates through tissue to the site of inflammation. Once the neutrophil has migrated out of the blood it cannot return to the circulation. It remains in tissues for several hours, then dies and is finally cleared by macrophages.

After arrival at the site of inflammation, the neutrophil must attach to the microorganism and phagocytose it. This is facilitated by neutrophil surface receptors for C3b and the Fc portion of immunoglobulin molecules which are opsonins coating the bacteria (Unkeless and Wright 1984). Some bacteria attach to neutrophil surfaces without this opsonization. Once attachment occurs pseudopodia extend around the micro-organism and completely engulf it in a phagosome (Bainton 1973). A novel type of engulfment termed "coiling phagocytosis," in which cytoplasmic pseudopods coil around the bacterium has been observed by Horwitz (1984) with engulfment of Legionella pneumophila.

After engulfment, neutrophil lysosomes fuse with phagosomes to form phagolysosomes. Using different pH indicator dyes, Jensen and Bainton (1973) observed a fall in pH to 6.5 within 3 min and 4.0 within 15 min of phagocytosis in rat neutrophils, whereas Cech and Lehrer (1984) observed a pH of 7.4 after 30 min and 5.7 after 60 min in human neutrophils. Ultrastructurally, alkaline phosphatase reactive secondary granules of rabbit heterophils appear to fuse first with the phagosome followed by fusion with peroxidase- and acid hydrolase-containing primary granules (Bainton 1973). Thus, the drop in pH within phagolysosomes appears to correspond with sequential fusion of granule constituents.

Degranulation into the extracellular space may occur in vivo and in vitro. Cytochalasin B, f-Met-Leu-Phe, and phorbal-myristate acetate are capable of inducing secretion (White and Estensen 1974; Hoffstein and Weissman 1978; Bentwood and Henson 1980). Both ultrastructural and biochemical studies indicate sequential degranulation with release of secondary granules followed by primary granules (White and Estensen 1974; Bentwood and Hensen 1980).

A respiratory burst occurs during phagocytosis and generates oxidants which participate in bacterial killing (Fantone and Ward 1982; Babior 1984). Initially a membrane oxidase oxidizes NADPH or NADH to form an O_2 radical and NADP or NAD. The superoxide reacts with hydrogen ions spontaneously or with superoxide dismutase to form hydrogen peroxide. NADP and H_2O_2 further stimulate the hexose monophosphate shunt. Detoxification is accomplished by the following reactions involving reduced (GSH) and oxidized (GSSG) glutathione:

$$2GSH + H_2O_2 \xrightarrow{\text{glutathione peroxidase}} GSSG + H_2O$$

$$GSSG + NADPH \xrightarrow{\text{glutathione reductase}} 2GSH + NADP$$

Catalase may also contribute to detoxification by catalyzing divalent reduction of H_2O_2 to water and oxygen. Oxygen dependent killing occurs when myeloperoxidase catalyzes the formation of additional highly reactive oxidizing radicals from H_2O_2 and a halide such as chloride. Similarly, superoxide radicals can contribute to formation of hydroxyl or iron hydroxide radicals which are active in bacterial killing. The ability to generate superoxide anion and subsequently H_2O_2 can be evaluated at the light microscope level in viable neutrophils by observing a colorimetric change with reduction of nitroblue tetrazolium. NADH oxidase has been localized at the ultrastructural level on the plasmalemma and in phagocytic vacuoles using cerous ions to precipitate H_2O_2 formed in the enzymatic reaction (Briggs et al. 1975; Ohno et al. 1982).

Bacterial killing may also occur as a result of anaerobic mechanisms (Klebanoff 1975). Lysozyme destroys muramic acid in bacterial cell walls. Cationic proteins and lactoferrin also contribute to killing.

However, many of the enzymes within phagosomes contribute primarily in completing digestion of dead bacteria.

Eosinophilic leucocytes

Eosinophil migration and phagocytosis have many similarities to that observed for neutrophils. Although eosinophils phagocytose and kill bacteria, they do so less avidly than neutrophils (Baehner and Johnson 1971) and appear to have little importance in bacterial infection. On the other hand, eosinophils play a significant role in metazoan infections (McLaren 1982). They also often appear in great numbers at sites which accumulate antigen-antibody complexes (Litt 1964) or, in sites of allergic reaction, particularly after mast cell degranulation (Goetzl 1976).

Eosinophils are particularly effective in attacking larvae in helminth infections. This generally requires that the host be immunized against the larvae and that the larvae are coated with IgG and/or IgE. The attachment of eosinophils to the larvae appears to be mediated by Fc and complement receptors. The larvae cannot be phagocytosed, however, degranulation of eosinophil lysosomes on the larva surface does occur and may result in substantial damage to the larvae, although ultrastructural observations fail to demonstate significant morphologic disruption of the parasite membrane (McLaren et al. 1977). The helminth is killed after 18 - 24 hr exposure to the degranulated eosinophil content which remains adherent to the parasite surface (Caulfield et al. 1985). Eosinophils may also be induced to degranulate in vitro with an ionophore as a result of calcium dependent activation of phospholipase A2 and production of arachidonic acid metabolites (Henderson et al. 1983). Eosinophils lack lysozyme but contain more peroxidase, acid hydrolases, basic protein, and phospholipase than do neutrophils (Goetzl 1976). The basic protein appears to be particularly larvocidal for such organisms as Schistosoma and Trichinella. The injury sustained by larvae after attack by eosinophils may kill the organism or facilitate the organism's destruction by other phagocytic cells.

Both a large crystalloid containing lysosome and a small non-crystalloid containing lysosome with distinct acid phosphatase activities can be seen in eosinophils (Parmley and Spicer 1974). The crystalloid granules are readily involved in macroendocytosis of bacteria during which acid phosphatase activity is unmasked (Bass et al. 1981). Eosinophil small granules are involved in microendocytic uptake of particulate material such as colloidal gold (Komiyama and Spicer 1975b).

Eosinophils frequently migrate into areas of allergic inflammation where mast cells and basophils degranulate releasing among other products, a specific eosinophil chemotactic factor (Goetzl 1976). Once immobilized in these sites eosinophils presumably participate in degradation of a variety of toxic mediators such as histamine, slow reacting substance of anaphylaxis (leukotrienes) and platelet activating factor. This is accomplished utilizing the eosinophil granule constituents histaminase, aryl sulfatase B, and phospholipase (Goetzl 1976; Zeiger and Colten 1977).

Basophilic leucocytes and mast cells
    Although basophils and mast cells differ in morphology,
distribution, and life span and probably origin, as outlined in previous
sections, their functions overlap significantly. This is related in
part to their content of histamine and basophilic sulphated
glycosaminoglycans, which in mast cells appears to be predominantly
heparin (Uvnas 1974). The number of mast cells and basophils vary
widely among species (Spicer 1963). Basophils are rarely if ever
observed in rats whereas mast cells are abundant in connective tissues
and body fluids of these animals. Rabbits have readily identifiable
circulating basophils but few, if any tissue mast cells. Other species
such as humans and guinea pigs have relatively moderate numbers of
basophils and mast cells.

Both mast cells and basophils have surface receptors for IgE (Sullivan
et al. 1971; Dvorak et al. 1983). An allergic reaction is initiated
when an antigen bridges two IgE molecules. Degranulation ensues and can
progress to a systemic anaphylactic reaction. In vitro degranulation
has been studied in detail and can be induced by antigens, Concanavalin
A, compound 48/80 and the calcium ionophore A 23187 (Kessler and Kuhn
1975; Lichtenstein 1975; Dvorak et al 1982a, 1983). The reactions
appear to be mediated by calcium influx and/or cyclic AMP. A variety of
inflammatory mediators are released by these cells and include
histamine, serotonin, heparin, chondroitin sulphate, leukotrienes or
slow reacting substances of anaphylaxis, platelet activating factor,
proteases and acid hydrolases (Sampson and Archer 1978; Dvorak et al.
1983). These substances promote smooth muscle contraction, increased
vascular permeability, anticoagulation, platelet aggregation and
chemotaxis of other leucocytes into the inflammed area (Lewis et al.
1975; Dvorak et al. 1977, 1983). Although the physiologic benefit of
such a fulminant reaction is not apparent, a more confined reaction may
prove beneficial to the host in promoting influx of other leucocytes
needed to overcome a specific host insult. Following degranulation,
mast cells synthesize new granules, whereas basophils do not (Dvorak et
al. 1983).

Basophils and mast cells also appear in delayed hypersensitivity
reactions, where their functions are not clearly understood. Dvorak et
al. (1976) have proposed a more refrained type of basophil degranulation
in allergic contact dermatitis, termed "piecemeal degranulation". This
reaction presumably involves transport of granule content to the cell
surface by cytoplasmic vesicles. The reaction proceeds over several
hours with 10% of basophils degranulating by 24 hr and more than 60% by
72 hr.

Basophils and mast cells are capable of endocytosis. Macroendocytosis
of red cells by mast cells has been described by Spicer et al. (1975).
Ultrastructural studies have also demonstrated endocytosis of thorium
dioxide by mast cells and retention of this material for as long as 10
months (Padawer 1969, 1974). Collidal gold is microendocytosed into
small vesicles and transported to specific mast cell granules with
subsequent unmasking of acid phosphatase within these granules (Komiyama

and Spicer 1975a). Similarly, circulating basophils are capable of
endocytosing horseradish peroxidase (Dvorak et al. 1972). Both the
involvement of granules in endocytic activities as well as their content
of acid hydrolases (Komiyama and Spicer 1974, 1975a; Parmley et al.
1979b), proteases, esterases (Dvorak et al. 1977) and peroxidase
(Ackerman and Clark 1971; Christie and Stoward 1978) identify the
granules as specialized lysosomes rather than just secretory granules.
Furthermore, these granules are involved in lysosomal enlargement in the
Chediak-Higashi syndrome similar to that seen for lysosomes in other
myeloid cells (Spicer et al. 1981).

Monocyte-macrophages

Monocytes and their progeny are phagocytic cells that engulf
and degrade microorganisms, effete cells, cell debris and even some
tumour cells. They also secrete substances which regulate immune
responses and myelopoiesis (Cline et al. 1978).

Monocytes and macrophages phagocytose and destroy microorganisms
utilizing aerobic and anaerobic pathways similar to but less efficient
than that described for neutrophils. Activated or primed macrophages
phagocytose and kill microorganisms more rapidly than do non-activated
cells (Carr and Daems 1980; Johnston and Kitagawa 1985). Alveolar
macrophages appear to more selectively utilize mitochondria and aerobic
metabolism as an energy source (Simon et al. 1977). Morphologically,
mitochondria are more prominent in monocytes and macrophages, compared
to neutrophils. Macrophages have more granules and cytoplasm than
monocytes, consequently, the phagocytic capability of macrophages is
greater than monocytes. Macrophages also can synthesize a variety of
different lysosomal components depending upon the need or extent of
activation, unlike neutrophils. The granules may contain acid
phosphatase, peroxidase, lysozyme, proteases and esterases. Macrophage
granules may also contain enzymes such as dipeptidyl aminopeptidase II
not found in monocytes (Sannes et al. 1977). The transformation of
monocytes to various types of macrophages occurs within the tissues and
is associated with the development of a variety of heterophagic granules
or secondary lysosomes (Figure 44). Primary lysosomes in macrophages
may also be exocytosed, whereas secondary lysosomes or residual bodies
are retained even after treatment with cation ionophores (Spicer et al.
1979). The enzyme content of resident macrophages appears to vary
specifically depending upon tissue site and function (Carr and Daems
1980). For example, osteoclasts appear to secrete material which
results in degradation of calcified bone and cartilage at the ruffled
border (Takagi et al. 1982) and splenic macrophages are more involved in
red cell turnover and iron metabolism.

Macrophages appear to engulf a variety of antigens which are degraded or
processed and then presented to lymphoid cells for initiation of cell
mediated or humoral responses (Cline et al. 1978). Antigens presented
in this way are apparently much more immunogenic than unprocessed
antigens.

Macrophages secrete interleukin 1, a substance that stimulates T lymphocyte production of T cell growth factor or interleukin 2 (Cline et al. 1978). Prostaglandins such as PGE_2, are also elaborated by macrophages and appear to inhibit activation of lymphocytes. Monocytes and macrophages also produce colony stimulating factor, a glycoprotein hormone which stimulates neutrophil production. Similarly interferon, an antiviral compound, is released by macrophages (Cline et al. 1978).

T lymphocytes secrete a variety of substances, or lymphokines, which enhance macrophage killing of microorganisms in intracellular phagocytic vacuoles (Cline et al. 1978; Rocklin et al. 1980). These substances also promote macrophage destruction of tumour cells through direct cytolytic surface contact which can be independent or specific for tumour antigens. Lymphokines accomplish these functions through macrophage activation which results in transformation of the resting macrophage to an activated macrophage with increased cytoplasm, cytoplasmic granules, and cytoplasmic projections, as well as increased metabolic activities (i.e. hexose monophosphate shunt).

Macrophages lining blood vessels or sinuses remove old cells with modified surface or decreased surface charge as occurs with effete red cells (Crosby 1959). Similarly, antibody coated cells modified by viral infection or malignant transformation are attacked and removed by macrophages (Carr and Daems 1980).

A variety of toxic substances are also removed by macrophages and retained or degraded in residual bodies. These substances include asbestos fibres, carbon particles, and lead. Some substances, including iron are both inactivated and stored by macrophages. Haemosiderin is present in many macrophages of iron-replete animals and virtually all macrophages of iron-loaded animals. The ability to store iron protects parenchymal cells from free radical damage which occurs when iron storage or binding capabilities are exceeded.

Platelets
A cascade of platelet and coagulation reactions results in haemostasis. Platelets are particularly important in filling gaps in endothlial cells. These gaps occur as a result of injury, inflammation or develop spontaneously. The platelet responds sequentially with 1) a change in shape, 2) adhesion to the endothelial gap, 3) primary aggregation, 4) release of a variety of mediators, and 5) secondary aggregation.

Platelets transform from a disk to a spiked sphere shape without a change in volume prior to adhesion (Born 1970). This change may be induced by a variety of substances including collagen, thrombin, and adenosine diphosphate (ADP). Transmission electron microscopy demonstrates movement of microtubules towards the interior of the platelet, although contraction of platelets appears somewhat independent of the location of the microtubules (White 1983). Removal of calcium with EDTA does not prevent the change of shape although adhesion and aggregation

are inhibited. The change in platelet shape to a spiked sphere
presumably promotes adhesion.

Following the change of shape, platelets adhere to a variety of substances including collagen, microfibrils, basement membrane components,
and glass. Adhesion to collagen is not calcium dependent, whereas
adhesion to glass and microfibrils does depend on calcium ions (Jamieson
1974). Factor VIII on the platelet surface, particularly the von
Willebrand portion, also appears necessary for proper adhesion (Weiss et
al. 1973).

Platelet aggregation in vivo occurs at the site of initial adhesion
(Hirsh and Doery 1971). In vitro both primary or reversible and
secondary or irreversible aggregation can be identified depending upon
the inducing agent. The latter phenomenon represents a second wave in
which additional platelet ADP is released promoting further
aggregation. The reactions are dependent on calcium and can be measured
in vitro using collagen, ADP, epinephrine, and thrombin.
The release reaction occurs as aggregation proceeds. The released
products include the dense granule contents: ADP, 5-hydroxytryptamine
(serotonin), and prostaglandin metabolites as well as alpha granule
enzymes such as acid phosphatase (Hirsh and Doery 1971). Potassium is
also released, although the platelet membrane appears relatively
intact. Ultrastructurally, the platelet plug filling the endothelial
gap appears degranulated and relatively free of organelles.

In addition to exocytosis, platelets are also capable of limited endocytosis. White (1968) has demonstrated incorporation of thorium
particles into platelet granules from plasma.

Lymphocytes
Lymphocytes may be divided into T and B classes which differ
in development and function although a significant overlap in morphology
exists (Kay et al. 1979). A minority of lymphocytes are null cells and
lack T or B function. These cells presumably are non-committed cells
capable of acquiring either T or B cell characteristics.

Several functional populations of T lymphocytes have been identified and
exert their effects by cell contact and/or secretion of chemical
mediators. Effector T cells are primarily involved in delayed hypersensitivity reactions. Helper T cells promote antibody production by B
lymphocytes and have Fc receptors (Moretta et al. 1977). Suppressor T
cells inhibit B cell antibody production and have surface receptors for
the Fc portion of IgG (Steinberg and Klassen 1977). Natural killer
cells are large granular lymphocytes which are capable of attacking
virally infected and tumour cells and exerting a cytolytic effect
without prior sensitization (Roder and Pross 1982; Carpen et al. 1982).

The subclassification and separation of lymphocyte subsets has been
greatly enhanced by the availability of monoclonal antibodies (Table V,
Hsu et al. 1983; Bernard et al. 1984; Knowles et al. 1985). Helper
cells react with the monoclonal antibodies T4 and Leu 3 and have the CD4

cluster of differentiation antigen, whereas suppressor cells are T8 and Leu 2 positive and have been assigned CD8 based on the new nomenclature for cluster of differentiation antigens (Bernard et al. 1984). Similarly, the extent of differentiation of thymocytes can be clarified using monoclonal antibodies with immature thymocytes expressing T9 and 10 antigens, not found in mature thymocytes.

Activated or sensitized T lymphocytes secrete a variety of chemical mediators or lymphokines (Rocklin et al. 1980). Lymphocytotoxin has a cytolytic effect on target cells and may be important in tumour immunity. Migration inhibition factor is also released by activated lymphocytes and results in accumulation of macrophages around micro-organisms and antigens. Transfer factor is another secretory product capable of transferring delayed hypersensitivity reactions to specific antigens. Activated lymphocytes are also capable of inducing non-specific blast transformation of other lymphocytes through secretion of lymphocyte transforming factor.

B lymphocytes are primarily responsible for humoral immunity and can be identified by immunoglobulin receptors on the membrane (Cooper 1981). These can be receptors for IgM, IgG, IgA, IgD or IgE. Fully mature B cells secrete specific antibody of one immunoglobulin class. The immunoglobulin molecule contains two identical light chains termed kappa or lambda, and two heavy chains designated gamma, alpha, mu, delta and epsilon for IgG, IgA, IgM, IgD and IgE respectively (Spiegelberg 1974). The antigen binding sites are in the N terminal end of the light and heavy chains, termed the variable region, which forms the idiotype unique to each clone of antibodies. Papain splits the immunoglobulin molecule into two Fab fragments containing one variable region each and an Fc fragment responsible for binding to cell surface Fc receptors. Serum IgM is a polymer of 5 immunoglobulin molecules connected by a polypeptide, the J chain. Secretory IgA on the other hand is a polymer which also contains a secretory piece produced by epithelial cells that secrete IgA.

IgG is the most abundant and smallest immunoglobulin with a molecular weight of 160,000, a sedimentation coefficient of 7S, and a biologic half life of 23 days. At least four subclasses may be identified. IgG_1 and IgG_3 are capable of activating complement, whereas the other subclasses fail to do so or do so less actively.

Although IgA is the second most abundant immunoglobulin in the serum, its activity appears most critical in its secretory form and location. It is predominantly synthesized by plasma cells in the submucosa and is subsequently processed and secreted by mucoid epithelial cells and salivary glands into the intestinal tract. Here it is capable of neutralizing a variety of microorganisms.

IgM is the largest immunoglobulin with a molecular weight of 900,000, a sedimentatin coefficient of 19S, and a biologic half life of 5 days. It is the third most common antibody in the serum and is the first antibody synthesized after antigen challenge. It fixes complement and may facilitate production of IgG.

IgE is the predominant antibody involved in allergic reactions and is found on the surface of basophils and mast cells (Sullivan et al. 1971). IgD is also a minor antibody, however its function is not fully understood.

SUMMARY AND CONCLUDING REMARKS
The development of new technologies has resulted in a better understanding of the variety and complexities of blood cells. Morphological, cytochemical and immunological studies have allowed identification of distinct blood cell types to which specific functions have been assigned. This review has attempted to provide an overview to correlate these observations for mammalian blood cells. Future studies and refinement will result in identification of additional blood cell subtypes and functions.

FIGURES
For transmission electron microscopy, morphological specimens were stained with uranyl acetate and lead citrate. Cytochemical preparations were not counterstained unless otherwise specified. Diaminobenzidine (DAB); High Iron Diamine (HID); Thiocarbohydrazide (TCH); Silver Proteinate (SP); Periodate (PA). All specimens were fixed in glutaraldehyde and post-fixed in OsO_4 unless otherwise specified. HID-TCH-SP and PA-TCH-SP stained specimens were not post-fixed in OsO_4.

REFERENCES

Abrahamson, D.R. & Fearon, D.T. (1983). Endocytosis of the C3b receptor of complement within coated pits in human polymorphonuclear leukocytes and monocytes. Lab. Invest., 48, 162-8.

Ackerman, G.A. (1968). Ultrastructure and cytochemistry of the developing neutrophil. Lab. Invest., 290-302.

Ackerman, G.A. (1971). The human neutrophilic promyelocyte. Z. Zellforsch. 118, 467-81.

Ackerman, G.A. (1975). Surface differentiation of hemopoietic cells demonstrated ultrastructurally with cationized ferritin. Cell Tiss. Res., 159, 23-37.

Ackerman, G.A. & Clark, M.A. (1971a). Ultrastructural localization of peroxidase activity in human basophil leukocytes. Acta Haemat., 45, 280-4.

Ackerman, G.A. & Clark, M.A. (1971b). Ultrastructural localization of peroxidase activity in normal human bone marrow cells. Z. Zellforsch. 117, 463-75.

Ackerman, S.K. & Douglas, S.D. (1980). Monocytes. In The Reticuloendothelial System: A Comprehensive Treatise (Vol. I), ed. H. Friedman, M. Escobar, & S. Richard, pp. 297-327, chap. 7. New York: Plenum Press.

Amsel, S. & Dell, E.S. (1972). Bone formation by hemopoietic tissue: Separation of preosteoblast from hemopoietic stem cell function in the rat. Blood, 39, 267.

Avila, J.L. & Convit, J. (1976). Physiochemical characteristics of the glycosaminoglycan-lysosomal enzyme interaction in vitro. A model of control of leucocytic lysosomal activity. Biochem. J., 160, 129-36.

Babior, B.M. (1984). Oxidants from phagocytes: Agents of defense and destruction. Blood, 64, 959-66.

Baggiolini, M., Horisberger, U., Gennaro, R. & Dewald, B. (1985). Identification of three types of granules in neutrophils of ruminants. Lab. Invest. 52, 151-8.

Baehner, R.L. & Johnston, R.B., Jr. (1971). Metabolic and bactericidal activities of human eosinophils. Br. J. Haematol. 20, 277-85.

Bainton, D.F. & Farquhar, M.G. (1966). Origin of granules in polymorphonuclear leukocytes. Two types derived from opposite faces of the Golgi complex in developing granulocytes. J. Cell Biol., 28, 277-301.

Bainton, D.F. & Farquhar, M.G. (1968). Differences in enzyme content of azurophil and specific granules of polymorphonuclear leukocytes. I. Histochemical staining of bone marrow smears. J. Cell Biol. 39, 286-98.

Bainton, D.F. & Farquhar, M.G. (1970). Segregation and packaging of granule enzymes in eosinophilic leukocytes. J. Cell Biol., 45, 54-73.

Bainton, D.F., Ullyot, J.L. & Farquhar, M.G. (1971). The development of of neutrophil polymorphonuclear leukocytes in human bone marrow. J. Exp. Med., 134, 907-34.

Bainton, D.F. (1973). Sequential degranulation of the two types of polymorphonuclear leukocyte granules during phagocytosis of microorganisms. J. Cell Biol., 58, 249-64.

Bass, D.A., Lewis, J.C., Szejda, P., Cowley, L. & McCall, C.E. (1981). Activation of lysosomal acid phoSphatase of eosinophil leukocytes. Lab. Invest., 44, 403-9.

Bearman, R.M., Levine, G.D. & Bensch, K.G. (1978). The ultrastructure of the normal human thymus: A study of 36 cases. Anat. Rec., 190, 755-82.

Beckstead, J.H., Halverson, P.S., Ries, C.A. & Bainton, D.F. (1981). Enzyme histochemistry and immunohistochemistry on biopsy specimens of pathologic human bone marrow. Blood, 57, 1088-98.

Behnke, 0. (1968). Electron microscopial observations on the surface coating of human blood platelets. J. Ultrastruct. Res., 24, 51-69.

Benesch, R., Benesch, R.E. & Yu, C.I. (1968). Reciprocal binding of oxygen and diphosphoglycerate by human hemoglobin. Proc. Natl. Acad. Sci. (USA), 59, 526-32.

Bentwood, B.J. & Henson, P.M. (1980). The sequential release of granule constituents from human neutrophils. J. Immunol. 124, 855-62.

Bernard, A., Boumsell, L., Dausset, J., Milstein, C. & Schlossman, S.F. (1984). In Leukocyte Typing. New York: Springer-Verlag.

Bessis, M. (1973). Red cell shapes: An illustrated classification and its rationale. In Red Cell Shape: Physiology Pathology Ultrastructure, ed. M. Bessis, R.I. Weed, P.F. LeBlond, pp. 1-25. New York: Springer-Verlag.

Bessis, M.C. & Breton-Gorius, J. (1959). Ferritin and ferruginous micelles in normal erythroblasts and hypochromic hypersideremic anemias. Blood, 14, 423-32.

Beutler, E. & Yeh, M.K.Y. (1963). Erythrocyte glutathione reductase. Blood, 21, 573-83.

Bodel, P.T., Nichols, B.A. & Bainton, D.F. (1978). Differences in peroxidase localization of rabbit peritoneal macrophages after surface adherence. Am. J. Pathol., 91, 107-18.

Boggs, D.R. (1967). The kinetics of neutrophilic leukocytes in health and in disease. Semin. Hematol., 4, 359-86.

Borgers, M., Thone, F., De Cree, J. & De Cock, W. (1978). Alkaline phosphatase activity in human polymorphonuclear leukocytes. Histochem. J., 10, 31-43.

Born, G.V.R. (1970). Observations on the change in shape of blood platelets brought about by adenosine diphosphate. J. Physiol., 209, 487-511.

Borregaard, N., Heiple, J.M., Simons, E.R. & Clark, R.A. (1983). Subcellular localization of the b-cytochrome component of the human neutrophil microbicidal oxidase: Translocation during activation. J. Cell Biol., 97, 52-61.

Borregaard, N. & Tauber, A.I. (1984). Subcellular localization of
    the human neutrophil NADPH oxidase b-cytochrome and
    associated flavoprotein. J. Biol. Chem., 259, 47-52.
Brederoo, P. & Daems, W.T. (1977). A new type of primary granule in
    guinea pig heterophil granulocytes. Cell Biol. Internat.
    Reports, 1, 363-8.
Brederoo, P., van der Meulen, J. & Mommaas-Kienhuis, A.M. (1983).
    Development of the granule population in neutro-
    phil granulocytes from human bone marrow. Cell Tissue
    Res., 234, 469-96.
Breton-Gorius, J. & Guichard, J. (1972). Ultrastructural locali-
    zation of peroxidase activity in human platelets and
    megakaryocytes. Am. J. Pathol., 66, 277-86.
Breton-Gorius, J. & Guichard, J. (1975). Fine structural locali-
    zation of peroxidase activity in human platelets
    and megakaryocytes. Am. J. Pathol., 79, 523-36.
Bretz, U. & Baggiolini, M. (1974). Biochemical and morphological
    characterization of azurophil and specific granules of
    human neutrophil polymorphonuclear leukocytes. J. Cell
    Biol., 63, 251-69.
Briggs, R.T., Drath, D.B., Karnovsky, M.L. & Karnovsky, M.J.
    (1975). Localization of NADH oxidase on the surface of
    human polymorphonuclear leukocytes by a new cytochemical
    method. J. Cell Biol., 67, 566-86.
Bruns, R.R. & Palade, G.E. (1968). Studies on blood capillaries.
    I. General organization of blood capillaries in
    muscle. J. Cell Biol., 37, 244-99.
Calamai, E.G. & Spitznagel, J.K. (1982). Characterization of rat
    polymorphonuclear leukocyte subcellular granules. Lab.
    Invest., 46, 597-604.
Campbell, F.R. (1972). Ultrastructural studies of transmural
    migration of blood cells in the bone marrow of rats, mice
    and guinea pigs. Am. J. Anat., 135, 521-36.
Carpen, O., Virtanen, I. & Saksela, E. (1982). Ultrastructure of
    human natural killer cells: Nature of the cytolytic
    contacts in relation to cellular secretion. J. Immunol.,
    128, 2691-7.
Carr, I. & Daems, W.T. (1980). The macrophage: A bird's-eye
    view. In The Reticuloendothelial System: A
    Comprehensive Treatise (Vol. I), ed. H. Friedman, M.
    Escobar & S. Richard, pp. 1-17. New York: Plenum Press.
Catini, C., Gheri, G. & Miliani A. (1977). Cytochemical detection
    of histamine in the human granulopoietic series of
    healthy subjects and of patients affected by chronic
    myeloid leukaemia. A spectrophofluorimetric test for
    checking OPT-induced fluorescence in isolated cells.
    Histochem. J., 9, 141-51.
Caulfield, J.P., Lenzi, H.L., Elsas, P. & Dessein, A.J. (1985).
    Ultrastructure of the attack of eosinophils stimulated by
    blood mononuclear cell products on schistosomula of
    Schistosoma mansoni. Am. J. Pathol., 120, 380-90.

Cech, P. & Lehrer, R.I. (1984). Phagolysosomal pH of human neutrophils. Blood, 63, 88-95.

Christie, K.N. & Stoward, P.J. (1978). Endogenous peroxidase in mast cells localized with a semipermeable membrane technique. Histochem. J., 10, 425-33.

Clark, M.A. & Ackerman, A.G. (1971). Osmium-Zinc iodide reactivity in human blood and bone marrow cells. Anat. Rec. 170, 81-96.

Clark, J.M., Vaughan, D.W., Aiken, B.M. & Kagan, H.M. (1980). Elastase-like enzymes in human neutrophils localized by ultrastructural cytochemistry. J. Cell Biol. 84, 102-19.

Clementi, F. & Palade, G.E. (1969). Intestinal capillaries. I. Permeability to peroxidase and ferritin. J. Cell Biol., 41, 33-58.

Cline, M.J. Lehrer, R.I., Territo, M.C. & Golde, D.W. (1978). Monocytes and macrophages: Functions and disease. Ann. Intern. Med., 88, 78-88.

Cochrane, C.G. (1984). Mechanisms coupling stimulation and function in leukocytes. Fed. Proc., 43, 2729-31.

Coimbra, A. & Lopes-Vas, A. (1971). The presence of lipid droplets and the absence of stable sudanophilia in osmium-fixed human leukocytes[1]. J. Histochem. Cytochem., 19, 551-7.

Combs, J.W. (1966). Maturation rat mast cells. An electron microscopic study. J. Cell Biol., 31, 563-75.

Combs, J.W., Lagunoff, D. & Benditt, E.P. (1965). Differentiation and proliferation of embryonic mast cells of the rat. J. Cell Biol., 25, 577-92.

Cooper, M.D. (1981). Pre-B cells: Normal and abnormal development. J. Clin. Immunol., 1, 81-9.

Corash, L., Tan, H. & Gralnick, H.R. (1977). Heterogeneity of human whole blood platelet subpopulations. I. Relationship between buoyant density, cell volume, and ultrastructure. Blood, 49, 71-87.

Cramer, E.B. & Gallin, J.I. (1979). Localization of submembranous cations to the leading end of human neutrophils during chemotaxis. J. Cell Biol., 82, 369-79.

Cronkite, E.P., Fliedner, T.M., Bond, V.P., & Robertson, J.S. (1959). Anatomic and physiologic facts and hypotheses about hemopoietic proliferating systems. In Kinetics of Cellular Proliferation. ed. F. Stohlman, Jr., pp. 1-18. New York: Grune & Stratton, Inc.

Crosby, W.H. (1959). Normal functions of the spleen relative to red blood cells: A review. Blood, 14, 399.

Czarnetzki, B.M., Figdor, C.G., Kolde, G., Vroom, T., Aalberse, R. & de Vries, J.E. (1984). Development of human connective tissue mast cells from purified blood monocytes. Immunology, 51, 549-54.

Davis, W.C., Spicer, S.S., Greene, W.B. & Padgett, G.A. (1971a).
          Ultrastructure of bone marrow granulocytes in normal mink
          and mink with the homologue of the Chediak-Higashi trait
          of humans. I. Origin of the abnormal granules present
          in the neutrophils of mink with the C-HS trait. Lab.
          Invest., 24, 303-17.
Davis, W.C., Spicer, S.S., Greene, W.B. & Padgett, G.A. (1971b).
          Ultrastructure of cells in bone marrow and peripheral
          blood of normal mink and mink with the homologue of the
          Chediak-Higashi trait of humans. II. Cytoplasmic
          granules in eosinphils, basophils, mononuclear cells and
          platelets. Am. J. Pathol., 63, 411-32.
De Bruyn, P. (1981). Structural substrates of bone marrow
          function. Semin. Hemat. 18, 179-93.
Delaune, E. (1939). Observations on the bovine blood picture in
          health and under parasitism. Proc. Soc. Exp. Biol. Med.,
          41, 482-3.
Donohue, D.M., Gabrio, B.W., Finch, C.A. (1958). Quantitative
          measurement of hematopoietic cells of the marrow. J.
          Clin. Invest., 37, 1564-70.
Dunn, W.B., Hardin, J.H. & Spicer, S.S. (1968). Ultrastructural
          localization of myeloperoxidase in human neutrophil and
          rabbit heterophil and eosinophil leukocytes. Blood, 32,
          935-44.
Dunn, W.B. & Spicer, S.S. (1969). Histochemical demonstration of
          sulfated mucosubstances and cationic proteins in human
          granulocytes and platelets. J. Histochem. Cytochem., 17,
          668-74.
Dvorak, A.M., Dvorak, H.F. & Karnovsky, M.J. (1972). Uptake of
          horseradish peroxidase by guinea pig basophilic
          leukocytes. Lab. Invest., 26, 27-39.
Dvorak, A.M., Mihm, M.C., Jr. & Dvorak, H.F. (1976). Degranulation
          of basophilic leukocytes in allergic contact dermatitis
          reactions in man. J. Immunol., 116, 687-95.
Dvorak, H.F., Orenstein, N.S., Dvorak, A.M., Hammon, M.E., Roblin,
          R.O., Feder, J., Schott, C.F., Goodwin, J. & Morgan, E.
          (1977). Isolation of the cytoplasmic granules of guinea
          pig basophilic leukocytes: Identification of esterase
          and protease activities. J. Immunol., 199, 38-46.
Dvorak, A.M., Galli, S.J., Morgan, E., Galli, A.S., Hammond, M.E. &
          Dvorak, H.F. (1982a). Anaphylactic degranulation of
          guinea pig basophilic leukocytes. Lab. Ivest., 46, 461-
          75.
Dvorak, A.M., Nabel, G., Pyne, K., Cantor, H., Dvorak, H.F. & Galli,
          S.J. (1982b). Ultrastructural identification of the
          mouse basophil. Blood, 59, 1279-85.
Dvorak, A.M., Galli, S.J., Schulman, E.S., Lichtenstein, L.M. &
          Dvorak, H.F. (1983). Basophil and mast cell degranul-
          ation: Ultrastructural analysis of mechanisms of mediator
          release. Fed. Proc., 42, 2510-5.

Eguchi, M., Sannes, P.L. & Spicer, S.S. (1979). Peroxisomes of rat peritoneal macrophages during phagocytosis. Am. J. Pathol., 95, 281-8.

Eikelenboom, P., Dijkstra, C.D., Boorsma, D.M. & van Rooijen, N. (1985). Characterization of lymphoid and nonlymphoid cells in the white pulp of the spleen using immunohisto-peroxidase techniques and enzyme-histochemistry. Experientia, 41, 209-15.

Eylar, E.H., Madoff, M.A., Brody, O.V. & Oncley, J.L. (1962). The contribution of sialic acid to the surface charge of the erythrocyte. J. Biol. Chem., 237, 1992-2000.

Fantone, J.C. & Ward, P.A. (1982). Role of oxygen-derived free radicals and metabolites in leukocyte-dependent inflammatory reactions. Am. J. Ped., 107, 397-418.

Fittschen, C., Parmley, R.T., Austin, R.L. & Crist, W.M. (1983). Vicinal glycol-staining identifies secondary granules in human normal and Chediak-Higashi neutrophils. Anat. Rec., 205, 301-11.

Fujita, T., Kashimura, M. & Adachi, K. (1985). Scanning electron microscopy and terminal circulation. Experientia, 41, 167-79.

Gallin, J.I. & Seligmann, B.E. (1984). Neutrophil chemoattractant fMet-Leu-Phe receptor expression and ionic events following activation. In Contemporary Topics in Immunobiology, ed. R. Synderman, pp. 83-103, vol. 14, chap. 3. New York: Plenum Press.

Gennaro, R., Dewald, B., Horisberger, U., Gubler, H.U. & Baggiolini, M. (1983). A novel type of cytoplasmic granule in bovine neutrophils. J. Cell Biol., 96, 1651-61.

Gerrard, J.M., White, J.G., Rao, G.H.R. & Townsend, D. (1976). Localization of platelet prostaglandin production in the platelet dense tubular system. Am. J. Pathol., 83, 283-94.

Gilmour, J.R. (1941). Normal hematopoiesis in intra-uterine and neonatal life. J. Pathol. Bact., 52, 25-55.

Goetzl, E.J. (1976). Modulation of human eosinophil polymorpho-nuclear leukocyte migration and function. Am. J. Pathol., 85, 419-35.

Goetzl, E.J. (1983). Leukocyte recognition and metabolism of leukotrienes. Fed. Proc., 42, 3128-31.

Gothlin, G. & Ericsson, J.L.E. (1976). The osteoclast. Review of ultrastructure, origin, and structure-function relationship. Clin. Orthop., 120, 201-31.

Grossi, C.E., Cadoni, A., Zicca, A., Leprini, A. & Ferrarini, M. (1982). Large granular lymphocytes in human peripheral blood: Ultrastructural and cytochemical characterization of the granules. Blood, 59, 277-83.

Grossi, C.E. & Greaves, M.P. (1981). Normal lymphocytes. In Atlas of Blood Cells. Function and Pathology, ed. F.D. Zucker, M.F. Greaves, C.E. Grossi & A.M. Marmont, pp. 347-408. Philadephia, PA: Lea & Febiger.

Hardin, J.H., Spicer, S.S. & Greene, W.B. (1969). Ultrastructural localization of antimonate deposits in rabbit heterophil and human neutrophil leukocytes. Lab. Invest., 21, 214-24.

Hardin, J.H. & Spicer, S.S. (1970). An ultrastructural study of human eosinophil granules: Maturation stages and pyro-antimonate reactive cation. Am. J. Anat., 128, 283-310.

Hardin, J.H. & Spicer, S.S. (1971). Ultrastructural localization of dialyzed iron-reactive mucosubstance in rabbit hetero-phils, basophils, and eosinophils. J. Cell Biol., 48, 368-86.

Hartwig, H. & Hartwig, H.G. (1985). Structural characteristics of the mammalian spleen indicating storage and release of red blood cells. Aspects of evolutionary and environment demands. Experientia, 41, 159-63.

Hastie, R. & Chir, B. (1974). A study of the ultrastructure of human basophil leukocytes. Lab. Invest., 31, 223-31.

Henderson, W.R., Chi, E.Y., Jorg, A. & Klebanoff, S.J. (1983). Horse eosinophil degranulation induced by the ionophore A23187. Ultrastructure and role of phospholipase A_2. Am. J. Pathol., 111, 341-9.

Hirsh, J. & Doery, J.C.G. (1971). Platelet function in health and disease. Prog. Hematol., 7, 185-34.

Hoffstein, S. & Weissman, G. (1978). Microfilaments and micro-tubules in calcium ionophore-induced secretion of lyso-somal enzymes from human polymorphonuclear leukocytes. J. Cell Biol., 78, 769-81.

Horwitz, M. (1984). Phagocytosis of the Legionnaires' disease bacterium (Legionella pneumophila) occurs by a novel mechanisms: engulfment with a pseudopod coil. Cell, 36, 27-33.

Hsu, S.M., Cossman, J. & Jaffe, E.S. (1983). Lymphocyte subsets in normal human lymphoid tissues. Am. J. Clin. Path., 80, 21-30.

Hudson, G., Cin, K.N. & Maxwell, M.H. (1976). Ultrastructure of Simian eosinophils. J. Anat., 122, 231-9.

Hudson, G. (1967). Eosinophil granules and uranyl acetate an electron microscope study of guinea-pig bone marrow. Exp. Cell Res., 46, 121-8.

Hung, K.S. (1972). Electron microscopic observations on eosinophil leukocyte granules in dog blood. Anat. Rec., 174, 165-74.

Ishizaka, T., Adachi, T., Chang, T.H. & Ishizaka, K. (1977). Development of mast cells in vitro. Biologic function of cultured mast cells. J. Immunol., 118, 211-7.

Jackson, C.W. (1973). Cholinesterase as a possible marker for early cells of the megakaryocytic series. Blood, 42, 413-21.

Jacob, H.S. & Jandl, J.H. (1962). Effects of sulfhydryl inhibition on red blood cells. I. Mechanism of hemolysis. J. Clin. Invest., 41, 779-92.

Jaffe, E.R. (1964). Metabolic processes involved in the formation
          and reduction of methemoglobin in human erythrocytes.
          In The Red Blood Cell, ed. C. Bishop & D.M. Surgnor, pp.
          397-422, chapt. 11. New York: Academic Press.
Jamieson, G.A. (1974). Interaction of platelets and collagen. In
          Platelets: Production, Function, Transfusion, and
          Storage, ed. M.G. Baldini & S. Ebbe, pp. 157-70. New
          York: Grune & Stratton, Inc.
Jensen, M.S. & Bainton, D.F. (1973). Temporal changes in pH within
          the phagocytic vacuole of the polymorphonuclear
          neutrophilic leukocyte. J. Cell Biol., 56, 379-88.
Johnston, R.B., Jr. & Kitagawa, S. (1985). Molecular basis for the
          enhanced respiratory burst of activated macrophages.
          Fed. Proc., 44, 2927-32.
Kay, H.E., Ackerman, S.R. & Douglas, S. (1979). Anatomy of the
          immune system. Semin. Hematol., 16, 252-82.
Kendall, M.D. (1981). The Thymus Gland. London: Academic Press.
Kessler, S. & Kuhn, C. (1975). Scanning electron microscopy of mast
          cell degranulation. Lab. Invest., 32, 71-7.
Kincade, P.W. & Phillips, R.A. (1985). Minisymposium summary: B
          lymphocyte development. Fed. Proc., 44, 2874-80.
Klebanoff, S.J. (1975). Antimicrobial mechanisms in neutrophilic
          polymorphonuclear leukocytes. Semin. Hemat., 12, 117-42.
Knisely, M.H. (1936). Spleen studies. I. Microscopic observations
          of the circulatory system of living unstimulated
          mammalian spleen. Anat. Rec., 65, 23-50.
Knowles, D.M., Bruke, J.S., Wang, J.M., Bonetti, F., Pellicci, P.G.,
          Flug, F., Dalia-Favera, R. & Wang, C.Y. (1985). SIg-E-
          ("Null-Cell") Non-Hodgkin's lymphomas: Multiparametric
          determination of their B- or T-Cell lineage. Am. J.
          Pathol., 120, 356-70.
Komiyama, A. & Spicer, S.S. (1974). Ultrastructural localization of
          a characteristic acid phosphatase in granules of rabbit
          basophils. J. Histochem. Cytochem., 22, 1092-104.
Komiyama, A. & Spicer, S.S. (1975a). Acid phosphatase demonstrated
          ultrastructurally in mast cell granules altered by
          pinocytosis. Lab. Invest., 32, 485-91.
Komiyama, A. & Spicer, S.S. (1975b). Microendocytosis in eosino-
          philic leukocytes. J. Cell Biol., 64, 622-35.
Krivit, W. (1977). Overwhelming postsplenectomy infection. Am. J.
          Hematol., 2, 193-201.
Landsberg, J.W. (1940). The blood picture of normal cats. Folia
          Haematologica, 64, 169-73.
Laurie, G.W., Leblond, C.P. & Martin, G.R. (1982). Localization of
          type IV collagen, laminin, heparan sulfate proteoglycan,
          and fibronectin to the basal lamina of basement
          membranes. J. Cell Biol., 95, 340-4.
Leary, A.G. & Ogawa, M. (1984). Identification of pure and mixed
          basophil colonies in culture of human peripheral blood
          and marrow cells. Blood, 64, 78-83.

Levine, R.F. (1982). The origin, development and regulation of magakaryocytes. Br. J. Haematol., 52, 173-80.

Lewis, R.A., Goetzl, E.J., Wasserman, S.I., Valone, F.H., Rubin, R.H. & Austen, K.F. (1975). The release of four mediators of immediate hypersensitivity from human leukemic basophils. J. Immunol., 114, 87-92.

Lewis, D.M., Lewis, J.C., Loegering, D.A. & Gleich, G.J. (1978). Localization of the guinea pig eosinophil major basic protein to the core of the granule. J. Cell Biol., 77, 702-13.

Lichtenstein, L.M. (1975). The mechanism of basophil histamine release induced by antigen and by the calcium ionophore A23187. J. Immunol., 114, 1692-9.

Litt, M. (1964). Eosinophils and antigen-antibody reactions. Ann. N.Y. Acad. Sci., 116, 964-85.

Lucht, U. (1980). Osteoclasts ultrastructure and functions. In The Reticuloendothelial System: A Comprehensive Treatise (Vol. I) Morphology, ed. H. Friedman, M. Escobar & S.M. Richard, pp. 705-34. New York: Plenum Press.

Malech, H.L., Root, R.K. & Gallin, J.I. (1977). Structural analysis of human neutrophil migration. Centriole, microtubule, and microfilament orientation and function during chemotaxis. J. Cell Biol., 75, 666-93.

Marikovsky, Y. & Danon, D. (1969). Electron microscope analysis of young and old red blood cells stained with colloidal iron for surface charge evaluation. J. Cell Biol., 43, 1-7.

Mason, D.Y., Farrell, C. & Taylor, C.R. (1975). The detection of intracellular antigens in human leucocytes by immuno-peroxidase staining. Br. J. Haematol., 31, 361-70.

Mayerson, H.S. (1930). The blood cytology of dogs. Anat. Rec., 47, 239-50.

Mazia, D. (1974). The cell cycle. What happens in the living cell between the time it is born in the division of another cell and the time it divides agains? New methods of investigation reveal four phases in the cycle. Sci. Am., 230, 55-64.

McLaren, D.J. (1982). The role of eosinophils in tropical disease. Semin. Hematol., 19, 100-6.

McLaren, D.J., Mackenzie, C.D. & Ramalho-Pinto, F.J. (1977). Ultra-structural observations on the in vitro interaction between rat eosinophils and some parasitic helminths (Schistosoma mansoni, Trichinella spiralis and Nippostrongylus brasiliensis). Clin. Exp. Immunol., 30, 105-18.

Mechanik, N. (1926). Untersuchungen uber das Gewicht des knochenmarkes des menschen. Z. Anat. Ent., 79, 58-99.

Miller, F., De Harven, E. & Palade, G.E. (1966). The structure of eosinophil leukocyte granules in rodents and in man. J. Cell Biol., 31, 349-62.

Moretta, L., Webb, S.R., Grossi, C.E., Lydyard, P.M. & Cooper, M.D. (1977). Functional analysis of two human T-cell sub-populations: Help and suppression of B-cell responses by T cells bearing receptors for IgM or IgG. J. Exp. Med., 146, 184-200.

Movat, H.Z. & Fernando, N.V.P. (1962). The fine structure of connective tissue. II. The plasma cell. Exp. Mol. Pathol., 1, 535-53.

Munro, H.N. & Linder, M.C. (1978). Ferritin: Structure, biosynthesis, and role in iron metabolism. Am. Physiological Soc., 58, 317-96.

Murata, F., Hardin, J.H. & Spicer, S.S. (1973). Coexistence of acid phosphatase and acid mucosubstance in the nucleoid of human blood platelet granules. Histochemie, 35, 319-29.

Murata, F. & Spicer, S.S. (1973). Morphologic and cytochemical studies of rabbit heterophilic leukocytes. Evidence for tertiary granules. Lab. Invest., 29, 65-72.

Murata, F. & Spicer, S.S. (1974). Ultrastructural comparison of basophilic leukocytes and mast cells in the guinea pig. Am. J. Anat., 139, 335-51.

Murphy, J.R. (1960). Erythrocyte metabolism. I. The equilibration of glucose-C^{14} between serum and erythrocytes. J. Lab. & Clin. Med., 55, 281-5.

Nichols, B.A., Bainton, D.F. & Farquhar, M.G. (1971). Differentiation of monocytes. Origin, nature, and fate of their azurophil granules. J. Cell Biol., 50, 498.

Nichols, B.A. & Bainton, D.F. (1973). Differentiation of human monocytes in bone marrow and blood. Lab. Invest., 29, 27-40.

Nichols, B.A., Setzer, P.Y., & Bainton, D.F. (1984). Glucose-6-phosphatase as a cytochemical marker of endoplasmic reticulum in human leukocytes and platelets. J. Histochem. Cytochem., 32, 165-71.

Nirmalan, G., Nair, S.G. & Simon, K.J., (1967). Hematology of Indian elephant (Elephas maximus). Can. J. Physiol. Pharmacol., 45, 985-91.

Noseworthy, J., Smith, G.H., Himmelhock, S.R. & Evans, W.H. (1975). Protein and glycoprotein electrophoretic patterns of enriched fraction of primary and secondary granules from guinea pig, polymorphonuclear leucocytes. J. Cell Biol., 65, 577-86.

Odeberg, H., Olsson, I. & Venge, P. (1975). Cationic proteins of human granulocytes. IV. Esterase activity. Lab. Invest., 32, 86-90.

Ogawa, M., Parmley, R.T., Bank, H.L. & Spicer, S.S. (1976). Human marrow erythropoiesis in culture: I. Characterization of methylcellulose colony assay. Blood, 48, 407-17.

Ohlsson, K., Olsson, I. & Spitznagel, J.K. (1977). Localization of
    chymotrypsin-like cationic protein, collagenase and
    elastase in azurophil granules of human neutrophilic
    polymorphonuclear leukocytes. Z. Physiol. Chem., 358,
    361-6.
Ohno, Y., Hirai, K., Kanoh, T., Uchino, H. & Ogawa, K. (1982).
    Subcellular localization of hydrogen peroxide production
    in human polymorphonuclear leukocytes stimulated with
    lectins, phorbol myristate acetate, and digitonin: An
    electron microscopic study using $CeCl_3$. Blood, 60, 1195-
    1201.
Okun, D.B. & Tanaka, K.R. (1978). Leukocyte alkaline phosphatase.
    Am. J. Hematol., 4, 293-9.
Oliver, C. & Essner, E. (1975). Formation of anomalous lysosomes in
    monocytes, neutrophils, and eosinophils from bone marrow
    of mice with Chediak-Higashi syndrome. Lab. Invest., 32,
    17-27.
Olsson, I. (1969). The intracellular transport of glycosamino-
    glycans (mucopolysaccharides) in human leukocytes. Exp.
    Cell Res., 54, 318-25.
Orenstein, N.S., Galli, S.J., Dvorak, A.M., Silbert, J.E. & Dvorak,
    H.F. (1978). Sulfated glycosaminoglycans of guinea pig
    basophilic leukocytes. J. Immunol., 121, 586-92.
Oski, F.A. (1972). Fetal hemoglobin, the neonatal red cell, and 2,3
    diphosphoglycerate. Pediatr. Clin. North Am., 19, 907-
    17.
Padawer, J. (1969). Uptake of colloidal thorium dioxide by mast
    cells. J. Cell Biol., 40, 747-58.
Padawer, J. (1974). Mast cells: Extended lifespan and lack of
    granule turnover under normal in vivo conditions. Exp.
    Mol. Pathol., 20, 269-80.
Parmley, R.T., Martin, B.J. & Spicer, S.S. (1973). Staining of
    blood cell surfaces with a lectin-horseradish peroxidase
    method. J. Histochem. Cytochem., 21, 912-22.
Parmley, R.T. & Spicer, S.S. (1974). Cytochemical and ultra-
    structural identification of a small type granule in
    human late eosinophils. Lab. Invest., 30, 557-67.
Parmley, R.T., Spicer, S.S. & Wright, N.J. (1975). The ultra-
    structural identification of tissue basophils and mast
    cells in Hodgkin's disease. Lab. Invest., 32, 469-75.
Parmley, R.T., Dow, L.W. & Mauer, A.M. (1977a). Ultrastructural
    cell cycle-specific nuclear and nucleolar changes of
    human leukemic lymphoblasts. Cancer Res., 37, 4313-25.
Parmley, R.T., Spicer, S.S. & O'Dell, R.F. (1977b). Ultrastructural
    identification of acid complex carbohydrate in cyto-
    plasmic granules of normal and leukaemic human mono-
    cytes. Br. J. Haematol., 39, 33-9.
Parmley, R.T., Spicer, S.S. & Alvarez, C.J. (1978). Ultrastructural
    localization of nonheme cellular iron with ferro-
    cyanide. J. Histochem. Cytochem., 26, 729-41.

Parmley, R.T., Denys, F.R. & Alvarez, C.J. (1979a). Ferrocyanide enhancement of concanavalin A-ferritin and cationized ferritin staining blood cell surface glycoconjugates. Histochem. J., 11, 379-89.

Parmley, R.T., Ostroy, F., Gams, R.A. & DeLucas, L. (1979b). Ferrocyanide staining of transferrin and ferritin-conjugated antibody to transferrin. J. Histochem. Cytochem., 27, 681-5.

Parmley, R.T., Spicer, S.S., Komiyama, A., Dow, L.W. & Austin, R.L. (1979c). Ultrastructural cytochemistry of basophils in chronic myelocytic leukemia. Exp. Mol. Pathol., 30, 41-54.

Parmley, R.T., Eguchi, M., Spicer, S.S., Alvarez, C.J. & Austin, R.L. (1980). Ultrastructural cytochemistry and radioautography of complex carbohydrates in heterophil granulocytes from rabbit bone marrow. J. Histochem. Cytochem., 28, 1067-80.

Parmley, R.T., Takagi, M., Spicer, S.S., Thrasher, A. & Denys, F.R. (1982). Ultrastructural visualization of complex carbohydrates in eosinophilic leukocytes. Am. J. Anat., 165, 53-67.

Parmley, R.T., Hajdu, I. & Denys, F.R. (1983a). Ultrastructural localization of the transferrin receptor and transferrin on marrow cell surfaces. Br. J. Haematol., 54, 633-41.

Parmley, R.T., Hurst, R.E., Takagi, M., Spicer, S.S. & Austin, R.L. (1983b). Glycosaminoglycans in human neutrophils and leukemic myeloblasts: Ultrastructural, cytochemical immunologic, and biochemical characterization. Blood, 61, 257-66.

Parmley, R.T., Rahemtulla, F., Cooper, M.D. & Roden, L. (1985). Ultrastructural and biochemical characterization of glycosaminoglycans in HNK-1 positive large granular lymphocytes. Blood, 66, 20-5.

Parmley, R.T., Doran, T. Boyd, L., Gilbert, C. & White, D. (1986). Unmasking of lysosomal glycosaminoglycans in phagocytic rabbit heterophils. J. Histochem. Cytochem. 34, 109.

Payne, D.N. & Ackerman, G.A. (1977). Ultrastructural autoradiographic study of the uptake and intracellular localization of ^{35}S-sulfate by developing human neutrophils. Blood, 50, 841-56.

Petri, S. (1983). Morphologie und zahl der Blutkorperchen bei 7--ca. 30 g. schweren normalen weissen Laboratoriumsmausen. Acta. Pathol. Microbiol. Scan., 10, 159.

Ponder, E., Yeager, J.F., Charipper, H.A. (1928). Studies in comparative haematology: I. Camelidae. Q. J. Exp. Physiol., 19, 115-26.

Presentey, B., Jerushalmy, Z., Ben-Bassat, M. & Perk, K. (1980). Genesis, ultrastructure and cytochemical study of the cat eosinophil. Anat. Rec., 196, 119-27.

Pryzwansky, K.B., Rausch, P.G., Spitznagel, J.K. & Herion, J.C. (1979). Immunocytochemical distinction between primary and secondary granule formation in developing human neutrophils: correlations with Romanowsky stains. Blood, 53, 179-85.

Quesenberry, P. & Levitt. L. (1979). Medical progress: Hematopoietic stem cells. N. Engl. J. Med. 301, 755-60.

Rausch, P.G., & Moore, T.G. (1975). Granule enzymes of polymorphonuclear neutrophils: a phylogenetic comparison. Blood, 46, 913-9.

Reed, C.E. & Bennett, J.M. (1975). N-Acetyl- -glucosaminidase activity in normal and malignant leukocytes. J. Histochem. Cytochem. 23, 752-57.

Rice, W.G., Kinkade, J.H., Jr., & Parmley, R.T. (1986). High resolution of heterogeneity among human neutrophil granules. Physical, biochemical, and ultrastructural properties of isolated fractions. Blood (in press).

Riggs, A. (1965). Functional properties of hemoglobins. Physiol. Rev. 45, 619-73.

Robertson, P.B., Ryel, R.B., Taylor, R.E., Shyu, K.W., & Fullmer, H.M. (1972). Collagenase: Localization in polymorphonuclear leukocyte granules in the rabbit. Science, 177, 64-5.

Rocklin, R.E., Bendtzen, K. & Grieneder, P. (1980). Mediators of immunity: Lymphokines and monokines. Adv. Immunol., 29, 56-136.

Roder, J.C. & Pross, H.F. (1982). The biology of the human natural killer cell. J. Clin. Immunol., 2, 249-63.

Sampson, D. & Archer, G.T. (1967). Release of histamine from human basophils. Blood, 29, 722-36.

Sannes, P.L., McDonald, J.K. & Spicer, S.S. (1977). Dipeptidyl aminopeptidase II in rat peritoneal wash cells. Cytochemical localization and biochemical characterization. Lab. Invest., 37, 243-53.

Sannes, P.L. & Spicer, S.S. (1979). The heterophagic granules of mast cells. Dipeptidyl aminopeptidase II activity and resistance to exocytosis. Am. J. Pathol., 94, 447-58.

Scarborough, R.A. (1931). The blood picture of normal laboratory animals. Yale J. Biol. Med., 3, 63-547.

Schaefer, H.E., Hubner, G. & Fischer, R. (1973). Spezifische mikrogranula in eosinophilen. Acta. Haematol., 50, 92-104.

Scott, R.E. & Horn, R.G. (1970a). Ultrastructural aspects of neutrophil granulocyte development in humans. Lab Invest., 23, 202-15.

Scott, R.E. & Horn, R.G. (1970b). Fine structural features of eosinophil granulocyte development in human bone marrow. J. Ultrastruct. Res., 33, 16-28.

Seifert, M.F. & Marks, S.C., Jr. (1985). The regulation of hemopoiesis in the spleen. Experientia 41, 192-9.

Shannon, A.W. & Zellmer, D.M. (1982). Heterogeneity in polymorpho-nuclear leukocyte neutrophil granules. Histochem. J., 14, 847-50.

Sheehan, H.L. & Storey, G.W. (1947). An improved method of staining leucocyte granules with Sudan black B. J. Pathol. Bacteriol., 59, 336-37.

Simionescu, M., Simionescu, N., & Palade, G.E. (1976). Segmental differentiations of cell junctions in the vascular endothelium. Arteries and veins. J. Cell Biol., 68, 705-23.

Simionescu, N. & Simionescu, M. (1983). The cardiovascular system. In Histology: Cell & Tissue Biology, ed. L. Weiss, pp. 371-433, 5th ed. New York: Elsevier Biochemical.

Simon, L.M., Robin, E.D., Phillips, J.R., Acevedo, J., Axline, S.G. & Theodore, J. (1977). Enzymatic basis for bioenergetic differences of alveolar versus peritoneal macrophages and enzyme regulation by molecular O_2. J. Clin. Invest., 59, 443-8.

Sixma, J.J., van den Berg, A., Hasilik, A., von Figura, K. & Geuze, H.J. (1985). Immuno-electron microscopical demon-stration of lysosomes in human blood platelets and mega-karyocytes using anti-cathepsin D. Blood, 65, 1287-91.

Song, S.H., & Groom, A.C. (1971). Storage of blood cells in the spleen of the cat. Am. J. Physiol., 220, 779-84.

Spicer, S.S. (1963). Histochemical properties of mucopolysaccharide and basic protein in mast cells. Ann. N.Y. Acad. Science, 103, 322-33.

Spicer, S.S., Green, W.B., & Hardin, J.H. (1969). Ultrastructural localization of acid mucosubstance and antimonate-precipitable cation in human and rabbit platelets and megakaryocytes. J. Histochem. Cytochem., 17, 781-92.

Spicer, S.S., Simson, J.A.V. & Farrington, J.E. (1975). Mast cell phagocytosis of red blood cells. Am. J. Pathol., 80, 481-98.

Spicer, S.S., Sannes, P.L., Eguchi, M. & McKeever, P.E. (1979). Fine structural and cytochemical aspects of the development of macrophages and of their endocytic and secretory activity. J. Reticuloendothel. Soc., 26, 49-65.

Spicer, S.S., Sato, A., Vicent, R., Eguchi, M. & Poon, M.C. (1981). Lysosome enlargement in the Chediak-Higashi syndrome. Fed. Proc., 40, 1451-5.

Spiegelberg, H.L. (1974). Biological activities of immunoglobulin of different classes and subclasses. Adv. Immun., 19, 259-94.

Spitznagel, J.K., Dalldorf, F.G., Leffell, M.S., Folds, J.D. (1974). Character of azurophil and specific granules purified from human polymorphonuclear leukocytes. Lab. Invest., 30, 774-85.

Steinberg, A.D., & Klassen, L.W. (1977). Role of suppressor T cells in lymphopoietic disorders. Clin. Haematol., 6, 439-78.

Stolinski, C. & Brethnach, A.S. (1975). Freeze-Fracture Replication of Biology Tissues. pp. 74-80 and 132-5. London: Academic Press.

Sullivan, A.L., Grimley, P.M. & Metzger, H. (1971). Electron microscopic localization of immunoglobulin E on the surface membrane of human basophils. J. Exp. Med., 134, 1403-16.

Sullivan, A.L., Grasso, J.A. & Weintraub, L.R. (1976). Micro-pinocytosis of transferrin by developing red cells: An electron-microscopic study utilizing ferritin-conjugated transferrin and ferritin-conjugated antibodies to transferrin. Blood, 47, 133-143.

Synderman, R. & Pike, M.C. (1984). Transductional mechanisms of chemoattractant receptors on leukocytes. In Contemporary Topics in Immunobiology, ed. R. Synderman, pp. 1-24, vol. 14, chap. 1. New York: Plenum Press.

Takagi, M., Parmley, R.T., Toda, Y. & Denys, F.R. (1982). Extra-cellular and intracellular digestion of complex carbo-hydrates by osteoclasts. Lab. Invest., 46, 288-97.

Thomas, H.M. (1974). The oxyhemoglobin dissociation curve in health and disease; role of 2,3-diphosphoglycerate. Am. J. Med., 57, 331-48.

Thurlow, P.J., Kerrigan, L., Harris, R.A. & McKenzie, I.F.C. (1985). Analysis of human bone marrow with monoclonal antibodies. J. Histochem. Cytochem., 33, 1183-9.

Till, J.E., McCulloch, E.A. & Simnovitch, L. (1964). A stochastic model of stem cell proliferation based on the growth of spleen colony forming cells. Proc. Natl. Acad. Sci. (USA.), 51, 29-36.

Tokuyasu, K., Madden, S.C. & Zeldis, L.J. (1968). Fine structural alterations of interphase nuclei of lymphocytes stimu-lated to growth activity in vitro. J. Cell Biol., 39, 630-60.

Unkeless, J.C. & Wright, S.D. (1984). Structure and modulation of Fc and complement receptors. In Contemporary Topics in Immunobiology, ed. R. Synderman, pp. 171-82, vol. 14, chap. 6. New York: Plenum Press.

Uvnas, B. (1974). The isolation of secretory granules from mast cells. In Methods of Enzymology, 31: Biomembranes (Part A), ed. S. Fleischer, & L. Packer, Pp. 395-402. New York: Academic Press.

van Ewijk, W. & Nieuwenhuis, P. (1985). Compartments, domains and migration pathways of lymphoid cells in the splenic pulp. Experientia, 41, 199-208.

Valone, F.H. (1984). Regulation of human leukocyte function by lipoxygenase products of arachidonic acid. In Contemporary Topics in Immunobiology, ed. R. Synderman, pp. 155-8, vol. 14, chap. 5. New York: Plenum Press.

Ward, J.M., Wright, J.F. & Wharran, G.H. (1972). Ultrastructure of granulocytes in the peripheral blood of the cat. J. Ultrastruct. Res., 39, 389-96.

Weiss, H.J., Rogers, J. & Brand, H. (1973). Defective ristocetin--induced platelet aggregation in von Willebrand's disease and its correction by factor VIII. J. Clin. Invest., 52, 2697-2707.

Weiss, L. (1983). Lymphatic vessels and lymph nodes. In Histology: Cell & Tissue Biology, pp. 527-43. 5th ed. New York: Elsevier Biomedical.

Weiss, L. (1985). New trends in spleen research: Conclusion. Experientia, 41, 243-8.

Weiss, L. & Tavassoli, M. (1970). Anatomical hazards to the passage of erythrocytes through the spleen. Semin. Hematol., 7, 372-80.

Weiss, L., Powell, R. & Schiffman, F.J. (1985). Terminating arterial vessels in red pulp of human spleen: A transmission electron microscopic study. Experientia, 41, 233-42.

West, B.C., Rosenthal, A.S., Gelb, N.A., & Kimball, H.R. (1974). Separation and characterization ofhuman neutrophil granules. Am. J. Pathol., 77, 41-61.

Wetzel, B.K., Horn, R.G. & Spicer, S.S. (1967a). Fine structural studies in the development of heterophil eosinophil and basophil granulocytes in rabbits. Lab. Invest., 16, 349-81.

Wetzel, B.K., Spicer, S.S. & Horn, R.G. (1967b). Fine structural localization of acid and alkaline phosphatases in cells of rabbit blood and bone marrow. J. Histochem. Cytochem., 15, 311-34.

White, J.G. (1968). The transfer of thorium particles from plasma to platelets and platelet granules. Am. J. Pathol., 53, 567-75.

White, J.G. (1983). Ultrastructural physiology of platelets with randomly dispersed rather than circumferential band microtubules. Am. J. Pathol., 110, 55-63.

White, J.G. & Estensen, R.D. (1974). Selective labilization of specific granules in polymorphonuclear leukocytes by phorbol myristate acetate. Am. J. Pathol., 75, 45-60.

White, J.G. & Gerrard, J.M. (1976). Ultrastructural features of abnormal blood platelets. A review. Am. J. Pathol., 83, 590-14.

White, J.G. & Clawson, C.C. (1980). The surface-connected canalicular system of blood platelets--a fenestrated membranes system. Am. J. Pathol., 101, 353-9.

Wilkinson, P.C. (1982). Chemotaxis and Inflammation, 2nd ed. New York: Churchill-Livingston, Inc.

Wintrobe, M.M. (1933). Variations in the size and hemoglobin content of erythrocytes in the blood of various vertebrates. Folia Haematol., 51, 32-49.

Wintrobe, M.M. (1974). Clinical Hematology, ed. M.M. Wintrobe, pp. 1791-23.

Wintrobe, M.M., Schumacker, H.B. & Schmidt, W.J. (1936). Values for number, size and hemoglobin content of erythrocytes in normal dogs, rabbits and rats. Am. J. Physiol., $\underline{114}$, 502-7.

Yam, L.T., Li, C.Y. & Crosby, W.H. (1971). Cytochemical identification of monocytes and granulocytes. Am. J. Clin. Pathol., $\underline{55}$, 283-90.

Yamada, E. & Yamauchi, R. (1966). Some observations on the cytochemistry and morphogenesis of the granulocytes in the rat bone marrow as revealed by electron microscopy. Acta Haemat. Jap., $\underline{29}$, 530-41.

Yokota, S., Tsuji, H. & Kato, K. (1984). Localization of lysosomal and peroxisomal enzymes in the specific granules of rat intestinal eosinophil leukocytes revealed by immuno-electron microscopic techniques. J. Histochem. Cytochem., $\underline{32}$, 267-74.

Zeiger, R.S. & Colten, H.R. (1977). Histaminase release from human eosinophils. J. Immunol., $\underline{118}$, 540-3.

Zucker-Franklin, D. (1967). Electron microscopic study of human basophils. Blood, $\underline{29}$, 878-90.

Figure 1. Several hundred cells arising from a CFU-C are
seen by phase contrast microscopy in this human granulocyte
colony grown in soft agar. Bar = 100 µm.

Figure 2. This scanning electron micrograph illustrates a
compact erythrocyte colony arising from a CFU-E. Human
marrow cells cultured in methyl cellulose. Bar = 10 µm.

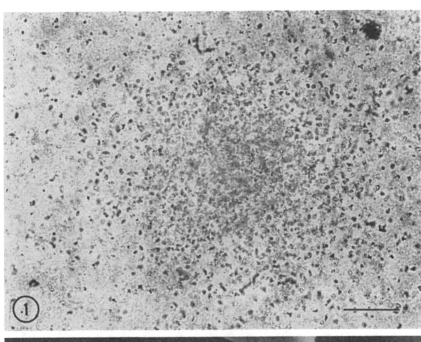

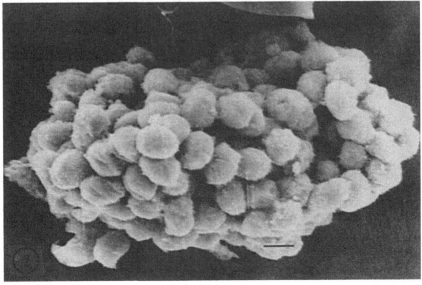

Figure 3. Cultured human erythroblasts at higher
magnification (cf Fig. 2). Bar = 10 μm.

Figure 4. Human proerythroblast containing abundant
ribosomes, sparse endoplasmic reticulum and a small Golgi
region. Bar = 1 μm.

Figure 5. Wright's stained human polychromatophilic (P) and
orthochromatophilic (O) erythroblasts. Bar = 10 μm.

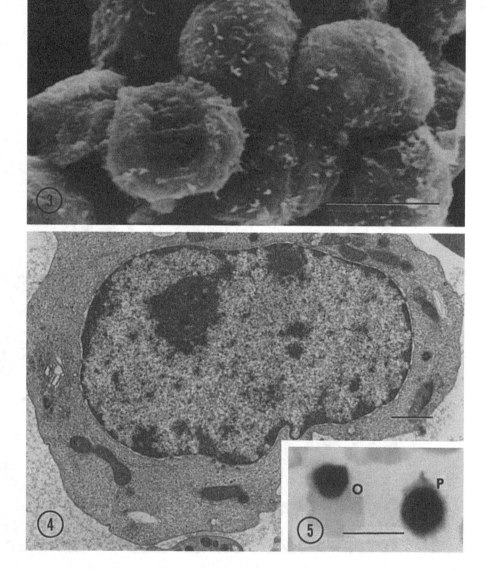

Figure 6. Human marrow erythroblasts surround this macrophage (M). Basophilic erythroblasts (B) contain dispersed nuclear chromatin and abundant polyribosomes. The nuclear chromatin is more condensed and fewer polyribosomes are evident in polychromatophilic erythroblasts (P). Several surface caveolae (arrows) are present and at higher magnification, these contain ferritin particles. Bar = 1 μm.

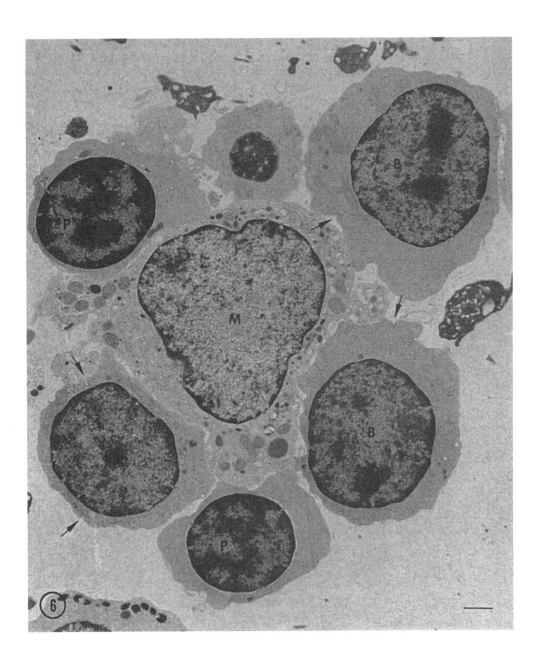

Figure 7. The surface of this human basophilic erythroblast
stains intensely for the transferrin receptor visualized by
sequential exposure to mouse monoclonal antibody to
transferrin receptor (OKT9) and horseradish peroxidase
conjugated goat antimouse antibody (Parmley et al. 1983a).
Two stained endocytic caveolae (E) and numerous intrinsic
ferritin particles are evident in the cytoplasm and on the
surface (arrows). Bar = 1 µm.

Figure 8. A polychromatophilic erythroblast stains for the
transferrin receptor (cf Fig. 7). Diffuse cytoplasmic heme
is DAB reactive. Cytoplasmic channels continuous with the
cell surface also stain (inset). Bar = 1 µm. Inset bar =
0.1 µm.

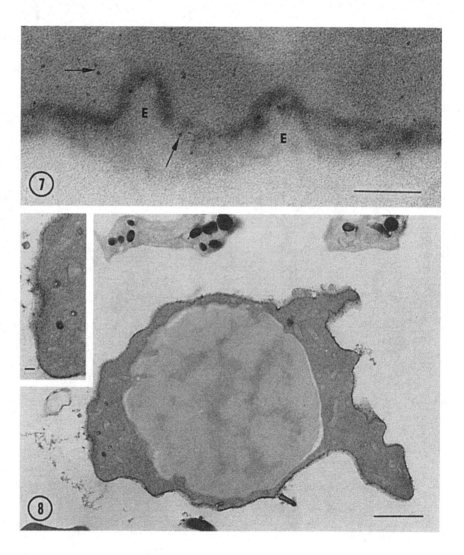

Figures 9-11. In this scanning electron micrograph of a
human erythrocyte (Fig. 9), only a few concanavalin A-
ferritin-ferrocyanide precipitates (arrows) localize hexoses
on the cell surface (Parmley et al. 1979a). In contrast to
human erythrocytes, rabbit erythrocytes contain abundant Con
A-ferritin-ferrocyanide reactive hexoses as seen by scanning
(Fig. 10) and transmission electron microscopy (Fig. 11).
Bar = 1 μm.

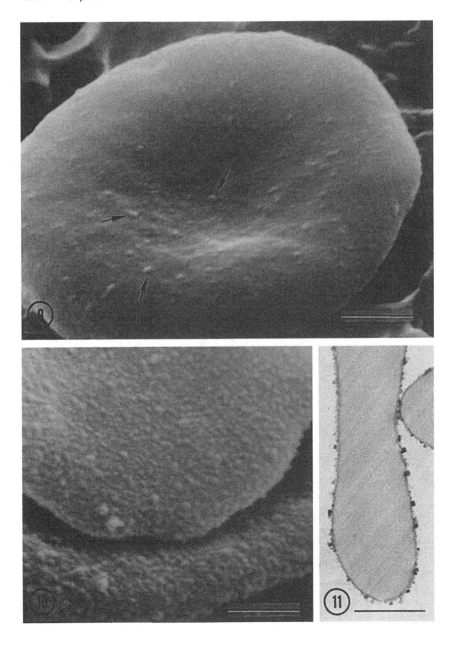

Figure 12. This schematic diagram illustrates development of early, mid and late myeloid cells in rabbit bone marrow. The myeloblast becomes committed to the heterophil, basophil or eosinophil series with synthesis of the respective unique granule types. Courtesy Dr. S.S. Spicer (Medical University of South Carolina, Charleston, S.C.).

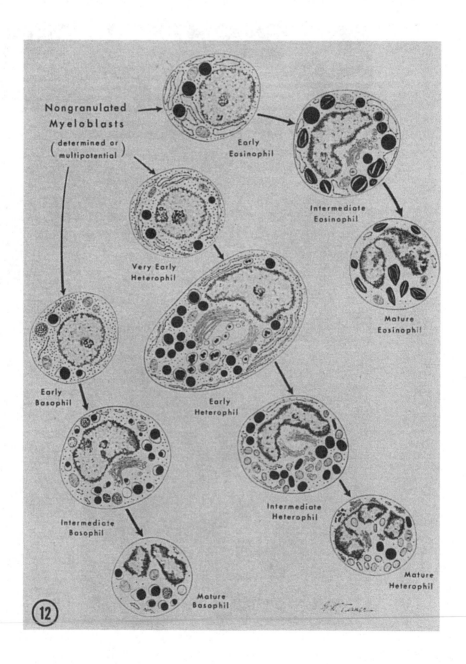

Figure 13. A human myelocyte and segmented neutrophil in a right's stained preparation. Bar = 10 μm.

Figure 14. A human marrow promyelocyte (below) contains dispersed nuclear chromatin, a moderate amount of rough endoplasmic reticulum and dense primary granules some of which contain lucent crystalloids (arrows). Distinct secondary granules are not evident in promyelocytes. The segmented neutrophil (above) contains two nuclear profiles and numerous primary (P) and smaller secondary (S) granules (enlarged in inset). Bar = 1 μm. Inset bar = 0.2 μm.

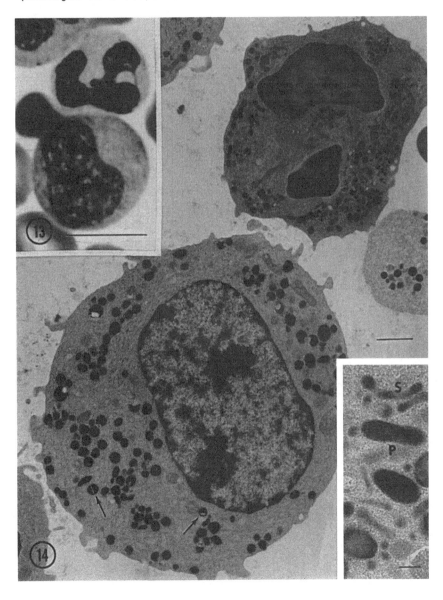

Figure 15. Dense particulate PA-TCH-SP staining is seen over glycogen doposits in this human segmented neutrophil. Nucleus (N). Bar = 1 μm.

Figure 16. After amylase digestion of en block specimens, PA-TCH-SP intensely stains glycoproteins in secondary granules (S) and Golgi lamellae (G, enlarged below). Mature primary granules (P) stain weakly. Nucleus (N). Bars = 1 μm.

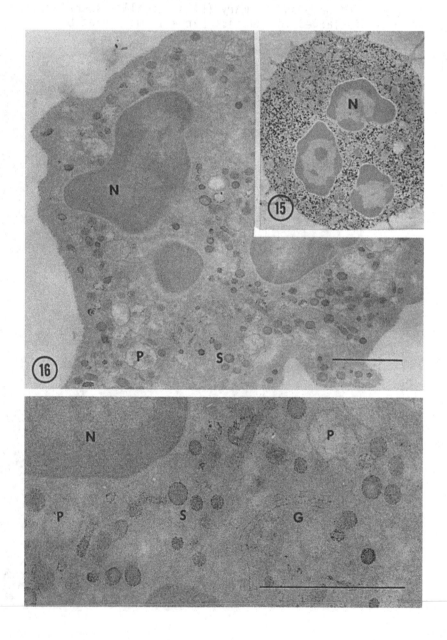

Figure 17. In late rabbit heterophils primary granules (P)
have variable density, frequently with a more intensely
stained periphery. Secondary granules (S) are smaller and
more homogeneous (enlarged in inset). Bar = 1 μm. Inset bar
= 0.2 μm.

Figure 18. Lead precipitate localizes alkaline phosphatase
reactivity in a crescent distribution within secondary
granules (S) of this rabbit late heterophil (enlarged in
inset). Primary granules (P) are unreactive, but demonstrate
variable peripheral osmiophilia. Lobes of nucleus (N). Bar
= 1 μm. Inset bar = 0.2 μm.

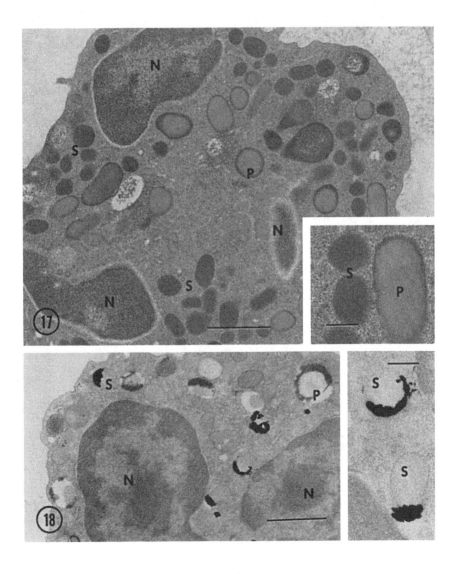

Figure 19. Human marrow promyelocyte (lower cell) containing
abundant DAB reactive peroxidase in endoplasmic reticulum
(ER), Golgi lamellae (G) and primary granules. The segmented
neutrophil (upper cell) contains peroxidase-positive primary
granules and peroxidase-negative secondary granules (arrows,
enlarged in inset). Bars = 1 μm.

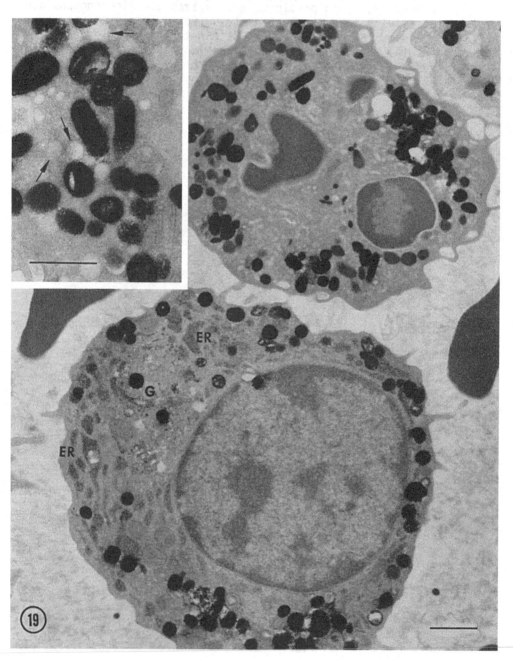

Figure 20. This cow blood neutrophil contains three nuclear lobes with at least three granules evident in this DAB stained preparation. The first granule type (G1) contains peroxidase activity and is smaller than the other granule types, unlike human neutrophils (cf Fig. 19) and rabbit heterophils (cf Fig. 17). A second granule type (G2) is minimally dense, whereas the third and largest granule type (G3) is electron lucent. Thin section counterstained with lead citrate. Bar = 1 μm.

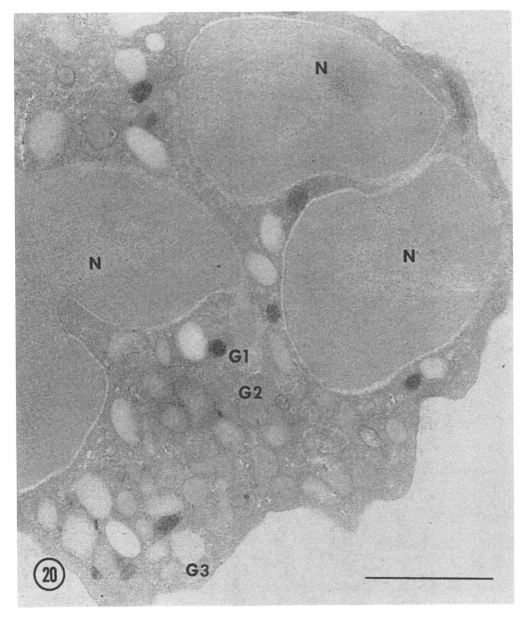

Figure 21. This human marrow promyelocyte contains acid phosphatase activity in Golgi lamellae (G) and immature primary granules (P1), whereas staining is absent or masked in mature primary granules (P2). Nucleus (N). Bar = 1 μm.

Figure 22. This human segmented neutrophil demonstrates acid phosphatase activity in Golgi lamellae (G), presumably involved in tertiary granule synthesis. Staining is masked in most, if not all, mature primary granules. Nucleus (N). Bar = 1 μm.

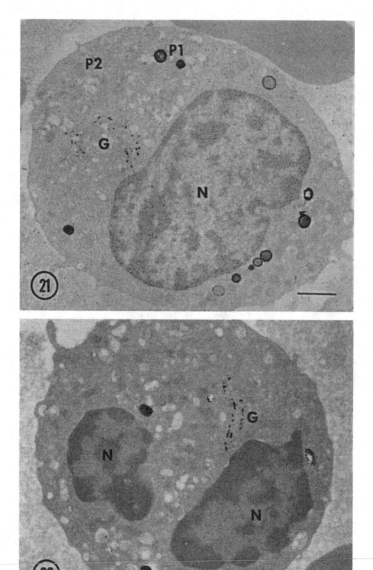

Figure 23. HID-TCH-SP localizes sulphated glycosaminoglycans in intensely stained immature primary granules (P1) in human promyelocytes (enlarged at right) whereas mature primary granules (P2) lack staining. Nucleus (N). Bars = 1 μm.

Figure 24. Tertiary granules (T), a presumed immature primary granule (P1) and some Golgi (G) associated vesicles (arrows, enlarged at right) demonstrate HID-TCH-SP staining in this human segmented neutrophil. Nucleus (N). Bars = 1 μm.

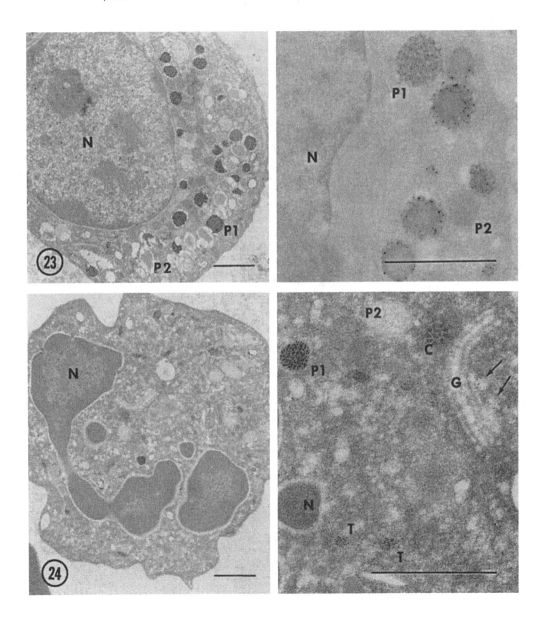

Figure 25. Wright's stained human blood eosinophil with two
nuclear lobes. Bar = 10 μm.

Figure 26. This human eosinophilic myelocyte contains
immature (non crystalloid containing) granules and mature
granules with distinct crystalloids (C). Bar = 1 μm.

Figure 27. Note the needle like crystalloids and larger
granule size in this rabbit eosinophilic myelocyte (enlarged
in inset). Immature granules lack crystalloids. Golgi
derived spherule (arrow). Bars = 1 μm.

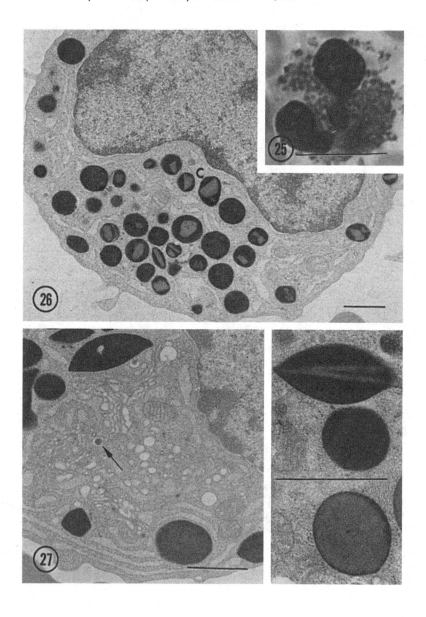

Figure 28. HID intensely stains sulphated glycoconjugates in immature granules (I) in this human eosinophilic myelocyte, however, mature crystalloid granules (C) lack staining. Bar = 1 μm.

Figures 29, 30. Rabbit eosinophilic myelocytes. HID stains sulphate in some Golgi derived vesicles and spherules (arrows) as well as immature granules (I) (Figure. 29). HID-TCH-SP stains the rim of these immature eosinohil granules (I) whereas mature crystalloid granules (C) lack staining (Fig. 30). Bars = 1 μm.

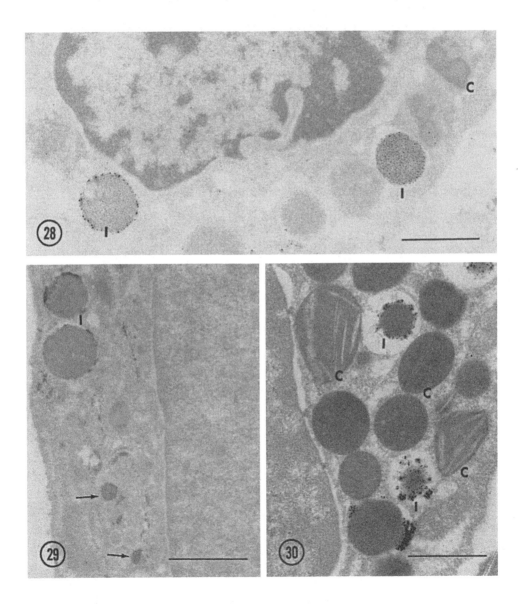

Figures 31, 32. PA-TCH-SP intensely stains vicinal glycol
containing glycoconjugates in Golgi vesicles (arrows, Fig.
31) of this human eosinophilic myelocyte after diastase
digestion of glycogen. Several stages of human eosinophil
granule genesis can be identified with PA-TCH-SP (Fig. 32).
The matrix and membrane of the most immature granules (T1)
stain intensely with a progressive decrease in staining noted
in maturing granules (T2, T3) and crystalloid granules (C).
Tubulovesicular structures (arrow) stain intensely. Bars =
1 μm and 0.1 μm.

Figure 33. This rat late eosinophil demonstrates particulate
PA-TCH-SP staining of glycogen and moderate staining of
numerous tubulovesicular structures (arrows). Crystalloid
granules (C). Bar = 1 μm.

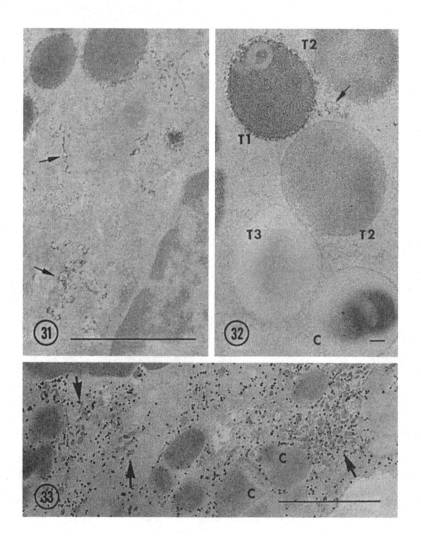

Figure 34. Acid phosphatase staining is intense in crystalloid granules (C), small granules (S) and Golgi vesicles (G) of this human blood eosinophil. Nucleus (N). Bar = 1 μm.

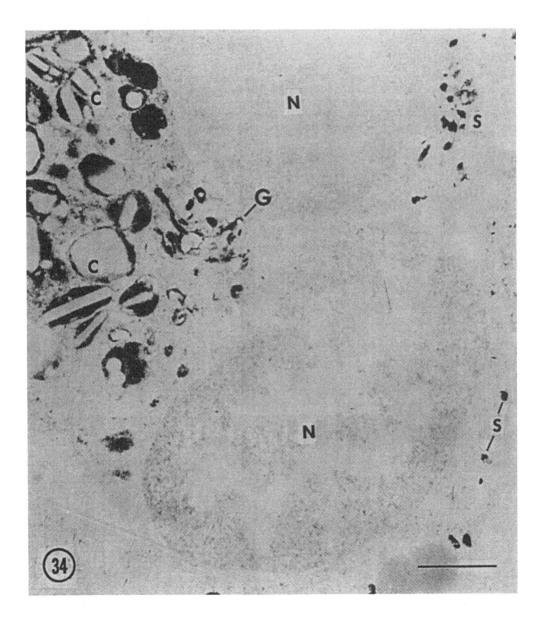

Figure 35. This cat blood eosinophil contains unique circular or cylindrical crystalloid profiles (C). Periodicity is evident in crystalloids at higher magnification (enlarged in inset). Tubulovesicular structures (T) are numerous. The crystalloid granules are much larger than primary granules (P) in the partially illustrated adjacent neutrophil. Bars = 1 μm.

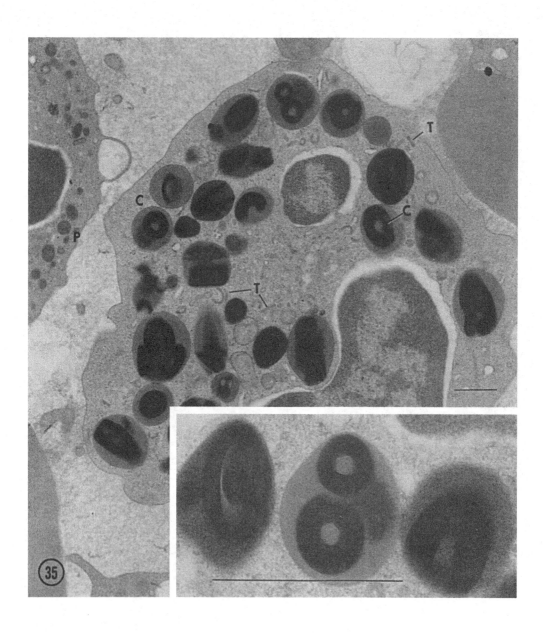

Figure 36. Intensely basophilic granules obscure the nuclear
lobes in this Wright's stained human blood basophil. Bar =
10 μm.

Figure 37. This human late basophil contains many specific
granules (G) with a coiled thread distribution of particulate
material (enlarged in inset). Some granules appear
extracted. The cytoplasm also contains numerous small
granular and/or tubulovesicular structures (arrow). Nucleus
(N). Bar = 1 μm. Bar in inset = 0.5 μm.

Figure 38. This cat blood basophil contains several specific
granules (G), however, they lack the coiled thread appearance
seen in human (cf Fig. 37) and rabbit specimens. Nucleus
(N). Bar = 1 μm.

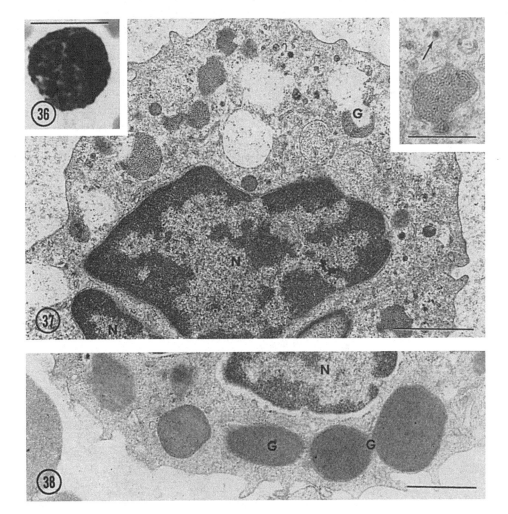

Figure 39. Dialyzed iron stains acidic glycoconjugates (carboxylated and sulphated) in specific granules of this human basophil. Staining is also evident on the plasmalemma and in mitochondrial membranes. The granule staining varies within some granules (enlarged in inset) possibly as a result of extraction of stained material or failure of en block stain penetration. The sialic acid rich erythrocyte (E) surface also stains intensely. Bars = 1μm.

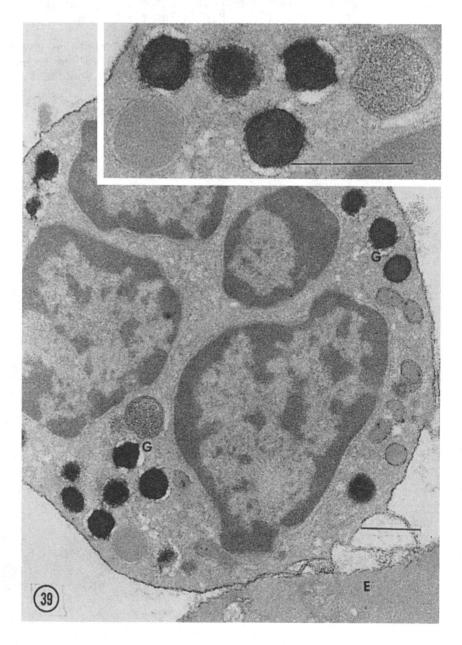

Figure 40. HID intensely stains sulphated glycosaminoglycans
in specific granules (G) of this human basophil.
Sialoglycoproteins on the basophil and erythrocyte (E) cell
surface lack staining seen with the dialyzed iron method (cf
Fig. 39), further attesting to the specificity of HID
staining for sulphate. Nuclear lobes (N). Bar = 1 μm.

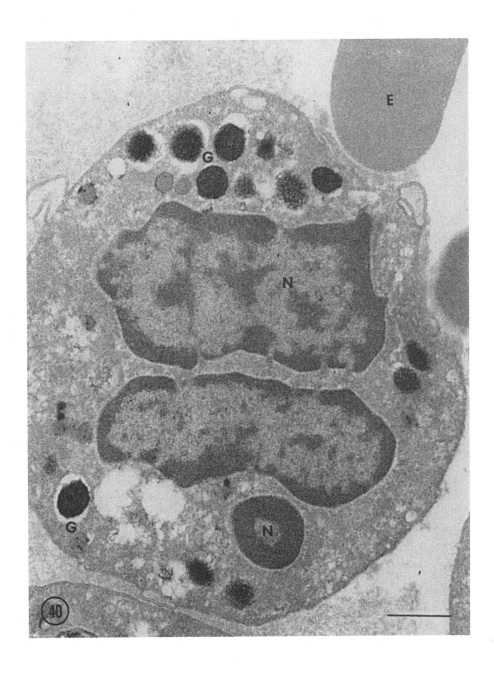

Figure 41. A Wright's stained human monocyte with a folded
or indented nucleus. Bar = 10 μm.

Figure 42. This human monocyte contains numerous pleomorphic
small granules and a central Golgi complex (G). The
cytoplasm contains more ribosomes, scattered small segments
of endoplasmic reticulum, larger mitochondria (M) and little,
if any glycogen, in contrast to neutrophils. The folded
nucleus results in two nuclear profiles in this micrograph
with moderately dispersed chromatin. Bar = 1 μm.

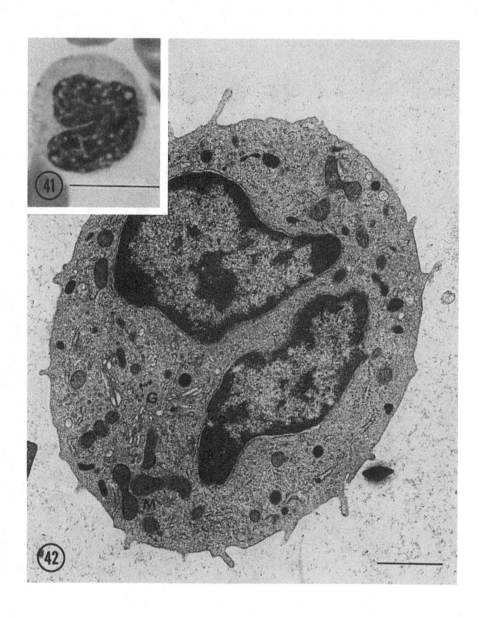

Figure 43. DAB localizes peroxidase activity in endoplasmic reticulum (ER), Golgi lamellae (G) and pleomorphic granules in this human marrow promonocyte. Note the smaller size of granules and less dilated endoplasmic reticulum when compared to promyelocytes (cf Fig. 19). The nucleus (N) has an irregular profile and contains dispersed nuclear chromatin with a prominent nucleolus. Mitochondria (M). Bar = 1 μm.

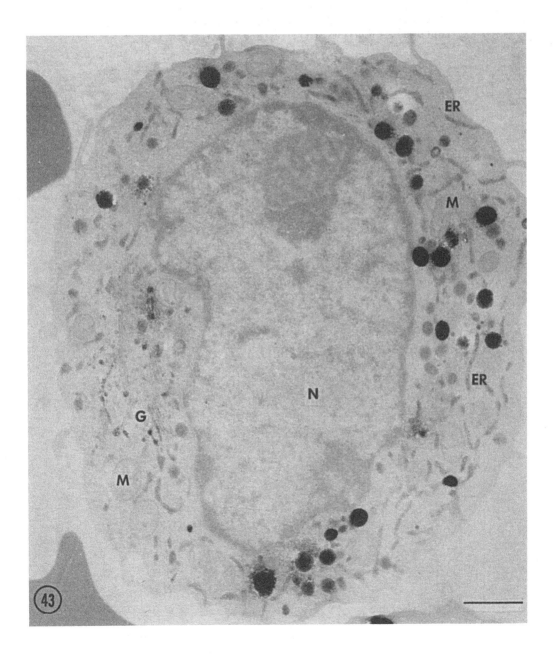

Figure 44. This human marrow macrophage has an irregular
nuclear profile with dispersed nuclear chromatin and numerous
heterophagic granules and vacuoles (H) containing digested
endocytosed material. There are scattered segments of
endoplasmic reticulum. Bar = 1 μm.

Figure 45. This guinea pig placental macrophage, or
Hoffbauer cell, contains a ferrocyanide reactive, haemosiderin laden, heterophagosome (H). Ferric iron in
cytoplasmic ferritin is also ferrocyanide reactive
(arrows). Nucleus (N). Bar = 1 μm.

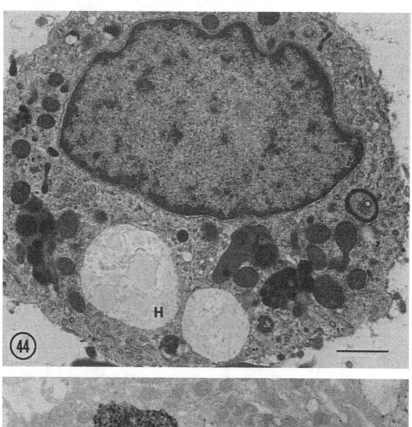

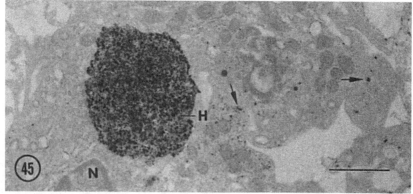

Figure 46. A multinucleated granular megakaryocyte in a
Wright's stained human marrow preparation. Bar = 10 μm.

Figure 47. A human marrow megakaryocyte with abundant alpha
granules containing denser nucleoids (A, enlarged in
inset). There are few, if any, dense bodies. this granular
megakaryocyte has not yet fully matured and is not actively
releasing platelets although demarcation membranes (DM) can
be seen. Nucleus (N). Bar = 1 μm. Inset bar = 1 μm.

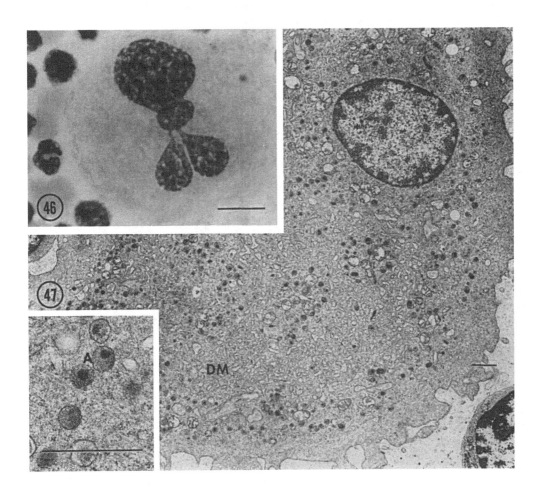

Figure 48. Dense bodies (D) containing serotonin and calcium
are prominent in these cat blood platelets. Numerous alpha
granules (A), several mitochondria (M) and dilated segments
of the open canalicular system (arrows) are evident. Bar = 1
μm.

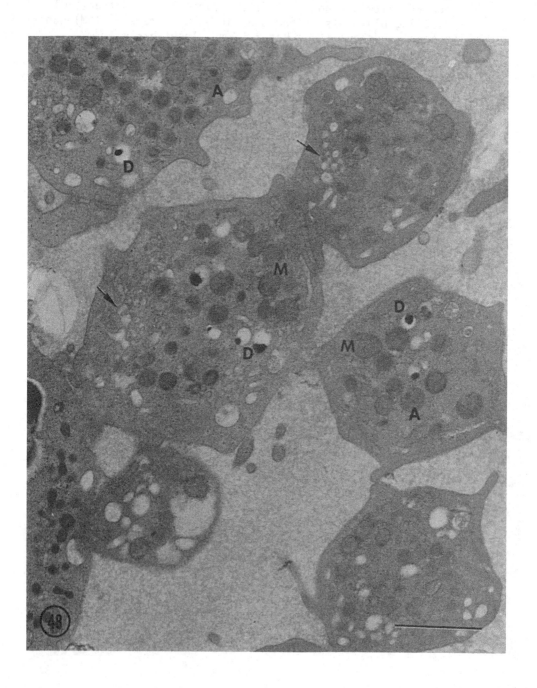

Figure 49. HID-TCH-SP stains sulphated glycosaminoglycans in alpha granules (A) of this human blood platelet. Non-sulphated acidic sialoglycoproteins present on the cell surface and membranes of the open canalicular system (O) fail to stain. Thin section not counterstained. Bar = 1 µm.

Figure 50. DAB stains platelet peroxidase activity in the dense tubular system (D) of this human blood platelet. Staining is optimal when specimens are fixed in formalin and/or tannic acid containing solutions. The open canalicular system (O) and alpha granules (A) do not stain. Bar = 1 µm.

Figure 51. Osmium-zinc iodide intensely stains the dense tubular system and mitochondria of these human blood platelets. Similar staining of a variety of proteins occurs in vesicular structures and endoplasmic reticulum of other blood cells. Alpha granules (A) and the open canalicular system (O) lack staining. Bar = 1 µm.

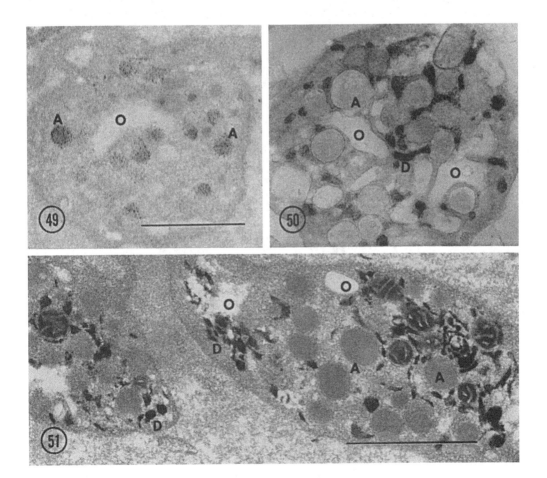

Figure 52. Two human lymphocytes with sparse cytoplasm in a Wright's stained preparation. Bar = 10 μm.

Figures 53-55. This human blood lymphocyte (Fig. 53) contains moderately clumped chromatin and sparse cytoplasm with few organelles and predominantly monoribosomes. The ring-shaped single nucleolus (enlarged at right) contains a segregated nucleolar organizing region (NO) surrounded by fibrillar dense material (F) and granular material (G). Pyroantimonate-osmium fixation localizes cations associated with clumped nuclear chromatin and the nucleolar organizing region (arrows) which is segregated in inactive cells (Fig. 54) and dispersed in S phase cells (Fig. 55). Bars = 1 μm.

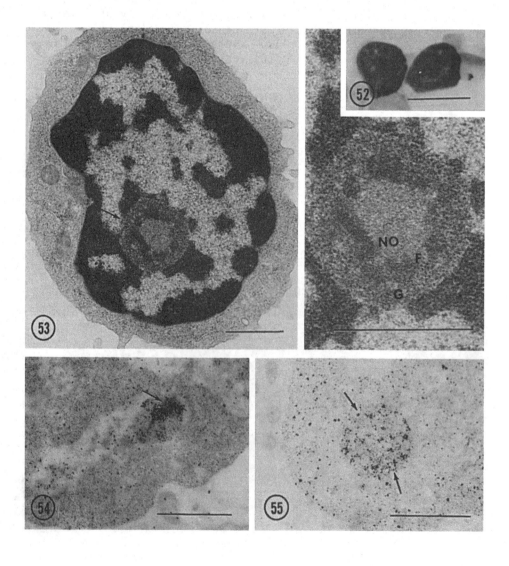

Figures 56-57. This human blood lymphocyte, 48h after stimulation with phytohaemagglutinin, has massively increased in size and readily incorporates ³HTdR (visualized by silver grains). The labelling of the peripheral condensed chromatin and nucleolus are characteristic of late S phase cells and contrasts with the dispersed chromatin labelling of early S phase (Parmley et al. 1977a). The nuclear chromatin is dispersed and the nucleolus is enlarged with scattered nucleolar organizing regions (enlarged in Figure 57). The cytoplasm contains predominantly polyribosomes and vesicular mitochondria. Autoradiograph developed with gold-elon-ascorbic acid. Emulsion removed with lead citrate and uranyl acetate staining. Bar = 1 μm.

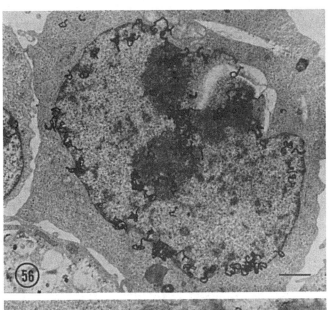

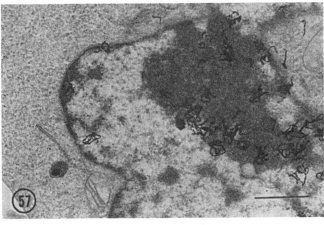

Figure 58. HID-TCH-SP stains sulphated glycosaminoglycans in granules (G) of these human large granular lymphocytes isolated with a fluorescent activated cell sorter using human natural killer (HNK-1) monoclonal antibody. The staining represents chondroitin and heparan sulphate and can be removed with chrondroitinase ABC digestion and nitrous acid (Parmley et al. 1985). Nucleus (N). Bar = 1 μm.

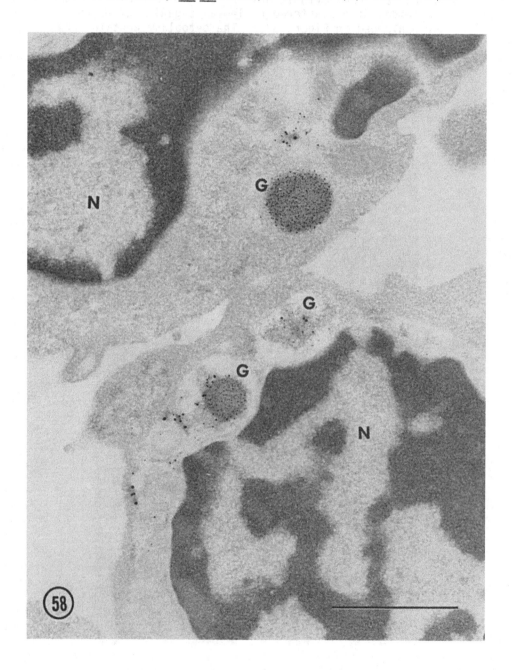

Figure 59. A human plasma cell is illustrated in this
Wright's stained marrow preparation. Note the abundant
basophilic cytoplasm, clear perinuclear Golgi region and
eccentric nucleus with clumped peripheral chromatin.
Bar = 10 μm.

Figure 60. This human marrow plasma cell contains a
prominent Golgi region (G) and abundant dilated and
elongated segments of rough endoplasmic reticulum (ER).
Occasional secondary lysosomes (L) with heterogeneous
material are present. Bar = 1 μm.

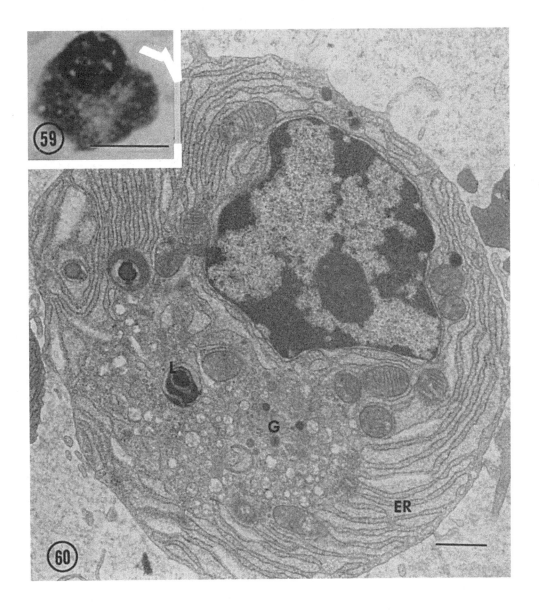

Figure 61. This human marrow plasma cell contains
numerous slightly azurophilic Russell bodies in this
Wright's stained specimen. Bar = 10 μm.

Figure 62. Russell bodies (R), representing precipitated
immunoglobulin, appear moderately dense within dilated
segments of rough endoplasmic reticulum. The Golgi
apparatus (G) is prominent and there are scattered
mitochondria (M). Nucleus (N). Bar = 1 μm.

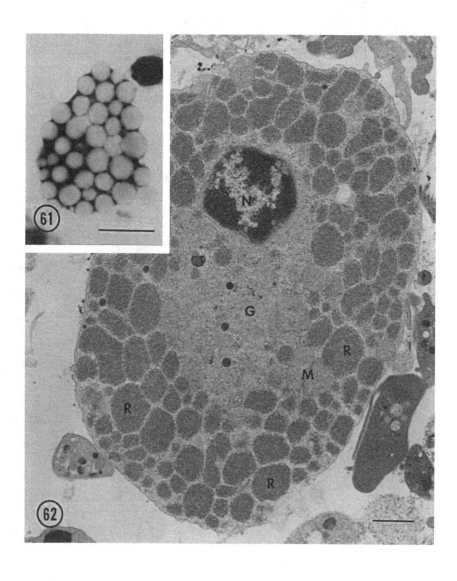

Figures 63-64. This human marrow mast cell contains a
single round nucleus with moderately dispersed chromatin. The granules appear homogeneously dense with
occasional crystalline structure (arrow, enlarged in Fig.
64.). These granules may have substantial variability in
morphology during degranulation (not shown). Bars = 1 μm.

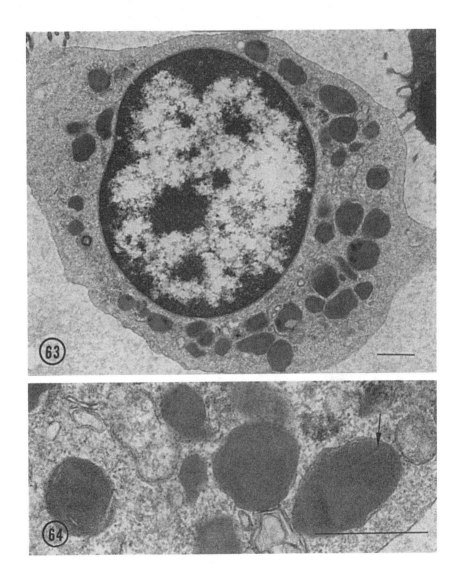

Figure 65. This Wright's stained human marrow osteoclast
contains several distinct round nuclei. Bar = 10 μm.

Figure 66. The ruffled (R) border of this rat osteoclast
contacts bone (B) and calcified cartilage (C). Sulphated
glycosaminoglycans stain heavily with HID-TCH-SP and the
precipitates progressively decrease in size near the ruffled border indicating extracellular digestion of this
material (enlarged above). Residual body staining enlarged in rght inset. Bars = 1 μm. Inset bar = 0.2 μm.

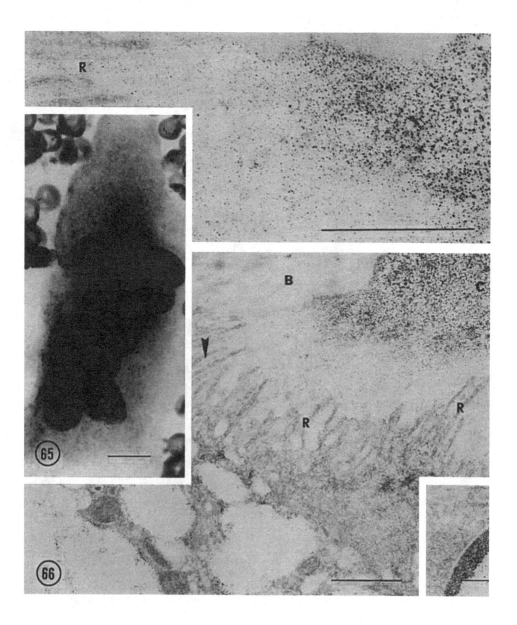

Taxonomic Index

A

B

C

Subject Index

E

Y